A TEXTBOOK OF

BIOTECHNOLOGY

VOLUME I:
GENETICS AND
MOLECULAR BIOLOGY

A TEXTBOOK OF
BIOTECHNOLOGY

VOLUME I: GENETICS AND MOLECULAR BIOLOGY

By

REHANA KHAN
Asstt. Prof. Zoology & Environment
Cambridge Girls P.G. College
Bhopal (M.P.)

UNIVERSITY SCIENCE PRESS
(An Imprint of Laxmi Publications Pvt. Ltd.)
An ISO 9001:2008 Company
BENGALURU • CHENNAI • COCHIN • GUWAHATI • HYDERABAD
JALANDHAR • KOLKATA • LUCKNOW • MUMBAI • RANCHI • NEW DELHI
BOSTON (USA) • NAIROBI (KENYA)

A TEXTBOOK OF BIOTECHNOLOGY

Printed and bound in India
Typeset at Kalyani Computer Services, Delhi
First Edition : 2007, Second Edition : 2010
ISBN 978-93-80386-60-7

Branches

Bengaluru	080-26 75 69 30	
Chennai	044-24 34 47 26,	24 35 95 07
Cochin	0484-237 70 04,	405 13 03
Guwahati	0361-254 36 69,	251 38 81
Hyderabad	040-27 55 53 83,	27 55 53 93
Jalandhar	0181-222 12 72	
Kolkata	033-22 27 43 84	
Lucknow	0522-220 99 16	
Mumbai	022-24 93 12 61	
Ranchi	0651-220 44 64	

PUBLISHED IN INDIA BY

UNIVERSITY SCIENCE PRESS
(An Imprint of Laxmi Publications Pvt.Ltd.)
An ISO 9001:2008 Company
113, GOLDEN HOUSE, DARYAGANJ,
NEW DELHI - 110002, INDIA
Telephone : 91-11-4353 2500, 4353 2501
Fax : 91-11-2325 2572, 4353 2528
www.laxmipublications.com info@laxmipublications.com

C—
Printed at: Repro Knowledgecast Limited, Thane

Dedicated
to
My Family
&
My Teacher

CONTENTS

PREFACE FOR SECOND EDITION

Much knowledge and skill have gathered momentum in biotechnology. After the publication of first edition of this book, new equipments, tools and techniques have added to usefulness for common man.

Government patronage and conducive policy for promoting, teaching, training and popularizing the benefits for the common man has helped a lot. The new generation find new occupation and carrier opportunities in biotechnology by which the fruit of well being will be shared at national and international level.

Though first edition got a welcome response by the beneficiaries still the stake hilder are expecting more with some edition to equipped them with new techniques in researches, to make them multiskilled and to take additional responsibilities with new vision.

The author express gratitude for the teachers, students and researchers dealing biotechnological problems who sent their valuable suggestions and prompted me to update and make it more useful and value added in its worth.

—AUTHOR

PREFACE FOR FIRST EDITION

The aim of this book is to present the subject of Biotechnology to the undergraduate B.Sc. IInd Year students of Biotechnology, M.P. Universities and students of various other Universities. This book provides an accurate and understandable matter on Biotechnology.

The author has been extra cautious in presenting the matter in easy and lucid language. The diagrams are especially designed for clarity and simplicity. It is comparable at international level and fully based on scientific research and teaching. Thus, in this book many topics which are not easily available have been incorporated for the convenience of the students.

I am very thankful to our principal Prof. Shakoor Khan for his precious suggestion, support and help.

I am very much thankful to my teacher Dr. Rabia Sultan, Prof. and Head, Saifia Science College, who inspired and guided me to bring this book in a readable form.

We all express our gratitude for the remarkable work of typing that has been done by Mr. Mirza Arif Baig.

We shall welcome comments and criticism from readers.

—AUTHOR

ACKNOWLEDGEMENT

The completion of the book has not at all been an easy task. It has been completed with the co-operation of so many hands. Hence, I acknowledge it heartily.

First and foremost I am thankful to my colleague Huda Zafar, who is Assistant Professor in the Department of Zoology and my students—Farheen, Versha, Avrti and M.A. Khan for the time and energy they have devoted in this job.

I cannot forget the co-operation and inspiration of my mother, my husband, Mr. Firdaus Khan without which it would have never been a success.

—AUTHOR

UNIT-1

1. Chromosomes : Chemical Composition
2. Structural Organisation of Chromatids
3. Special Chromosomes
4. Structural and Numerical Aberration Involving Chromosomes
5. Nature of Genetic Material : Nucleic Acids
6. Structure of DNA
7. DNA Replication in Prokaryotes and Eukaryotes

1 CHROMOSOMES: CHEMICAL COMPOSITION

INTRODUCTION

Chromosomes are the nuclear components of special organization, individuality and function.

They have the capability of self reproduction and play a vital role in the process of heredity, mutation, variation and evolutionary development.

It was *E. Strasburger* (1875), who observed thread-like structures during cell division. These thread-like structures were named as chromosomes (Gr. *Chrom* — colour, *soma* — body) by *Waldeyer* in 1888.

The dark staining network, which can readily be stained with basic dyes is termed as nuclear reticulum. The threads of this reticulum are made up of chromatin, which can be seen in the interphase nuclear stage during cell division. These chromatin threads or ribbon like structures are called CHROMOSOMES.

CHEMICAL COMPOSITION OF CHROMOSOMES

With the development of technology in biochemistry, the chemical composition of chromosomes has been carried out which revealed that major components of chromosomes

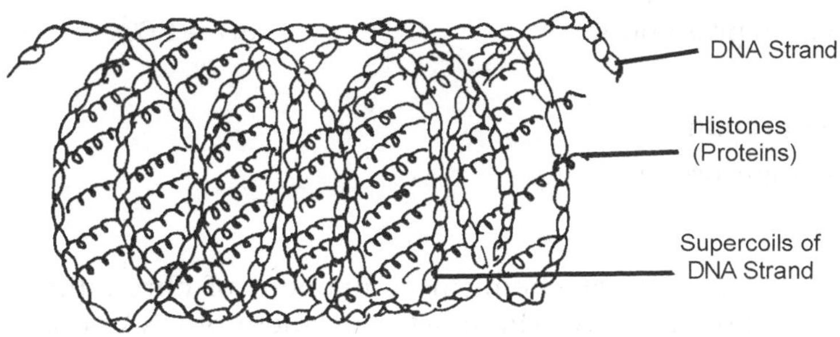

Fig. 1.1. Depicting possible relationship of histone protein and DNA double helix and chromonemata.

are DNA (Deoxyribose Nucleic Acid), RNA (Ribo Nucleic Acid), histones (basic proteins of low molecular weights) and acidic proteins (non histone proteins). Calcium has also been found associated in addition to these constituents but the exact way in which the nucleoprotein complex is assembled is still not very clear. It is thought by the cytologists that RNA is localised at certain regions of the chromosome but proteins are associated with DNA molecule along its entire length. RNA seems to be associated with DNA because some part or the other of chromatin is always being transcribed. Therefore, RNA is the transcription product of chromatin and probably not its integral part. According to *Ambrose & Easty* (1970), the electrical properties of nucleoprotein mainly determine the stability of chromosomes. Histones partly neutralize the negatively charged phosphate groups of DNA. Further neutralization is brought about by calcium bridges formed by the Ca^{++} ion between phosphate groups. On the basis of *Ambrose & Gopal Ayenger's* (1952) observations it has also been reported that any factor which reduces the negative charges on nucleoprotein will tend to lead to the formation of a more compact structure and conversely any factor, which increases the negative charges will cause repulsive forces to operate, thus expanding the chromosomes.

Table 1.1. Showing Variation in Chemical Composition of Chromatin from the Same Source

	Source	*% DNA*	*% RNA*	*% Histone*	*% Non-histone*
	Embryonic axis	39	10	40	11
Pea	Vegetable bud	40	4	52	4
	Growing cotyledon	43.5	6	34.5	16

The relative proportion of components suggests that, generally, there occurs more protein than DNA in chromosomes. Relative amount of different constituents of chromosome are given as follows:

DNA-histone complex	90 – 92%
Histone	55%
DNA	45%
Residual parts (protein) of chromosomes	8 – 10%
DNA	2 – 3%
RNA	12 – 14%
Acid proteins or protomers	83 – 86%

1. DNA
2. Histone protein
3. Non-histone protein

1. DNA

DNA is the most important chemical component of chromatin, since it plays the central role of controlling heredity. The most convenient of DNA is picogram (10^{-12}). DNA of chromatin represents the following two phenomena:

The C-value. The DNA in nuclei was stained using the feulgen reaction and the amount of stain in single nuclei was measured using a special microscope called cytophotometer. Both of these techniques demonstrated that nuclei contain a constant amount of DNA. Thus, all the cells in an organism contain the same DNA content (2C) provided that they are diploid. Gametes are haploid and, therefore, have half the DNA content (1C). Some tissues such as liver, contain occasional cells that are polyploid and their nuclei have a correspondingly higher DNA content (4C or 8C) (see Table 1.2).

Table 1.2. DNA Content and Chromosome Component (after *De Robertis* and *De Robertis Jr.*, 1987)

S.No.	*Cells*	*Mean DNA-feulgen Content*	*Pressured Chromosome Set*
1.	Spermatid	1.68 (1C)	Haploid (n)
2.	Liver	3.16 (2C)	Diploid ($2n$)
3.	Liver	6.30 (4C)	Tetraploid ($4n$)
4.	Liver	12.80 (8C)	Octaploid ($8n$)

Thus, each species has a characteristic content of DNA, which is constant in all individuals of that species. This is called the C-value.

The C-value Paradox. Eukaryotes vary greatly in DNA content but always contain much more DNA than prokaryotes. Lower eukaryotes in general have less DNA, such as nematode *Caenorhabditis elegans* which has only 20 times more DNA than *E. coli*, or the fruit fly (*D. melanogaster*) which has 40 times more DNA (*i.e.*, 0.18 pg or picogram per haploid genome). Vertebrates have greater DNA content (about 3 pg), in general about 700 times more than *E. coli*. One of the highest DNA content is that of the *Salamander amphiuma* which has 84 pg of DNA. Man has about 3 pg of DNA genome, or 3×10^9 base pairs, *i.e.*, the human genome could accommodate about 3 million average sized proteins if all the DNA were coding (or containing structural genes) and if this was true, salamanders would have 30 times more genes than human beings. This is called C-value paradox (*Gall*, 1981). It was detected quite early that there was little connection between the morphological complexity of eukaryotic organisms and their DNA content. For example, *E. coli* (containing 3,400,000 base pairs in its DNA) has about 3000 genes. Although it is difficult to estimate how many different genes exist in the human genome, there are probably not more than 20000 to 30000 genes. There is no reason to believe that salamanders should have more.

(**Note:** *According to a most recent estimate, there are 100000 genes in human genome, see Deviah, 1994*).

2. Histones

Histones are very basic proteins, basic, because they are enriched in the amino acids—arginine and lysine, to a level of about 24 mole per cent. Arginine and lysine at physiological pH are cationic and can interact electrostatically with anionic nucleic acids. Thus, being basic, histones bind tightly to DNA which is an acid. There are five types of histones in the eukaryotic chromosomes, namely H1, H2A, H2B, H3 and H4.

One of the important discoveries that has come from chemical studies is that the primary structures of histones have been highly conserved during evolutionary history. For example, histone H4 of calf and of garden pea contains only two amino acid differences in a protein of 102 residues (*DeLange*, 1969). Likewise, the sequence of histone H3 from rat differs only in two amino acids from that of pea, out of 102 total amino acid residues. These organisms are estimated to have an evolutionary history of at least 600 million years, during which they diverged structurally. This conservation of structure suggests that over the eras, histones have a very similar and crucial role in maintaining the structural and functional integrity of chromatin. Such an evolutionary conservation suggests that the functions of these two histones involve nearly all of their amino acids so that a change in any position is deleterious to the cell.

Histone H1 is the least rigidly conserved histone protein. It contains 210 to 220 amino acids and may be represented by a variety of forms even within a single tissue. H1 is present only once per 200 base pairs of DNA (in contrast to rest of the four types of histones, each of which is present twice) and is rather loosely associated with DNA. H1 histone is absent in yeast, *Saccharomyces cerevisiae*. Histones besides determining the structures of chromatin, play a regulatory role in the repression activity of genes.

3. Non-histones

In contrast to the modest population of histones in chromatin, non-histone proteins display more diversity. In various organisms, number of non-histones can vary from 12 to 20. Heterogeneity of these proteins is not conserved in evolution as the histones. These non-histones differ even between different tissues of the same organism suggesting that they regulate the activity of specific genes.

About 50 per cent non-histones of chromatin have been found to be structural proteins. They include such proteins as actin α and β-tubulins and myosin. Although for sometime these contractile proteins were thought to be contaminants, it is now believed that they are vital ingredients for the chromosome condensation and in the movement of chromosomes during mitosis and meiosis (*Thorpe*, 1984). Many of the remaining 50 per cent of non-histones include all the enzymes and factors that are involved in DNA replication, in transcription and in the regulation of transcription. These proteins are not very highly conserved among organisms, although they must carry out similar enzymatic activities. Apparently, they are not as important as the histones in maintaining chromosome integrity.

ISOLATION METHOD

Several methods have been developed over the years to prepare chromatin for microscopical examination. A relatively simple approval is to first prepare purified nuclei from thymus. Liver or any other desired source nuclei are then lysed with detergent and the chromatin is pelleted at 20000 rpm (about 50000 g) in a preparative centrifuge. Chromatin can also be isolated biochemically by purifying nuclei and then lysing them in hypotonic solution. When prepared in this way chromatin appears as a viscous gelatinous substance. Thus, chromatin which has been isolated from rat liver contains DNA, RNA and protein. Rat liver chromatin has been used as a model for chromatin. It possesses a histone to DNA ratio near 1 : 1, a non-histone protein to DNA ratio of 0.6 : 1 and RNA/DNA ratio of 0.1 : 1.

2 STRUCTURAL ORGANISATION OF CHROMATIDS

INTRODUCTION

While describing the structure of the chromosomes during various phases of cell cycle, cell biologists have introduced many terms for various components. It is thus necessary to be familiar with the following terms to understand the structure of the chromosome more clearly.

1. **Chromatid.** At mitotic metaphase each chromosome consists of two symmetrical structures, called *chromatids.* Each chromatid contains a single DNA molecule. Both chromatids are attached to each other only by the centromere. They get separated at the beginning of anaphase, when the sister chromatids of a chromosome migrate to the opposite poles.

2. **Chromonemata.** During mitotic prophase, the chromosomal material becomes visible as very thin filaments, called *chromonemata* (a term coined by *Vejdovsky* in 1912). A chromonema, represents a chromatid in the early stages of condensation. Therefore, 'chromatid' and 'chromonema' are two names for the same structure: a single linear DNA molecule with its associated proteins. The chromonemata form the gene-bearing portions of the chromosomes. According to old view, a chromosome may have more than one chromonemata which are embedded in the achromatic and amorphous substance, called matrix. The matrix is enclosed in a sheath or pellicle. Both matrix and pellicle are non-genetic materials. They appear only at metaphase when the nucleolus disappears. It is believed that nucleolar material and matrix are interchangeable, *i.e.*, when chromosomal matrix disappears, the nucleolus appears and *vice versa.* Electron microscopic observations, however, have questioned the occurrence of pellicle and matrix in them.

3. **Chromomeres.** The chromomeres are bead-like accumulations of chromatin material that are sometimes visible along interphase chromosomes. The chromomere-bearing chromatin has an appearance of a necklace in which several beads occur on a string. Chromomeres become especially clear in the polytene chromosomes, where they become aligned side by side, constituting the chromosome beads. At metaphase, the chromosomes are tightly coiled and the chromomeres are no longer visible. Chromomeres are regions of tightly folded DNA and of great interest for the cell biologists. They are believed to correspond to the units of genetic function in the chromosomes (see *De Robertis & De Robertis*, Jr. 1987). In fact, for a long time the most of the geneticists considered these chromomeres as genes, *i.e.*, the units of heredity.

4. **Centromere and Kinetochore.** Originally it was considered that the centromere consists of small granules or spherules. The centromere of the chromosome of *Trillium* has the diameter of 3 µm and the spherules have the diameter of 0.2 µm. The chromonemata remain connected with the spherules of the centromere. Currently it is held that centromere is the region of the chromosome to which are attached the fibres of mitotic spindle. The centromere (a term much preferred by the geneticists) lies within a thinner segment of chromosome, the primary constriction. The regions of chromosome flanking, the centromere—contain highly repetitive DNA and may stain more intensely with the basic dyes, *i.e.*, it is a constitutive heterochromatin. Centromeres are found to contain specific DNA sequences with special proteins bound to them, forming a disc shaped structure, called kinetochore (a term that is much preferred by the cytologists). Under the Electron Microscope, the kinetochore appears as a plate or cup-like disc, 0.20 to 0.25 nm, in diameter, situated upon the primary constriction or centromere. In thin electron microscopic sections, the kinetochore shows a trilaminar structure, *i.e.*, a 10 nm thick, dense outer proteinaceous layer, a middle layer of low density and a dense inner layer tightly bound to the centromere. The DNA of centromere does not exist in the form of nucleosome (*Ris & Witt*, 1981). Further, emanating from the convex surface of outer layer of kinetochore, in addition to the microtubules, a "corona" or "collar" of fine filaments has been observed. During mitosis, 4 to 40 microtubules of mitotic spindle are attached to the kinetochore and provide the force for chromosomal movement during anaphase. The main function of the kinetochore is to provide a centre of assembly for microtubules, *i.e.*, it serves as a nucleation centre for the polymerization of tubulin protein into microtubules (*Telzer et al.*, 1975).

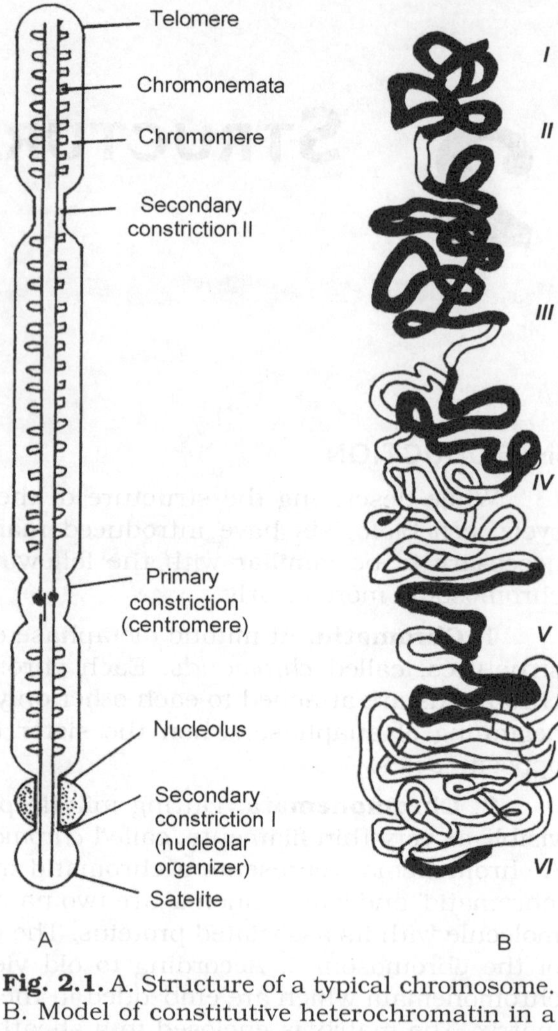

Fig. 2.1. A. Structure of a typical chromosome. B. Model of constitutive heterochromatin in a mammalian metaphase chromosome. (*I*). Constitutive heterochromatin, (*II*). Secondary constriction I or nucleolar organizer, (*III*). Primary constriction or centromere, (*IV*). Euchromatin, (*V*). Secondary constriction II possible site of 5S rRNA cistrons, (*VI*). Telomere (after *De Robertis et al.*, 1975).

The chromosomes of most organisms contain only one centromere and are known as monocentric chromosomes. Some species have diffuse centromeres, with microtubules attached along the length of the chromosome, which are called holocentric chromosomes. The chromosomes of the *Ascaris megalocephala* and hemipterans have diffused type of the centromere. In some chromosomal abnormality (induced for example by X-rays), chromosomes may break and fuse with other, producing chromosomes without centromere (acentric chromosomes) or with two centromeres (dicentric chromosomes). Both types of these chromosomal aberrations are unstable. The acentric chromosomes cannot attach to the mitotic spindle and remain in the cytoplasm. The dicentric chromosomes lead to fragmentation, since, two centromeres tend to migrate to opposite poles.

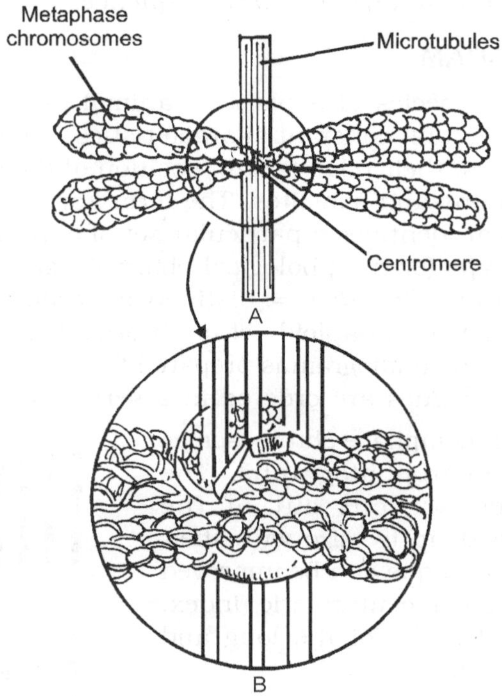

Fig. 2.2. A. A metaphase chromosome showing the folded fibre structure and the centromere with implanted microtubules. B. An insect at higher magnification showing the convex electronic dense layer and the fibrillar material forming the "corona" of the centromere (after *De Robertis et al.*, 1975).

5. **Telomeres.** The chromosome terminates at either end in a fine structure usually called as telomere (Gr; *telo* = fat), a term used by *Muller* (1938) to indicate the uniqueness of this portion or they are the specialized ends of the chromosomes which exhibit physiological differentiation and polarity. Sometimes the broken ends of the chromosomes may fuse to assume the behaviour but never mixes with the telomeres. It means telomere possesses polarity and checks the union of other fragments. In meiotic prophase the telomeres are often attracted to centriole and seen to migrate towards the nuclear membranes. This behaviour is called as Bouquet stage.

6. **Secondary Constriction.** The chromosomes besides having the primary constriction or the centromere possess secondary constriction at any point of the chromosome in a set. Secondary constrictions can be distinguished from primary constriction or centromere as chromosomes bend or exhibit angular deviation only at the position during anaphase.

7. **Nuclear Organizers.** These areas contain secondary constrictions with the genes coding for 5.8S, 10S and 28S ribosomal RNA and that induce the formation of nuclei. The secondary constriction may arise because the RNA genes are transcribed very actively and, thus, interfering with chromosomal condensation. In human beings, the nucleolar organizers are located in the secondary constriction of chromosomes 13, 14, 15, 21 and 22, all of which are acrocentric and have satellites.

8. **Satellite.** Sometimes the chromosomes bear round elongated or knob-like appendages known as satellites. The satellite remains connected with the rest of the chromosome by a thin chromatin filament. The chromosomes with the satellite are designated as the sat chromosomes. The shape and size of the satellite remain constant.

Chromosome satellites are a morphological entity and should not be confused with satellite DNAs which are highly repeated DNA sequence.

KARYOTYPE AND IDIOGRAM

All the members of a species of a plant or animal are characterized by a set of chromosomes which have certain constant characteristics. These characteristics include the number of chromosomes, their relative size, position of the centromere, length of the arms, secondary constrictions and satellites. The term karyotype has been given to the group of characteristics that identifies a particular set of chromosomes. A diagrammatic representation of a karyotype (or morphological characteristics of the chromosomes) of a species is called idiogram (Gr. *idios* = distinctive; *gramma* = something written). Commonly, the chromosomes of a haploid set of an organism are ordered in a series of decreasing length. Sometimes an idiogram is prepared for the diploid set of chromosomes, in which the pairs of homologues are ordered in a series of decreasing size.

A karyotype of human metaphase chromosomes is obtained from their microphotographs. The individual chromosomes are cut out of the microphotographs and lined up by size with their respective partners. The technique can be improved by determining the so called centromeric index, which is the ratio of the lengths of the long and short arms of the chromosome.

Some species may have special characteristics in their karyotypes; for example, the mouse has acrocentric chromosomes, many amphibians have only metacentric chromosomes and plants frequently have heterochromatic regions at the telomeres.

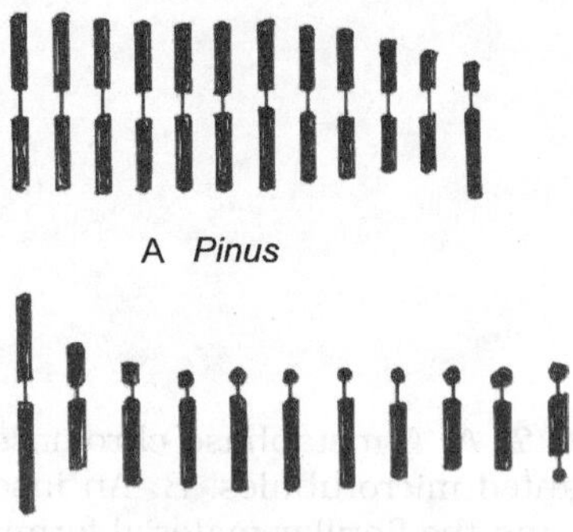

A *Pinus*

B *Ginkgo biloba*

Fig. 2.3. Karyotypes: A. Symmetric; B. Asymmetric.

CHROMATIN

The nucleoplasm contains a number of coiled structures which readily take basic stains such as basic fuchsin. Such chromatin fibres are visible only in the interphase nucleus. During cell division, they become thick, ribbon like structures called the chromosomes. Chemically DNA, RNA, proteins and lipids are the principal components of a nucleus. In the chromosomes, DNA and histone, a basic protein, are bounded to form a deoxironucleoprotein. When DNA and histone are extracted from the chromosome using the enzymes D-nase and protease respectively, the substrate of the chromosome is observed as a fine thread made up mostly of residual protein, RNA, and a small amount of DNA. This substrate is called the **residual chromosome.** The residual protein is an acidic protein containing large amounts of amino acids such as tryptophan and tyrosine. Proteins, serve as the skeletal structures of chromosomes to which the nucleic acid is attached. RNA rarely accounts for more than 5% of the total chromatic present. Histones are constituent of the chromatin of all eukaryotes except fungi.

MATERIAL OF THE CHROMOSOMES

As mentioned above, material of the chromosomes is known as chromatin. Depending on their staining properties, the following two types of chromatin may be distinguished in the interphase nucleus:

1. **Euchromatin.** Portions of chromosomes that stain lightly are only partially condensed and do not exhibit heterophycrosis. This chromatin is termed as euchromatin. It represents most of the chromatin that disperses after mitosis has been completed. Euchromatin contains structural genes which replicate and transcribe during G1-and S-phase of interphase. The euchromatin is considered genetically active chromatin, since it has a role in the phenotype expression of the genes. In euchromatin, DNA is found packed in 3 to 8 nm fibres.

2. **Heterochromatin.** Certain areas of chromatin stain darker than others. Such regions are called **heterochromatin.** Such a difference in staining is called heteropyenosis, which is characteristic of heterochromatin. In 1928, *Heitz* defined heterochromatin as *those regions of chromosomes that remain condensed during interphase and early prophase and form the, so-called chromocenter.* Heterochromatin is characterized by its especially high content of repititive DNA sequences and contains very few, if any, structural genes (*i.e.*, genes that encode proteins). It is late replicating (*i.e.*, it is replicated when the bulk of DNA already has been replicated) and is not transcribed.

The heterochromatic regions have a higher RNA content as compared to the euchromatic regions.

TYPES OF HETEROCHROMATIN

In an interphase nucleus, usually there is some condensed chromatin around the nucleolus, called perinucleolar chromatin, and some inside the nucleolus, called intranucleolar chromatin. Both types of heterochromatin appear to be connected and together, they are referred to as nucleolar chromatin.

Dense clumps of deeply staining chromatin often occur in close contact with the inner membrane of the nuclear envelope (*i.e.*, with the nuclear lamina) and is called condensed peripheral chromatin. Between the peripheral heterochromatin and the nucleolar heterochromatin are regions of lightly staining chromatin, called dispersed chromatin. In the condensed chromosomes, the heterochromatic region can be visualized as regions that stain more strongly or more weakly than the euchromatic regions, showing the so called positive or negative heteropyknosis of the chromosomes (Gr. *Hetero* = different; *pyknosis* = staining). Heterochromatin has been further classified into the following types:

1. **Constitutive Heterochromatin.** In such a heterochromatin, the DNA is permanently inactive and remains in the condensed state throughout the cell cycle. This most common type of heterochromatin occurs around the centromere, in the telomeres and in the C-bands of the chromosomes. In *Drosophila virilis*, constitutive heterochromatin exists around the centromeres and such pericentromeric heterochromatin occupies 40 per cent of the chromosomes. In many species, entire chromosomes become heterochromatic and are called B chromosomes, satellite chromosomes or accessory chromosomes and contain very minor biological roles. Such chromosomes comprising wholly constitutive heterochromatin occur in corn, many phytoparasitic insects and salamanders. In the fly *Sciara*, large metacentric heterochromatic chromosomes are found in the cells, but are absent in somatic cells. Entire Y chromosome of male *Drosophila* is heterochrom, even though containing six gene loci which are necessary for male fertility (see *Suzuki, et al.*, 1985).

Constitutive heterochromatin contains short repeated sequences of DNA, called satellite DNA. This DNA is called satellite DNA because upon ultracentrifugation, it separates from the main component of DNA. Satellite DNA may have a higher or lower G+C content than the main fraction. For example, the mouse satellite DNA is a 240 base pair sequence that is repeated about 1000,000 or 10^6 times in the mouse genome, constituting 10 per cent of the total mouse DNA. The exact significance of constitutive heterochromatin is still unexplained.

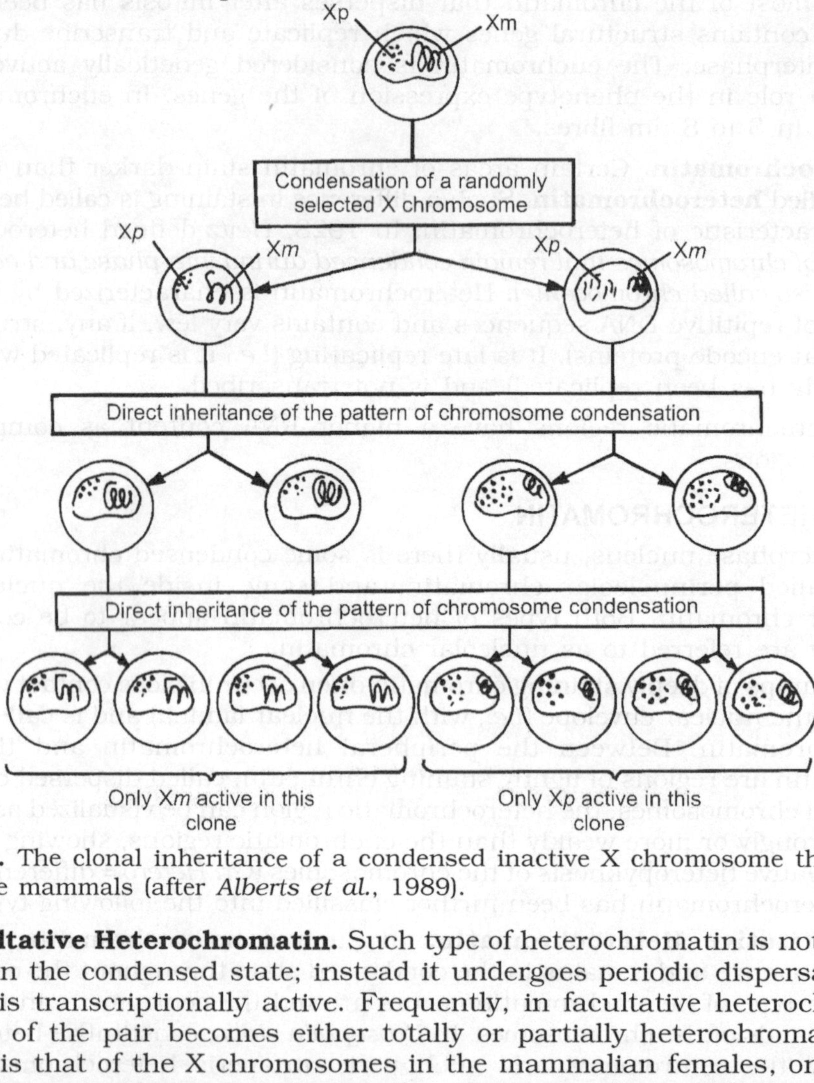

Fig. 2.4. The clonal inheritance of a condensed inactive X chromosome that occurs in female mammals (after *Alberts et al.*, 1989).

2. **Facultative Heterochromatin.** Such type of heterochromatin is not permanently maintained in the condensed state; instead it undergoes periodic dispersal and during this time it is transcriptionally active. Frequently, in facultative heterochromatin one chromosome of the pair becomes either totally or partially heterochromatic. The best known case is that of the X chromosomes in the mammalian females, one of which is active and remains euchromatic, whereas the other is inactive and forms at interphase, the sex chromatin or Barr body (named after its discoverer, Canadian cytologist, *Murray L. Barr*). Barr body contains DNA which is not transcribed and is not found in males. Indeed, the number of Barr bodies is always one less than the number of X chromosomes (*i.e.*, in humans, XXX female has two Barr bodies and XXXX female has three Barr bodies, M.L. Barr, 1959).

DOSAGE COMPENSATION AND LYONIZATION

In mammals all female cells contain two X chromosomes, while male cells contain one X and one Y chromosome. Presumably because a double dose of X chromosome products would be lethal, the female cells have evolved a mechanism for permanently inactivating one of the two X chromosomes. This inactivation is often termed lyonization after the name of British cytogeneticist *Mary Lyons*. In mice, this occurs between the third and the sixth day of development, when one or the other of the two X chromosomes in each cell is chosen at random and condensed into heterochromatin (or Barr body). Because the inactive chromosome is faithfully inherited, every female is a mosaic composed of clonal groups of cells, in which only the paternally inherited X chromosome (X*p*) is active and a roughly equal number of groups of cells in which only the maternally inherited X chromosomes (X*m*) is active (Fig. 2.4).

NUCLEOSOMES

Nucleosomes are the fundamental packing units of chromatin and give chromatin the 'beads on a string' appearance. Each nucleosome is disc-shaped about 11 nm in diameter. It consists of a core particle and a smaller spacer or linker DNA.

1. **Core Particle.** The core particle consists of the octamer of histones, having two copies each of H2A, H2B, H3 and H4. It is about 11 nm in diameter and 6 nm in height. A-strand of DNA having 146 base pairs, is tightly wrapped around this core forming two circles (73 nucleotides per turn).

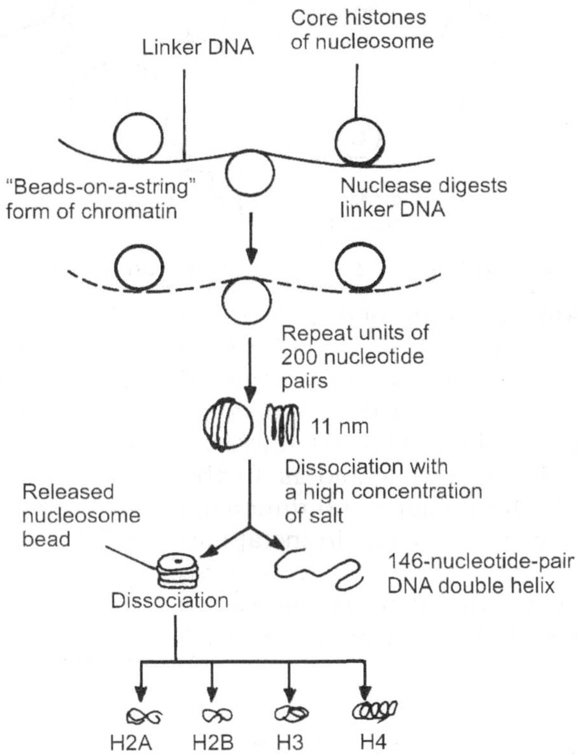

Fig. 2.5. Results of dissociation of nucleoprotein fibre.

2. **Spacer DNA or Linker DNA.** It is a small segment of DNA having just four base pairs. One unit of histone H1 is associated with it. There is considerable variation in the length of spacer region in different species, varying from one base pair to about 80 base pairs in sea-urchin sperm.

On an average, nucleosomes repeat at intervals of about 200 nucleotide pairs. Thus, each human cell with 6×10^9 DNA nucleotide pairs contains 3×10^7 nucleosomes.

3. **Nucleosome Packing or Formation of Chromatin Fibre.** Nucleosomes are packed upon one another to form a chromatin fibre or nucleoprotein fibre with a diameter of about 30 nm. It appears to have a beaded appearance. The beads are nucleosomes. These are connected by linker DNA.

The 11 nm nucleoprotein fibre represents the first level of organisation of chromatin. It is seen in interphase nucleus. The thick chromatin fibre of 30 nm arises by spiral coiling of thin 11 nm chromatin fibre. It has a solenoid type of ultrastructure and has 6–7 nucleosomes per turn.

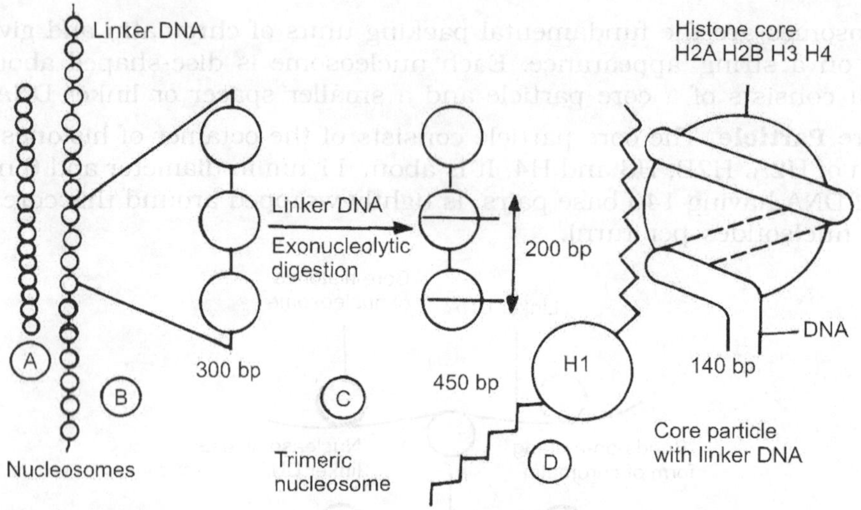

Fig. 2.6. Nucleoprotein fibre showing beads on a string appearance.

This type of packing requires one molecule of histone H1 per nucleosome. H1 gives polarity to chromatin fibre. The 30 nm fibre is called a solenoid. Inside the cells, this solenoid is further folded and packed into the supersolenoid of 300 nm diameter. In mitotic chromosomes, the supersolenoid is further condensed to produce the final shape and dimension of metaphase or anaphase chromosome. In metaphase chromosome, the two daughter DNA molecules are separately folded to produce two sister chromosomes called sister chromatids, held together by a single centromere.

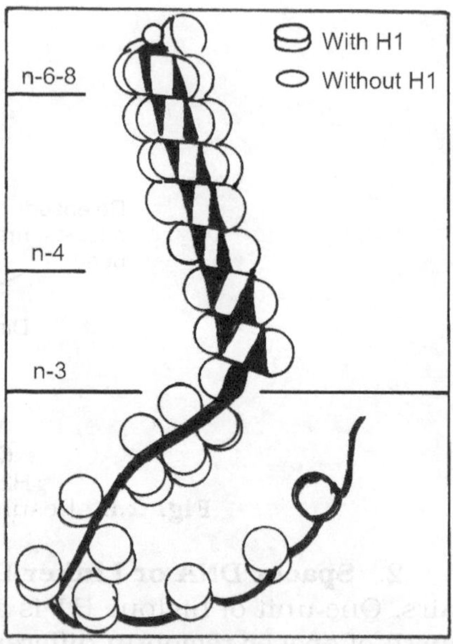

Fig. 2.7. Coiling of nucleosome fibre (Solenoid structure).

The nucleus in human cells is typically about 5 μm (5×10^{-4} cm) in diameter. The 30 nm chromatin fibre of a typical human chromo-some measures about 0.1 cm in length and could span the nucleus more than 100 times. Thus, 30 nm fibre exhibits further folding and forms a series of looped domains. In a typical human chromosome, their number may be 2000, when each loop contains about 20000 to 1000000 nucleotides (*i.e.*, about 0.5 μm of 30 nm fibre).

3 SPECIAL CHROMOSOMES

INTRODUCTION

The function of chromosomes is to carry the genetic information from one cell generation to another. DNA being the only permanent component of chromosome structure is the sole genetic material of eukaryotes. Some cells at certain particular stages contain large nuclei with giant or large sized chromosomes.

The following types of giant or large sized chromosomes are as follows:

POLYTENE CHROMOSOMES OR SALIVARY CHROMOSOMES

E.G. Balbiani (1881), an Italian cytologist was first to observe chromosomes very large, peculiar structures in the cells of certain secretory cells (salivary glands) of midge, chironomes (Diptera). These structures were long and sausage shaped and marked by swellings cross striation, but he did not recognize them as chromosomes. It was in 1933, that *Theophilus Paiter, Erust, Heitz* and *H. Baurr* rediscovered them in *Drosophila* and recognized them as the chromosomes. Since these chromosomes were discovered in the salivary gland cells, they were called *salivary gland chromosomes.*

Due to the occurrence of many chromonemata (DNA) in them, *Kollar* termed them as *polytene chromosomes.*

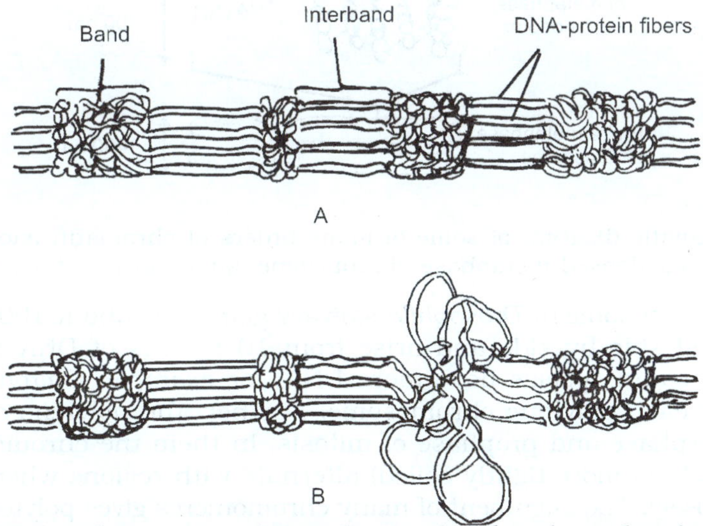

Fig. 3.1. A. Model of *Dupraw* and *Rac* showing band and interband structure in giant chromosome. B. The same model showing a possible method of "puffing".

Thus, some cells of the larvae of dipteran insects such as flies (*Drosophila*), mosquitoes and midge (chironomes) become very large having high DNA content. These cells are unable to undergo mitosis and are destined to die during metamorphosis. (Those cells of larva, which are destined to produce the adult structures after metamorphosis *i.e.*, imagnal disease remain diploid). Such polytenic cells are located most prominently in the salivary gland but also occur in Malpighian tubules, rectum, gut footpads, fat bodies, ovarian nurse cells etc.

Polyteny of giant chromosomes is achieved by replication of the chromosomal DNA several times without nuclear division (endomitosis). The resulting daughter chromatids do not separate but remain aligned side by side. In the process of endomitosis, the nuclear envelop does not rupture and no spindle formation takes place. In fact, polyteny differs from polyploidy in such a way that there is an excess of DNA per nucleus and the new chromosomes are separated from each other.

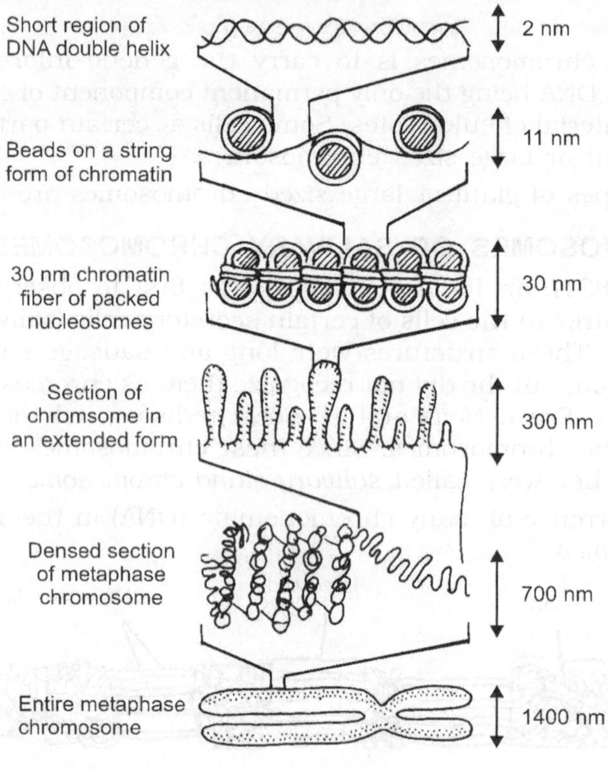

Fig. 3.2. Schematic diagram of some of many orders of chromatin packing which may give the highly condensed metaphase chromosome (after *Alberts et al.*, 1989).

A polytene chromosome of *Drosophila* salivary gland has about 1000 DNA molecules which are arranged side by side and arise from 10 rounds of DNA replication (2^{10} = 1024). Other dipter and species have more DNA, for example, *Chironomus* has 16000 DNA molecules in each polytene chromosome. Further, the polytene chromosomes are visible during interphase and prophase of mitosis. In them the chromomere (regions in which the chromatin is more tightly coiled) alternate with regions where the DNA fibres are folded more loosely. The alignment of many chromomeres gives polytene chromosomes, their characteristic morphology, in which a series of dark transverse bands alternates with clear zones called interbands. About 85 per cent of the DNA in polytene chromosomes is in bands and rest 15 per cent is in interbands. The crossbanding pattern of each polytene chromosome is a constant characteristic within a species and helps in chromosome mapping during cytogenetic studies. For example, in *Drosophila melanogaster*, there are about 5000 bands and 5000 interbands per genome and each band and interband represents a set of 1024 identical DNA sequences, arranged in file. Another peculiar characteristic of the polytene chromosomes is that the maternal and paternal homologous chromosomes remain associated side by side. This phenomenon is called *somatic pairing*. Consequently, in the salivary gland cells, the chromosome number always appears to be half of the normal somatic cells, *e.g.*, *Drosophila melanogaster* has only 4 polytene chromosomes. In *Drosophila*, pericentromeric heterochromatin of all polytene chromosomes also coalesces in a chromocenter.

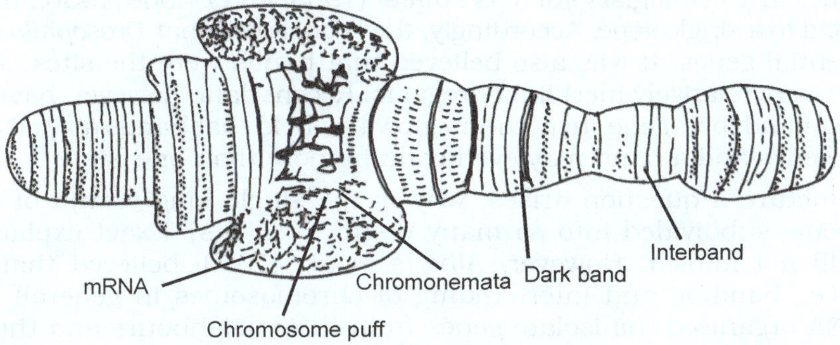

Fig. 3.3. Polytene chromosome of an insect, showing bands and interbands and a puff or Balbiani ring.

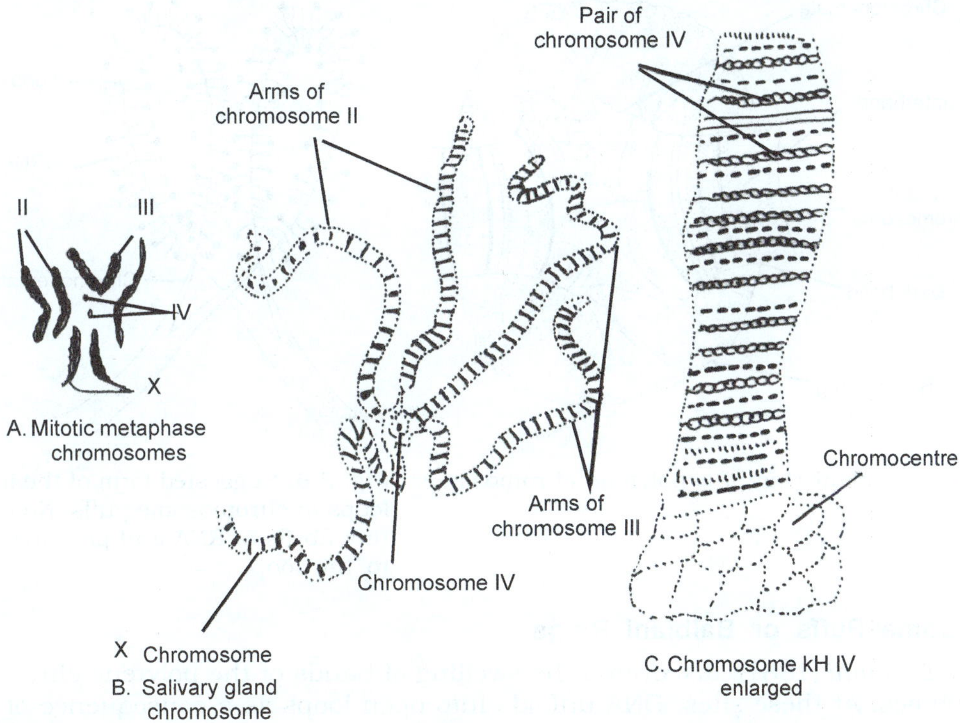

Fig. 3.4

The preparation of a slide of the polytene chromosomes of dipterans for light microscopy is rather easy. The larvae are taken at the third instar stage and the salivary glands are dissected out and squashed in aceto-carmine. In such preparations, these chromosomes in aggregate reach a length of as much as 2000 µm in *D. melanogaster.* In female *Drosophila*, the polytene chromosomes are found in the form of five long and one short strands, radiating from a single, more or less amorphous chromocenter. One long strand corresponds to the X chromosome and remaining four long strands are the left and right arms of II and III chromosomes containing maternal and paternal homologues in somatic pairing which lacks in the sex chromosomes of male fruit flies. Thus, in male *Drosophila*, X chromosome remains single and thin and Y chromosome exists indistinctly fused with the chromocentre.

One gene One band Hypothesis

The fixed pattern of bands and interbands in a *Drosophila* polytene chromosome suggested by the early cytologists such as *Painter* (1933) and *Bridge* (1936), that each band might correspond to a single gene. Accordingly, they concluded that *Drosophila* might contain only 5000 essential genes. It was also believed that bands were the sites of genes (DNA) and interbands were relatively inert linker regions. Recent data, however, have contradicted this simple "one-band, one-gene" hypothesis, now it is held that bands as well as interbands contain active genes and a band may even contain more than one genes.

At this juncture a question arises, why is the single long strand of chromatin in each chromosome subdivided into so many distinct regions? Exact explanation of this question is still not known. However, *Alberts et al.*, (1989) believed that this type of organization (*i.e.*, banding and interbanding of chromosomes in general) may help to: (*i*) keep the DNA organised, (*ii*) isolate genes from their neighbours and thereby prevent biological "crosstalk" or, (*iii*) regulate gene transcription for the cytodifferentiation, for example, constitutively expressed *"housekeeping genes"* could be located in interbands, whereas *cell-type specific genes* could be confined to the bands.

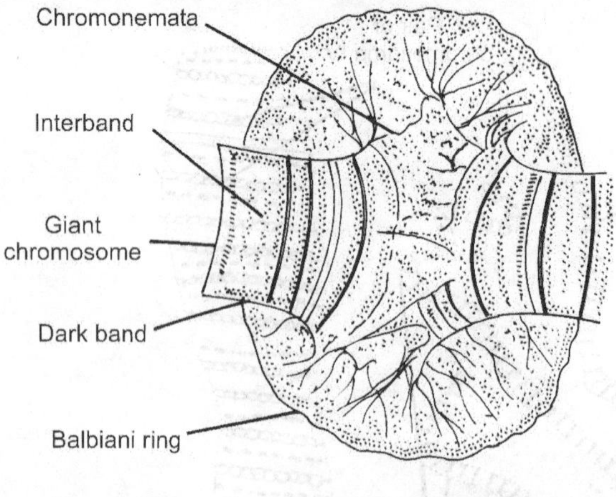

Fig. 3.5. A Balbiani ring of a polytene chromosome.

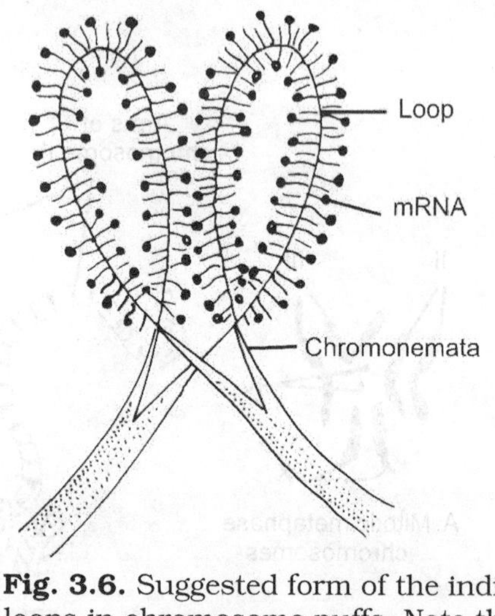

Fig. 3.6. Suggested form of the individual loops in chromosome puffs. Note that the tiny fibrils of RNA and proteins attached to the loop.

Chromosome Puffs or Balbiani Rings

E.G. Balbiani (1881), discovered the swelling of bands of the polytene chromosomes is chironomes. At these sites, DNA unfolds into open loops as a consequence of intense gene transcription (*e.g.*, mRNA formation). These rings appear fluffy or blunted under the electron microscope. In 1954, *W. Beerman* compared the polytene chromosomes of different tissues of Chironomus larvae and showed that although the pattern of bands and interbands was similar in all tissues, the distribution of puffs differed from one tissue to another. *Beerman & Bahr* (1954) had studied the fine structure of these puffs. According to them, these puffs represent regions where the tightly coiled chromosomal fibres open out to form many loops. In fact, puffing is a cyclic and reversible phenomenon. At definite times and in different tissues of the larvae, puffs may appear, grow and disappear. In salivary glands the appearance of some puffs has been correlated with the production of specific proteins, which are secreted in large amounts in the larval saliva (*Grossbach*, 1977). The process of puffing involves several processes such as the accumulation of acidic proteins, despiralization of DNA, accumulation of RNA polymerase II (an enzyme involved in the transcription of mRNA molecules), synthesis of mRNA and release of newly synthesized mRNA in the cytoplasm.

LAMPBRUSH CHROMOSOMES

The lampbrush chromosomes were first observed in salamander (amphibian) oocytes in 1882. Biologist coined the name 'Lampbrush' because the chromosomes look like the brushes which were used for cleaning the glass chimneys of old-fashioned paraffin or kerosene lamps. They were described in detail in shark oocytes by *R. Ruckert* in 1892. *Thorpe* (1984), *Burns & Bottino* (1989) preferred the term brush chromosomes for them. However, due to recent investigations of *Gall and Coworkers* (1962, 1983), the structure of these exceptionally large sized chromosomes has been interpreted in functional terms, *i.e.*, now they are merely visualized as means of "turning on and turning off" of the genes.

The lampbrush chromosomes occur at the diplotene stage of meiotic prophase in the primary oocytes of all animal species, both vertebrates and invertebrates. Thus, they have been described in *Sagitta* (Chactognatha), *Sepia* (mollusca), *Echinaster* (Echinodermata) and in several species of insects, shark, amphibians, reptiles, birds and mammals (humans). Lampbrush chromosomes are also found in spermatocytes of several species, giant nucleus of *Acetabularia* and even in plants (*Grun*. 1958). Generally, they are smaller and hairy" in invertebrates, than in vertebrates. Lampbrush chromosomes are best visualized in salamander oocytes because they have a high DNA content. For example, the largest chromosome having a length up to 1 mm has been observed in urodele amphibian. Thus, lampbrush chromosomes are much larger (longer) than the polytene chromosomes of insects.

Since the lampbrush chromosomes are found in the prolonged diplotene stage of meiotic prophase 1, they are present in the form of bivalents in which the maternal and paternal chromosomes are held together by chiasmata, at those sites where crossing over has previously occurred. The paired homologues are not condensed as usual chromosomes would be, instead they are very long and stretched out. Each bivalent has four chromatids, two in each homologue. The axis of each homologue consists of a row of granules or chromomeres from which lateral loops extend. The loops are always symmetrical, each chromosome having two of them, one for each chromatid. The loops can be categorized by size, thickness and other morphological characteristics. Each loop appears at a constant position in the chromosome. This fact helps in the chromosome mapping. There are about 10000 loops per chromosome set or haploid set (*e.g.*, oocytes of Salamander *triturus:* see *Grant*, 1978). Each loop has an axis, which is made of single

DNA molecule that is unfolded from the chromosome for the intense RNA synthesis. Thus, about 5 to 10 per cent of the DNA exists in the lateral loops, the rest being tightly condensed in the chromomeres, which are transcriptionally inactive. The centromeres of the chromosomes bear no loops. Each loop of lampbrush chromosomes is found to perform intense transcription of hnRNA or heterogeneous RNA molecules (*i.e.*, precursors of mRNA molecules for various ribosomal proteins and for five types of histone proteins). Electron microscopy of the loops has shown that the RNA polymerase enzyme molecules are attached to the principal axis (DNA) of the loop from which RNA fibrils of increasing length extend. As transcription continues along the DNA strand of loop, the fibrils of RNA (*i.e.*, hnRNA) lengthen. Proteins get associated with these RNA fibrils as they are formed and ultimately ribonucleoprotein product is released.

Thus, each lateral loop is covered by a matrix that consists of RNA. RNA transcripts with hnRNA binding proteins attached to them. Generally this matrix is asymmetrical, being thicker at one end of the loop than at the other. RNA synthesis starts at the thinner end and progresses towards the thicker end. Preparations spread for electron microscopy exhibit the typical 'christmas tree' images with nascent ribonucleoprotein chains attached perpendicularly to the DNA axis. Many of the loops correspond to a single transcriptional unit (or single gene) and they are transcribed continuously from end to end. They form a continuous thin-thick matrix. However, other loops contain several units of transcription (or many genes). Such loops include an extended section of chromatin that is not transcribed at all. For example, *Gall et al.*, (1981) found that in the American new *Notophthalmus viridescens*, clusters of the five histone genes are tandemly repeated in the genome but separated by about 50000 base pairs of repeats of 225 base pair satellite DNA.

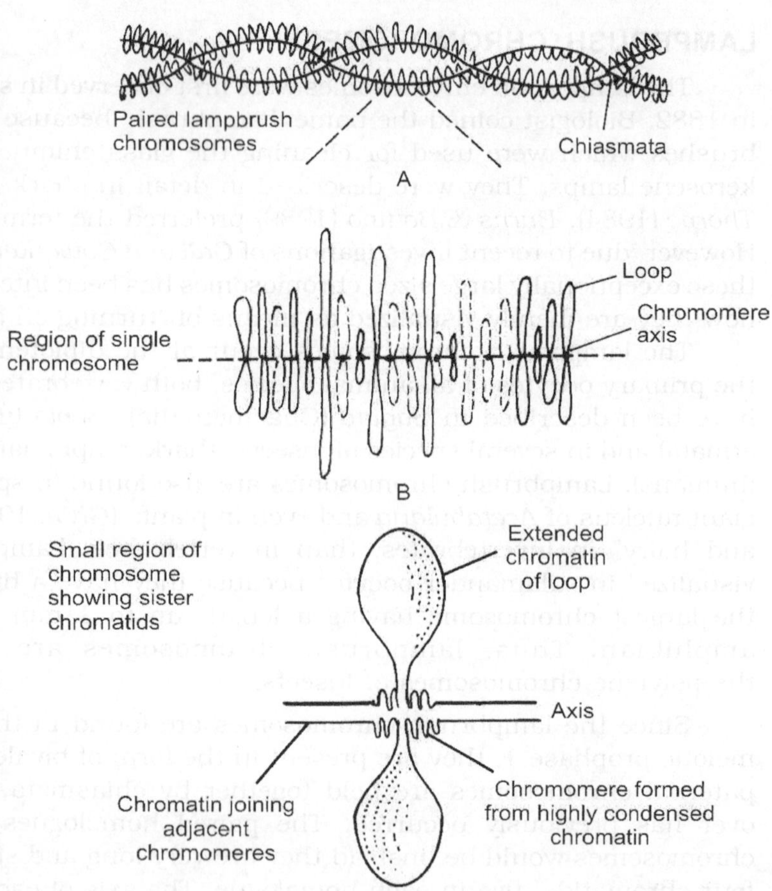

Fig. 3.7. Lampbrush chromosome structure. A. Bivalent or paired homologous chromosomes in pairing showing chiasmata, B. A part of one homologue showing paired loops given out by two chromatids, C. Single pair of loop (after *Alberts et al.*, 1989).

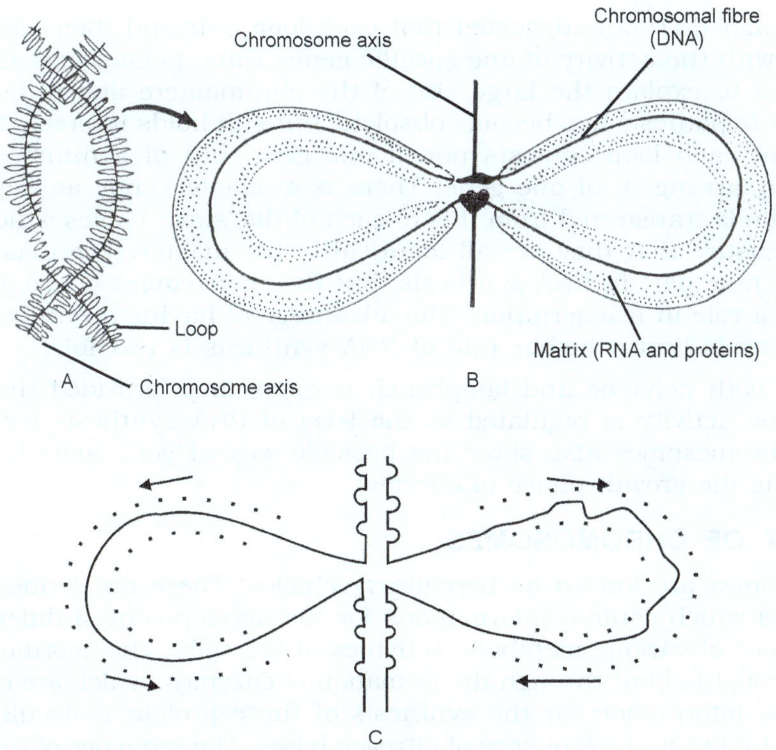

Fig. 3.8. Lampbrush chromosome. A. At low magnification. B. Loop magnified. C. RNA synthesis in a loop

Further, the number of pairs of loops gradually increases during meiosis till it reaches maximum in diplotene. Such a lampbrush stage may persist for months or years as the primary oocyte builds up a supply of mRNA molecules and other materials required for its ultimate development into a new individual. As meiosis proceeds further, number of loops gradually decreases and the loops ultimately disappear either due to disintegration or by reabsorption back into the chromosome. For example, the addition of histone proteins to the lampbrush chromosome stops the synthesis of RNA on the loops and causes the loops to retract into the chromosomes (see *Ambrose & Easty*, 1979).

Certain Hypothesis Regarding Nature of Loops

The loop of the lambrush chromosomes can be viewed in the following two ways:

1. It may be static, unchanging in length and constantly exposing the same stretch of chromosome fibre.
2. It may be dynamic, with new loop material spinning out of one side of a chromomere and returning to a condensed state on the other side. This is called spinning out and retraction hypothesis (*Gall & Callan*, 1962). It means that 100 per cent of the gene can be expressed during the lampbrush stage.

Recent DNA-RNA hybridization experiments have rejected the spinning out and retraction hypothesis (see *Grant*, 1978).

Master and Slave Hypothesis

Callan & Llyod (1960) suggested that each loop pair and their each chromomere is associated with the activity of one specific gene. Thus, master and slave hypothesis was postulated to explain the large size of the chromomere and of lampbrush loop. Presently, this hypothesis has become obsolete, but still holds interest. This hypothesis postulates that each loop consists not of one gene, but of a number of duplicated copies, linearly arranged, of one gene. There is a 'master' copy at each chromomere and information is transferred from this to each of the 'slave' copies which are matched against it to ensure that they are all identical to the master. The master copy of the gene does not take part in RNA synthesis, but the slave copies of the gene existing in the loop have a role in transcription. The advantage of having a number of duplicated copies of a gene is that a higher rate of RNA synthesis is possible.

Study of both polytene and lampbrush chromosomes provided the evidence that eukaryotic gene activity is regulated at the level of RNA synthesis (or transcription). Lampbrush chromosomes also show the possible way of gene amplification which is required during the growth phase of oocytes.

PHYSIOLOGY OF CHROMOSOMES

Chromosomes are known as hereditary vehicles. These are formed of strands of DNA molecules which contain informations for the development of different characters and performance of various metabolic activities of the cells. The coordinator of various functions is brought about through the formation of enzymes, which are complex protein molecules. The information for the synthesis of these protein molecules is contained in the DNA molecules in the sequence of nitrogen bases. The sequence of three nitrogenous bases codes for one molecule of a particular amino acid. It constitutes a triplet colour. DNA of a particular cistern transcripts its mRNA. It carries genetic code similar to its DNA and attached to the ribosome. The tRNA molecule picks up specific activated amino acid and recognizes the codons on the mRNA by their specific anti codons. The amino acid are thus linked together forming a polypeptide chain, which is finally changed into functional protein either single or by the association of several polypeptide chains.

4 Structural and Numerical Aberrations Involving Chromosomes

INTRODUCTION

During the present century, the science of cell genetics has made profound contributions to an understanding of the hereditary process. It was only in 1950's however that geneticist were able to establish the number of chromosomes in the human cell. These are the structures in the cell nucleus that encode the hereditary plan; the normal number in the human cell proved to be 46. Soon afterward came the discovery that mongolism is associated with the presence of 47 chromosomes. Similar gross abnormalities in the chromosome complement of the tissue cells have since been established in other diseases of man. To look through the microscope at the chromosomes of man is to see the very stuff that human life is made of.

The history of cell genetics can be said to date from 1968, when *Walter S. Sutton*, a graduate student at Columbia University in the U.S.A., who later became a surgeon, pointed out in a classic paper the striking parallelism between the inheritance of genetic traits and the behaviour of chromosomes in cells. Subsequent investigations by *Thomas Hunt Morgan* and *A.H. Sturtevant*, also at Columbia, showed that the units of heredity, the "factors" of *Gregor Johnn Mendel*, must be arranged in linear order on the chromosomes.

CYTOLOGY AND CYTOGENETICS

For many years before 1956, the chromosome number in man was recorded as 48. Suitable material for human chromosome studies was difficult to obtain, and the older techniques were not satisfactory for critical chromosome counts. Human chromosomes are small, and in earlier preparations, they overlapped on the microscope slide and were extremely difficult to distinguish. New methods of chemical treatment are now employed for fixing tissues with a minimum of distortion. Chemical treatment has also been developed for spreading chromosomes on the microscope slide, thus making it possible to identify each individual chromosome. Tissue culture methods have facilitated studies of cell division in living cells and have made metaphase stages available for chromosome counts. Recent developments in surgical treatment for cancer and other diseases have made more fresh human tissue available and modern techniques have provided better preparations for study. Testicular specimens have become particularly useful for chromosome investigations.

Types of Chromosome Changes

Cells of the same organism characteristically have the same number of chromosomes, with the exception of gemetes, which (in diploid organisms) have half as many chromosomes as somatic cells. Moreover, the organization and the number of genes on the chromosomes of an organism are the same from cell to cell. These characteristics of chromosome number and gene organization typically hold for all members of the same species. Exceptions to these normal conditions also occur and are known as chromosome aberrations. Chromosome aberrations may be defined as variations from the wild-type condition in either chromosome number or chromosome structure. Chromosome aberrations can arise spontaneously or can be induced experimentally by chemical or radiation mutagens, and in many instances they can be detected cytologically during mitosis and meiosis.

VARIATIONS IN CHROMOSOME NUMBER

Most eukaryotic organisms are diploid, meaning that they have two sets of chromosomes in their somatic cells. The gametes of these organisms are haploid since they have only one set of chromosomes.

Variations may occur from the wild type state in the total amount of genetic material in a cell. Such changes may occur spontaneously or may be experimentally induced. Variation in chromosome number, as opposed to variation in parts of chromosomes, gives rise to aneuploidy, monoploidy, and polyploidy.

CHANGES IN ONE OR A FEW CHROMOSOMES

Aneuploidy (Gk. *Aneu* = Uneven, *Ploid* = unit)

Aneuploidy describes the abnormal case in which one to a few whole chromosomes are lost from or added to the normal set of chromosomes (Fig. 4.1). In most cases, aneuploidy is lethal in animals and is detected mainly in aborted fetuses. Aneuploidy can occur, for example, from the loss of individual chromosomes in mitosis or meiosis; in this case nuclei are formed with less than the normal chromosome numbers. Aneuploidy may also result from non-disjunction, the irregular distribution to the cell poles of sister chromatids in mitosis or of homologous chromosomes in meiosis. In non-disjunction, nuclei with more and nuclei with less than the normal number of chromosomes are produced.

In the case of diploid organisms, aneuploid variations take four main forms:

1. **Nullisomy** (a nullisomic cell) involves a loss of one homologous chromosome pair; that is the cell is 2N–2.
2. **Monosomy** (a monosomic cell) involves a loss of a single chromosome; that is the cell is 2N–1.
3. **Trisomy** (a trisomic cell) involves a single extra chromosome; that is, the cell has three copies of one chromosome type and two copies of every other chromosome type. A trisomic cell is 2N+1.
4. **Tetrasomy** (a tetrasomic cell) involves an extra chromosome pair, resulting in the presence of four copies of one chromosome type and two copies of every other chromosome type. A tetrasomic cell is 2N+2.

Aneuploidy may involve the loss or the addition of more than one specific chromosome or a chromosome pair. Such situations are characterized, for example, by the term doubly monosomic (2N–1–1), doubly tetrasomic (2N+2+2), and so forth.

All forms of aneuploidy have serious consequences in meiosis. Monosomics, for example, will produce two kinds of haploid gametes, N and N–1. Alternatively, the odd, unpaired chromosomes in the 2N–1 cell may be lost during anaphase and not be included in either daughter nucleus, thereby producing two N–1 gametes. In the following sections we shall examine some examples of aneuploidy as they are found in the human population. In general, aneuploidy is more easily tolerated by plants than by animals.

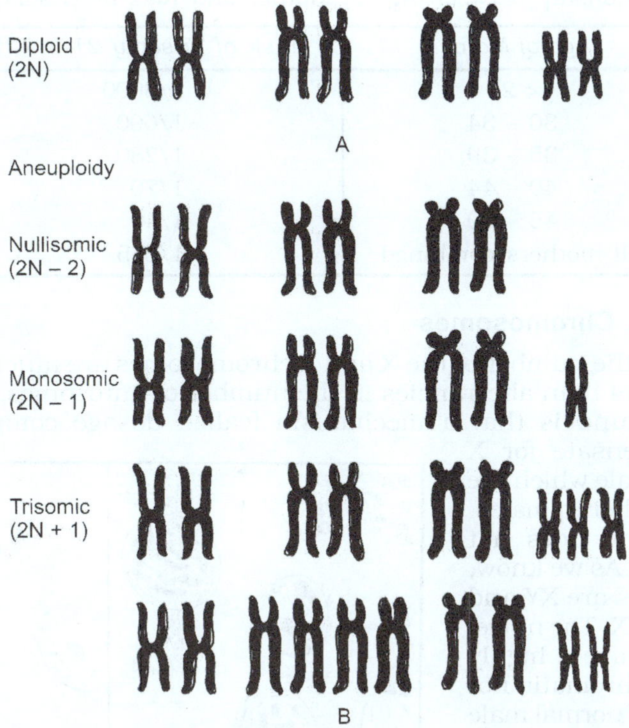

Fig. 4.1. A. Normal set of metaphase chromosomes in a diploid (2N) organism. B. Example of aneuploidy.

Trisomy–21

Trisomy–21 occurs when there are three copies of the chromosome 21, as shown in the karyotype. Individuals with trisomy–21 have down syndrome and exhibit such abnormalities as low IQ, epicanthal folds over the eyes, short and broad hands, and below average height. Fig. 4.2(B) is a photograph of a down syndrome individual. Trisomy of other chromosomes in humans is rarely found with the exception of the sex chromosomes, which we shall study in the following subsection. Trisomy–21 is probably more common in living humans than in other trisomics because chromosome 21 is a very small chromosome with fewer genes than most of the other chromosomes. The disruptions caused by the extra dose of chromosome 21 genes are less severe than would be the case with, say, the largest autosome (non-sex chromosome). Trisomics are produced for all autosomes since they are detected among fetuses that are aborted spontaneously, and it is from such fetuses that we have obtained knowledge about this sort of chromosome aberrancy. Such studies have also revealed that all monosomics in humans are lethal.

The relationship between the age of the mother and the probability of her having a trisomy–21 individual is indicated below. The correlation with the age of the father is much more tenuous. In females the eggs that are released monthly until menopause go through meiosis and arrest at diplotene at the age the individual becomes fertile. Apparently, the probability of nondisjunction occurring as an egg matures increases with the storage time of the egg in the ovary. It is important, then that older mothers to consider genetic counselling to determine whether the foetus has a normal complement of chromosomes. This test involves taking a sample of the amniotic fluid surrounding the foetus by inserting a needle through the abdominal wall (a procedure called aminocentesis) culturing the fetal cells contained therein, and performing a karyotype analysis.

Table 4.1. Relationship between Age of Mother and Risk of Trisomy–21 in Child.

Age of Mother	*Risk of Trisomy 21 in Child*
< 29	1/3000
30 – 34	1/600
35 – 39	1/280
40 – 44	1/70
45 – 49	1/40
All mothers combined	1/665

Aneuploidy in Sex Chromosomes

Aberrancies in the number of the X and Y chromosomes are much more readily found among living humans than aberrancies in the number of autosomes. The reason for they are more readily found is that a mechanism (called dosage compensation) exists in mammals to compensate for X chromosomes in female which are in excess of the number in males. Dosage compensation does not occur for autosomes. As we know, normal human males are XY and normal females are XX. The nuclei of normal females have a highly condensed mass of chromatin not found in the nuclei of normal male cells. This mass, which represents a cytologically condensed and inactivated X chromosome, was first discovered by Maurice Barr, and it has been named the Barr body. Therefore, XX individuals normally have two X chromosomes and one Barr body in their somatic cells, and XY individuals have one X chromosome, one Y chromosome, and no Barr bodies.

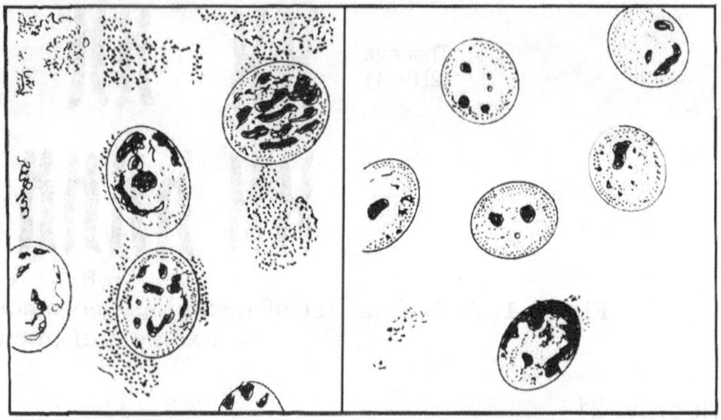

Fig. 4.2. Nucleus of A. Normal human female cells (XX) showing Barr bodies and B. Normal human male cells (XY) showing no Barr bodies.

In the somatic cells of individuals with a greater than normal number of X chromosomes, all but one of the X chromosomes become inactivated and produce Barr bodies. [A general formula for the number of Barr bodies is the number of X chromosomes minus one]. This concept was expanded in 1961 by *Mary Lyon* in what is called the Lyon hypothesis. Lyon proposed that the Barr body is an inactive (or mostly inactive) X chromosome, and her hypothesis provided an explanation for the survival of individuals with X chromosome aberrations. The components of the Lyon hypothesis are that the Barr body is a genetically inactive X chromosome (*i.e.*, it has become lyonized; the process is called lyonization), that the inactivation occurs at about the sixteenth day following fertilization, and that the X chromosomes to be inactivated is randomly chosen from the maternal and paternal X chromosomes in a process that is independent from cell to cell. Lyonization, then is a mechanism that, when it operates in cells with extra X chromosomes, serves to minimize the effects of multiple copies of X linked genes.

Individuals with Turner syndrome have two sets of autosomes but only one sex chromosome, an X, and hence are designated XO. The XO condition exemplifies monosomy of the X chromosome in humans. These people are females (because of the absence of the Y chromosome) and have a genomic complement of 45, X (indicating that they have a total of 45 chromosomes in contrast to the normal 46 and that the sex chromosome complement consists of one X chromosome). The XO individuals have no Barr bodies and exhibit few abnormalities until puberty, when they do not produce normal secondary sexual characteristics. Phenotypically, XO individuals tend to be short and have a weblike neck, poorly developed breasts, and immature internal sexual organs.

A number of human syndromes result from extra copies of the X and Y chromosomes. Klinefelter syndrome, for example, results from the presence of an extra X chromosome in a male, giving a 47, XXY male Table 4.2. The XXY individuals also have a single Barr body, which is not found in normal males.

All XXY males have poorly developed sexual characteristics and are often taller than the average male. Some have lower than average IQs. Individuals with similar phenotypes are also found with higher numbers of X chromosomes, such as XXXY, XXXXY, and so on. The number of Barr bodies in these cases follows the rule proposed earlier. In all cases nondisjunction is the likely abuse of the chromosome aberrancy, and survival of these individuals is ensured by the X inactivation mechanism.

Some people may have multiple Y chromosomes, sometimes with multiple X chromosomes also. Such individuals are males because of the presence of the Y chromosome, and tend to be taller than average. Occasionally, there are effects on fertility.

Finally, some females are trisomic for the X chromosomes. Their chromosomes complement is 47, XXX, and they have two Barr bodies per cell. These XXX individuals are sometimes mentally deficient and infertile.

The effects manifested in individuals with the X chromosome aberrancies are probably the result of the activities of the X chromosome genes during embryonic development until the time all but one of the X chromosomes are deactivated. Table 4.2, below presents a summary of various aneuploid abnormalities for autosomes and for sex chromosomes.

Table 4.2. Aneuploid Abnormalities in the Human Population.

Chromosomes	*Syndromes*	*Frequency of Birth*
Autosomes		
Trisomic 21	Down	1/700
Trisomic 13	Patau	1/5000
Trisomic 18	Edwards	1/10000
Sex Chromosomes Females		
XO, monosomic	Turner	1/5000
XXX, Trisomic		
XXXX, Tetrasomic		1/700
XXXXX, Pentasomic		
Sex Chromosomes Males		
XYY, Trisomic	Normal	1/1000
XXY, Trisomic		
XXYY, Tetrasomic		
XXXY, Tetrasomic	Klinefelter	1/500
XXXXY, Pentasomic		
XXXXXY, Hexasomic		

CHANGES IN COMPLETE SETS OF CHROMOSOMES

Monoploidy

Monoploidy and polyploidy involve variation from the normal state in the number of complete sets of chromosomes. Again considering an organism that is usually diploid, a monoploid individual has only one set of chromosomes (Fig. 4.3). Monoploidy is sometimes called haploidy, although haploidy is typically used to describe the chromosome complement of gametes.

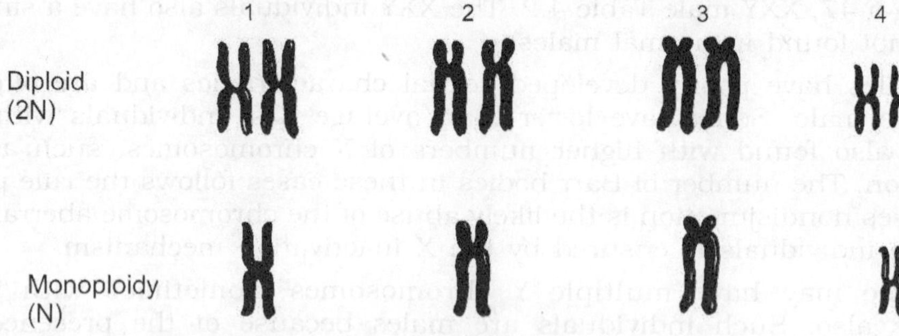

Fig. 4.3. Monoploidy (only one set of chromosomes).

Monoploidy occurs rarely. Owing to the presence of recessive lethal mutations in the chromosomes of many diploid eukaryotic organisms, which are usually counteracted by dominant alleles in heterozygous individuals, many monoploids probably do not survive and thus are undetected.

Polyploidy

Polyploidy describes the condition of a cell organism that has more than its normal number of sets of chromosomes. Animals appear to have evolved to the point that the consequences of more than two sets of chromosomes in a diploid organism are often severe or lethal. The molecular basis for the severe or lethal effects is not well understood. Thus, the evidence for polyploidy in animals often comes from studies of spontaneously aborted fetuses, so we know that polyploidy is one of the major bases for early, spontaneous abortions of human or other animal fetuses.

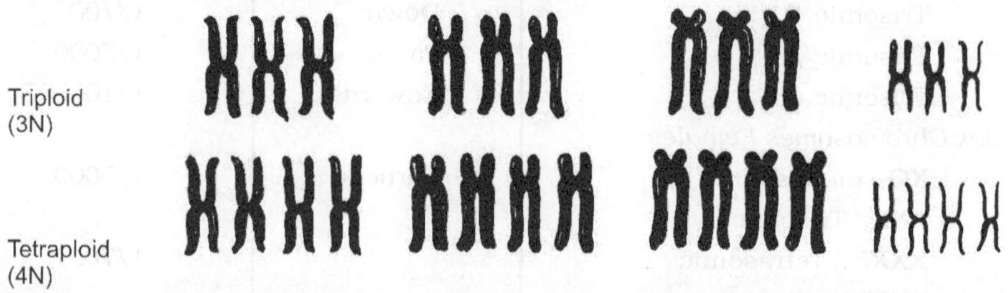

Fig. 4.4. Polyploidy (More than the normal number of sets of chromosomes).

There are two general classes of polyploids. Those that have an even number of chromosome sets and those that have an odd number of sets. Polyploids with an even number of chromosome sets have a better chance of being at least partially fertile since there is the potential for chromosomes to come together in pairs during meiosis. However, polyploids with an odd number of chromosome for each chromosome type, so that probability of a normal gamete is extremely low, and such organisms are usually sterile. An example of an important polyploid with an even number of chromosome sets is the cultivated bread wheat plant, *Triticum aestivum*, which is hexaploid (6N). This plant is fertile and produces gametes with 21 chromosomes. The cultivated banana is an example of a polyploid organism with an odd number of chromosome sets. It is triploid. As a result, the gametes have a variable number of chromosomes and few fertile seeds are set, thereby making the banana a highly palatable fruit to eat.

Animals appear to have evolved to the point that the consequences of more than two sets of chromosomes in a diploid organism are often severe or lethal. The molecular basis for the severe or lethal effects is not well understood. Thus, the evidence for polyploidy in animals often comes from studies of spontaneously aborted fetuses, so we know that polyploidy is one of the major basis for early, spontaneous abortions of human or other animal fetuses. In contrast, plants are much more likely to tolerate polyploidy although, as we have seen they may be sterile.

CHROMOSOMAL ABERRATIONS

Aberrations can occur in the structure of the chromosomes as well as in their number. The number of such variations is essentially infinite. chromosomal aberrations are found by fallowing processes:

Deletion

In deletions a piece of a chromosome break off and is lost. Such abnormalities, particularly if they involve sizeable amounts of the genetic material, could be lethal in many cases, and large deletions are in fact not commonly found.

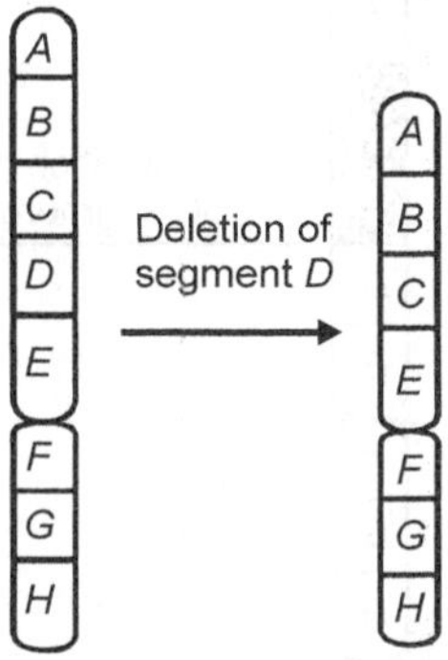

Fig. 4.5. Deletion with a chromosome segment there, (D) missing.

Duplication

A duplication is a chromosome aberration that results in the doubling of a segment of a chromosome. Duplications may or may not be lethal to an organism. The size of the duplicated segment may vary considerably, and duplicated segments may occur at different locations in the genome or in a tandem configuration (*i.e.*, adjacent to each other). When the order of genes in the duplicated segment is the opposite of the order of the original, it is a reverse tandem duplication; when the duplicated segments are tandemly arranged at the end of a chromosome, it is a terminal tandem duplication (Fig. 4.8). Cytologically, heterozygous duplications result in unpaird loops similar to those described for chromosome deletions.

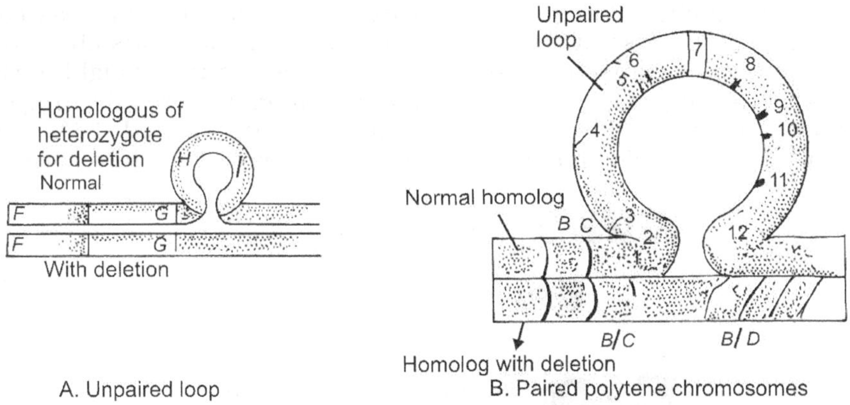

Fig. 4.6. Cytological effects at meiosis of heterozygosity for a deletion: A. formation of an unpaired loop when the homologous chromosomes pairs. B. paired *Drosophila* salivary gland polytene chromosomes in a strain with a heterozygous deletion. showing the unpaired region numbers refer to bands, which are presumed to indicate gene.

Evidence suggests that duplications of particular genetic regions can have unique phenotypic effects, as in the bar mutant on the X chromosome of *Drosophila melanogaster*. In strains carrying the bar mutation (not to be confused with the Barr body discussed earlier in the chapter), the number of faces of the compound eye is fewer than for the normal eye (giving the eye a bar shaped slit-like) rather than an oval appearance. Bar acts like an incomplete dominant mutation, since males heterozygous for bar have more faces, hence a somewhat larger bar-shaped eye than females homozygousfor bar. Males hemizygous for bar have very small eyes like those of homozygous bar females. Cytological examination of polytene chromosomes has shown that the trait is the result of a duplication of a small segment of the X chromosome. Some flies have been found three copies of that segment instead of the normal one copy, a condition called double bar. These flies have very small, slit-like eyes.

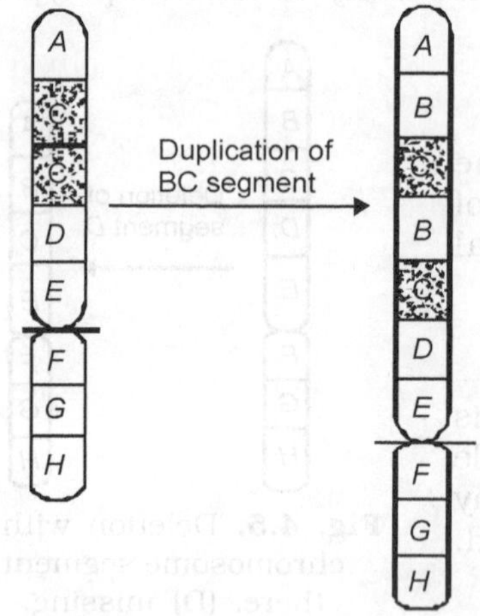

Fig. 4.7. Duplication with a chromosome segment (here, BC) repeated.

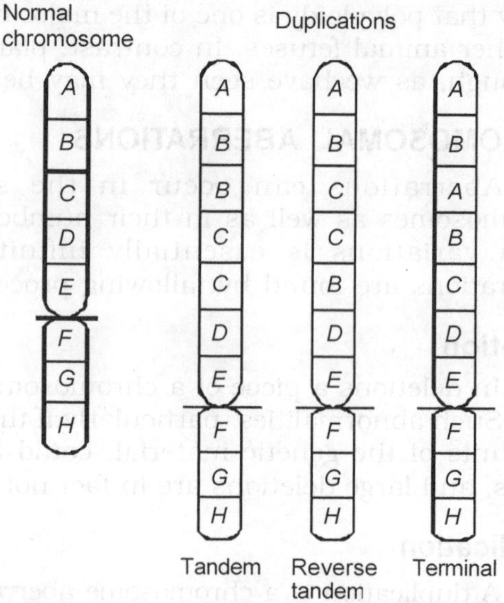

Fig. 4.8. Forms of chromosome duplications are tandem, reverse tandem, and terminal duplicalions.

A duplication such as bar probably arose as a result of a process called unequal crossing over. Unequal crossing over may result when homologous chromosomes paired inaccurately, perhaps because similar DNA sequences occur in neighbouring regions of chromosomes. Crossing over in the mispaired region will result in gametes with a duplication or a deletion. Fig. 4.9 shows, how the duplicated 16A segment of the X chromosome of *Drosophila* that is associated with bar could have arisen.

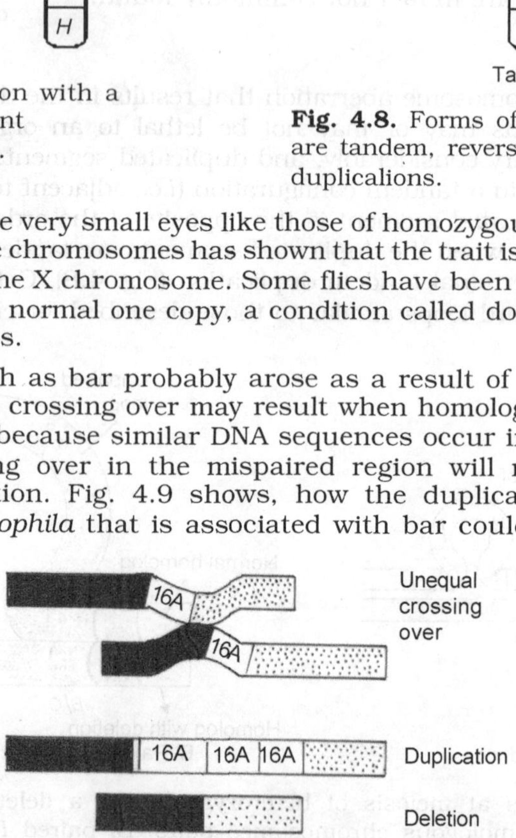

Fig. 4.9. Duplication and deletion of chromosome segments from unequal crossing over.

Inversion

An inversion is a chromosome mutation that results when a segment of a chromosome is excised and then reintegrated in an orientation 180° from the original orientation. When the inverted segment occurs on one chromosome arm and does not include the centromere, the inversion is called a *paracentric inversion.* When the inverted segment includes the part of both chromosome arms and, hence, includes the centromere, the inversion is called a *pericentric inversion.*

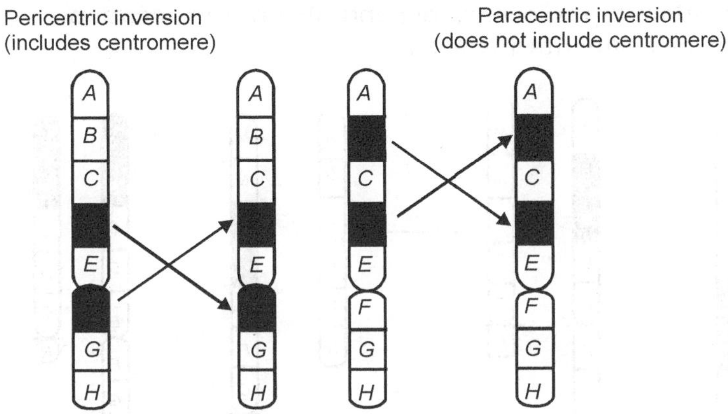

Fig. 4.10. Inversion.

Translocation

In translocation, breakage is followed by relocation of the fragments. A reciprocal translocation involves breaks in two nonhomologous chromosomes an thereafter a mutual exchange of the fragments. Depending on how the chromosomes behave during meiosis, some gamete will be formed with a normal, although displaced gene content—in other words, a balanced translocation. But translocation may also give rise to gametes that have duplications in others. Some offspring of gametes that carry translocations may therefore be normal and some may not. when the germ cells of even the apparently normal person who bears a balanced translocations later undergo meiosis, an unblanced gamete may results. In human pedigrees, a translocation often manifests itself when a disease is transmitted by unaffected indivduals whose collateral relatives show the disease.

The first recognizable translocation was found in 1959 by *Jerome Lejeune* of the University of Paris in a mentally deficient boy with congenital abnormalities of the backbone. The boy's cells had only 45 chromosomes, including one that was longer than its homologue. Lejeune suggested that part of the missing chromome had been delocated, thus explaining the extra length of the abnormal chromosome; the remaining fragment had been eliminated. some patients suffering from Down's syndrome and having a normal number of chromosomes may be explained by translocation. In these cases the extra chromosomal material, usually present as an extra "free" chromosome is translocated to another chromosome. The result is Down's syndrome with only 46 chromosomes. The disease would seem, therefore, to be related not so much to the number of chromosomes in the cell as to the extra chromosomal material.

A translocation (also called a transposition) is a chromosomal mutation in which a change in position of chromosome segments and the gene sequences they contain. Simple translocation may be classified as follows:

1. *Intrachromosomal* (within a chromosome) *translocations* (Fig. 4.11A) involve a change in position of a chromosome, either from one chromosome arm to the other or within the same chromosome arm.
2. *Interchromosomal* (between chromosomes) *translocations* involve the transfer of a chromosome segment from one chromosome into another chromosome (called a nonreciprocal translocation; see Fig. 4.11B) or the reciprocal exchange of segments between nonhomologous chromosomes (called a reciprocal translocation; see Fig. 11C). A variety of nonreciprocal and reciprocal translocation types occur, depending on the locations of the chromosome segment or segments involved.

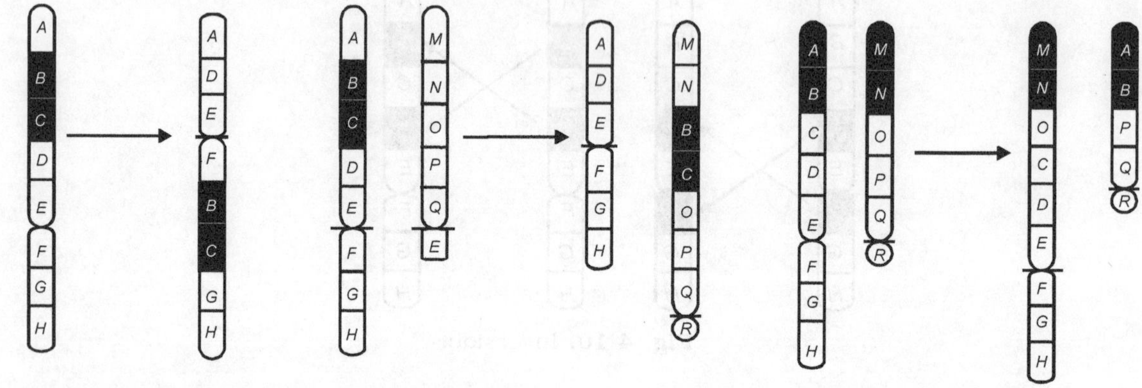

A. Nonreciprocal intrachromosomal translocation (transposition)

B. Nonreciprocal interchromosomal translocation

C. Reciprocal translocation

Fig. 4.11. Translocations: A. nonreciprocal intrachromosomal. B. nonreciprocal interchromosomal. C. reciprocal.

5 Nature of Genetic Material: Nucleic Acids

INTRODUCTION

The principal genetic material of living beings is Deoxyribonucleic acid and Ribonucleic acid which are chemically called as Nucleic acids. These are complex molecules composed of Carbon, Oxygen, Hydrogen, Nitrogen and Phosphate and are larger than most proteins. In 1869, a young Swiss physician had isolated it from puss cells of salmon sperm, a previously identified macromolecular substance, to which he gave the name—nuclein but he was unaware of the structure and function of nuclein. In 1889, with the help of his student Richard Altman, he recognised nuclein with higher molecular weight and was associated with basic protein to which he gave the name, protamine. *Emil Fischer*, in 1880 identified Purines and Pyrimidines. *Altmann*, in 1899 for the first time used the term Nucleic acid to describe the phosphorus containing nuclein. Kosser was awarded Nobel prize, in 1910 for demonstrating the presence of two types of pyrimidines (C and T) and two types of purines (A and G) in nucleic acids. *Ascoli Levine* and *Jones* discovered two kinds of nucleic acids—DNA and RNA. They are polynucleotides, which means a polymer consisting of nucleotides. Each nucleotide has three components. These are:

1. **Ring of 5 Carbon Sugar.** In case of ribonucleic acid, the sugar is ribose and in case of deoxyribonucleic acid, it is deoxyribose.

   DNA and RNA are differ only in one aspect that is the absence of a 2'-OH group in deoxyribose, a difference due to which DNA is chemically more stable than RNA.

2. **Nitrogenous Base.** There are two types of nitrogenous bases:

   (*i*) Pyrimidines are single ring compounds with nitrogen in 1 and 3 of 6 membered benzene ring. Most common types are cytosine, thymine and uracil.

   (*ii*) Purines, are double ring compounds. Their one molecule consists of a 5 membered imidazole ring joined to a pyrimidine ring at positions 4 and 5. Most common purines of nucleic acid are adenine and guanine.

   DNA and RNA both contain adenine, guanine and cytosine. Thymine is found only in DNA and uracil is found only in RNA. There are rare exceptions to this rule. Namely T is present in some tRNA molecules and there are a few phases whose DNA exclusively contains uracil. These bases are attached to carbon atom of sugar by an N-glycosidic bond.

3. **Phosphate Group.** Linked to the 5' carbon of the sugar by a phosphoester linkage. This phosphate is responsible for the strong negative charge of both nucleotides and nucleic acids:

Sugar	+	Base	=	Nucleoside		
Nucleoside	+	Phosphate	=	Nucleotide		

Thus

Nucleotide	=	Pentose sugar	+	Nitrogen base	+	Phosphate groups

DNA is the only genetic material of most living organisms, so it deserves priority in discussion.

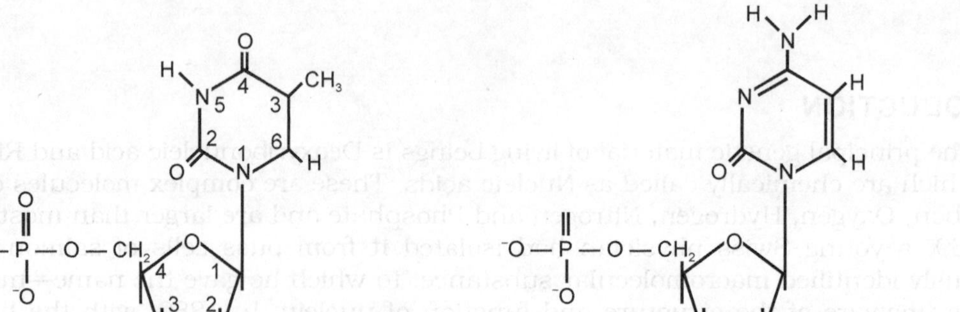

Fig. 5.1. Shows A. Pyrimidine Nucleotides B. Purine Nucleotides.

DEOXYRIBONUCLEIC ACID OR DNA

DNA is the macromolecule that makes up the major portion of the chromosomes. Its important task is to carry the genetic information for cell of an organism. The information from the DNA must be carried to the rest of the cell to function properly; which is accomplished by RNA.

There are four major nitrogenous bases are found in DNA, two are heterocyclic and another two are ringed purines. Two-ringed purines are adenine (A) and guanine (G), and one ringed pyrimidienes are cytosine (C) and thymine (T).

The part of each nucleotide, which contains a nitrogenous base and deoxyribose is called deoxyribonucleoside. The four kinds of deoxyribonucleosides and deoxyribonucleotides can be tabulated as follows :

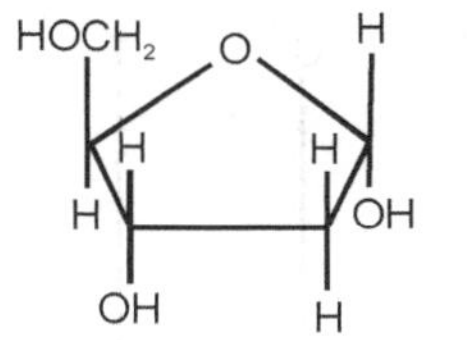

Fig. 5.2. Chemical formula of deoxyribose.

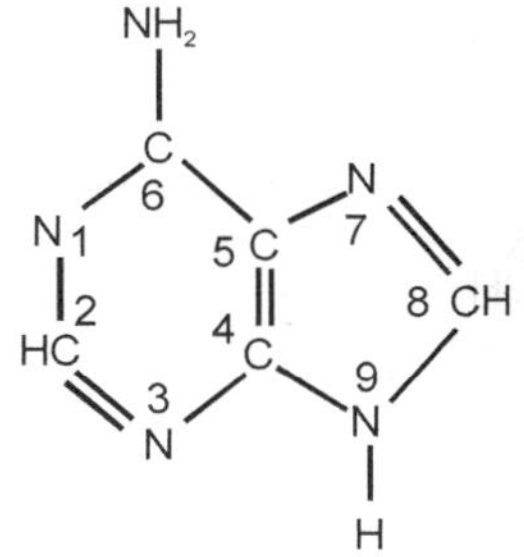

Fig. 5.3. Chemical formula of adenine (6-amino-purine, mw = 135.13 daltons).

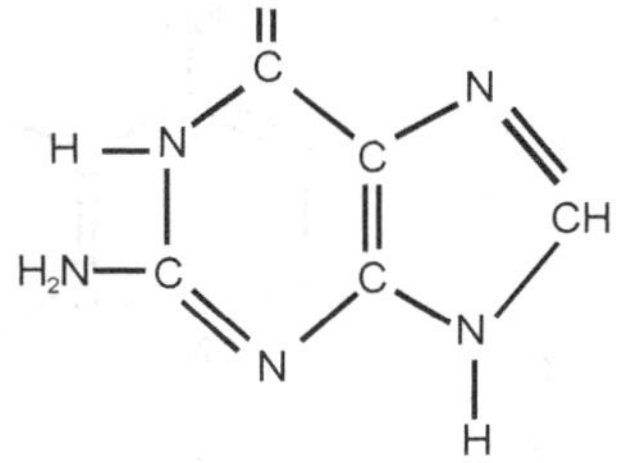

Fig. 5.4. Chemical formula of guanine (2-amino 6-hydroxy-purine, mw = 151.43 daltons).

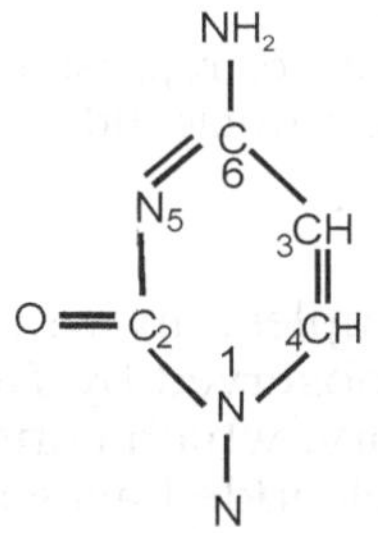

Fig. 5.5. Chemical formula of cytosine (2-hydroxy-6 aminopyrimidine mw = 111.10 daltons).

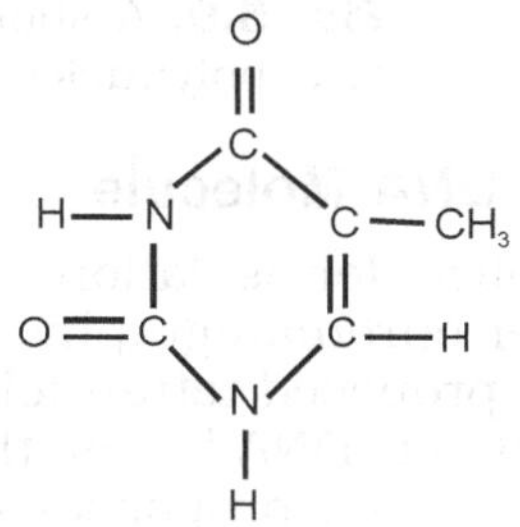

Fig. 5.6. Chemical formula of thymine (2,6-dihydroxy 5-methylpyrimidine, mw = 126.42 daltons).

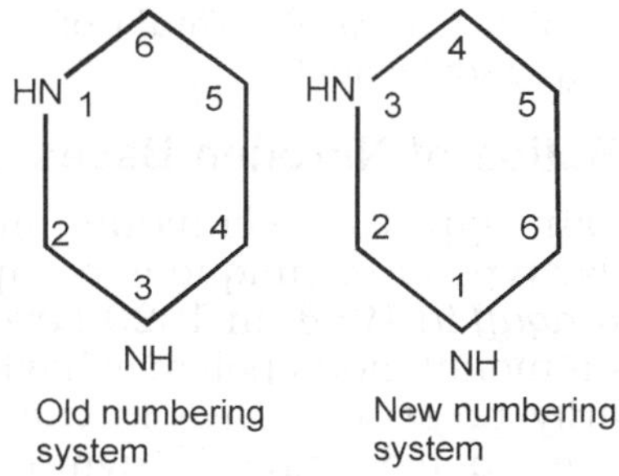

Fig. 5.7. Numbering of carbon atoms in a pyrimidine ring.

Table 5.1. Four Nitrogen Bases, Nucleosides and Nucleotides of DNA Molecule

Nitrogen Base	*Base + Deoxyribose = Deoxyribonucleoside*	*Deoxyribonucleoside + Phosphoric Acid = Deoxyribonucleotide*	*Abbreviation for Nucleotide*
1. Adenine (A)	Deoxyadenosine	Deoxyadenylic acid (Deoxyadenosine monophosphate)	dAMP
2. Guanine (G)	Deoxyguanosine	Deoxyguanylic acid (Deoxyguanosine monophosphate)	dGMP
3. Cytosine (C)	Deoxycytidine	Deoxycytidylic acid (Deoxycytidine monophosphate)	dCMP
4. Thymine (T)	Thymidine	Thymidylic acid (Thymidine monophosphate)	TMP

The four deoxyribonucleotides besides occurring in DNA molecule, also occuned in nucleoplasm and cytoplasm in their triphosphate forms such as deoxyadenosine triphosphate (dATP), deoxyguanosine triphosphate (dGTP), deoxycytidine triphosphate (dCTP) and thymidine triphosphate (TTP). The significance of occurrence of deoxyribonucleotides in triphosphate forms lies in the fact that during DNA replication, the DNA polymerase enzyme can act only on triphosphate or deoxyribonucleotides.

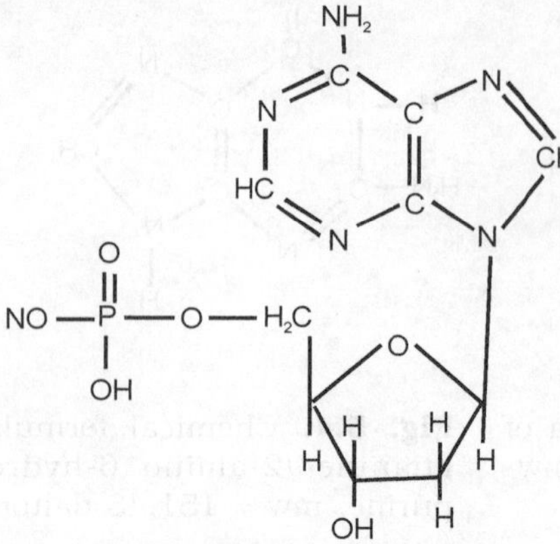

Fig. 5.8. Chemical formula of deoxycytidylic acid.

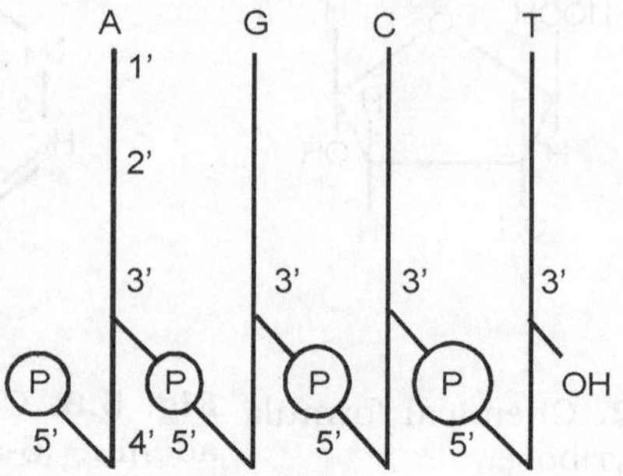

Fig. 5.9. A shorthand system of representation of a polynucleotide with four nucleotides.

Molar Ratios of Nitrogen Bases in DNA Molecule

Many types of techniques applied for isolation of DNA samples, in which one particular type of technique is a paper chromatography. First it is observed by *Hotchkis* and *charagaff* in 1948. In 1920 *Leven* proposed tetranucleotide theory, which considered DNA as a monotonous polymer having four DNA bases, the four nucleotide bases are not necessarily present in DNA in exactly equal proportions. Thus, Chargaff reported that the DNA extracted from calf thymus nuclei contains the four bases in the following molar proportions: 28% adenine, 24% guanine, 20% cytosine and 28% thymine. When he analyzed the DNA samples of varied animals, he found that the exact base composition of DNA differs according to its biological source. The relative amounts of purines and pyrimidines in samples of DNA of different living organisms are tabulated as follows:

Table 5.2. The Relative Amounts of Purines and Pyrimidines in Samples of DNA of Different Living Organisms

Sr. No.	*Source*	*Adenine* A	*Guanine* G	*Cytosine* C	*Thymine* T	$\frac{A+T}{G+C}$
1.	Beef sperm	28.7	22.2	22.0	27.2	1.26
2.	Human thymus	30.9	19.9	19.8	29.4	1.52
3.	Human liver	30.3	19.5	19.9	30.3	1.53
4.	Human sperm	30.7	19.3	18.8	31.2	1.62
5.	Hen red cells	28.8	20.5	21.5	29.2	1.38
6.	Rat bone marrow	28.6	21.4	21.5	28.4	1.33
7.	Herring sperm	27.8	22.2	22.6	27.5	1.23
8.	*Paracentrous lividus* (sea urchin) sperm	32.8	17.7	18.4	32.1	1.85
9.	Salmon	29.7	20.8	20.4	29.1	1.43
10.	Wheat germ	26.5	23.5	23.0	27.0	1.19
11.	Yeast	31.3	18.7	17.1	32.9	1.79
12.	*Diplococcus pneumoniae*	29.8	20.5	18.0	31.6	1.59

13.	K-12 *Escherichia coli*	26.0	24.9	25.2	23.9	1.00
14.	*Mycobacterium tuberculosis*	15.1	34.9	35.4	14.6	0.42
15.	Bacteriophage T_2	32.5	18.2	16.7	32.6	1.86

These different values of A, G, C and T in different samples of DNA suggests that DNA rather than being a monotonous polymer, carries genetic information in the form of specific nucleotide base sequences.

Physical, Molecular or Geometrical Organization of DNA

In 1938, *W. T. Astbury* was observed three dimensional structure of DNA by X-ray crystallographic study. The X-ray crystallographic studies of Astbury were continued by *Wilkins and his associates*, who managed to prepare highly oriented DNA fibre that allowed them to obtain an X-ray diffractional photograph. One of his female associates *Rosalind Franklin* obtained a superior X-ray diffraction photograph of DNA which confirmed Astbury's earlier inference of 3.4 internucleotide distance. Such studies demonstrated that DNA was a helical structure with a diameter of 20 Å and a pitch (one round) of about 34 Å. *Watson and Crick*, who were already engaged in constructing some suitable model for DNA structure, when observed Franklin's picture of DNA molecule, they immediately utilized that information in construction of a molecular model for DNA. In April 1953, Watson and Crick published their conclusion about the structure of the DNA in the same issue of '*Nature*', in which Wilkins and his colleagues presented the X-ray evidence for that structure.

Watson and Crick's Structural Model of DNA

In DNA molecule, the adjacent deoxyribonucleotides are joined in a chain by phosphodiester bridges or bonds which link the 5' carbon of the deoxyribose of one mononucleotide unit with the 3' carbon of the deoxyribose of the next mononucleotide unit. According to Watson and Crick, DNA molecule consists of two such polynucleotide chains wrapped helically around each other, with the sugar-phosphate chain on the outside (forming ribbon-like backbone of double helix) and purines and pyrimidines on the inside of the helix (projecting between two sugar phosphate backbone as transverse bars). The two polynucleotide strands are held together by hydrogen bonds between specific pairs of purines and pyrimidines.

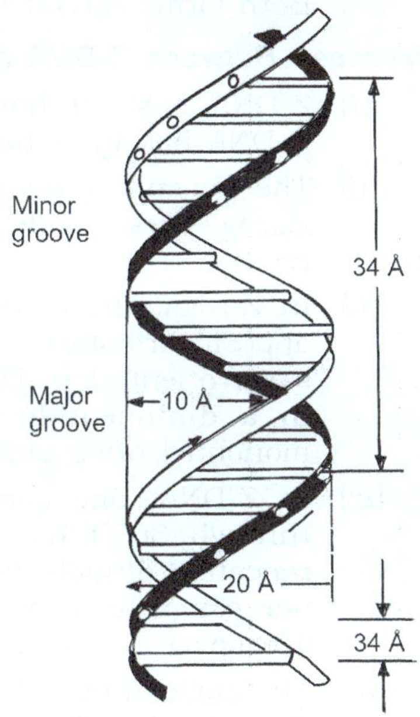

Fig. 5.10. Watson-Crick double helical DNA molecule.

The hydrogen bonds between purines and pyrimidines are such that adenine can bond only to thymine by two hydrogen bonds, and guanine can bond only to cytosine by three hydrogen bonds and no other alternative is possible between them. The specificity of the kind of hydrogen bonds that can be formed assures that for every adenine in one chain there will be thymine in the other. For every guanine in first chain there will be cytosine in the other and so on. Thus, the two chains are complementary to each other, that is, the sequence of nucleotides in one chain dictates the sequence of nucleotides in the other. Two strands run anti-parallely—that is, have opposite directions, one strand has phosphodiester linkage in 3' → 5' direction, while other strand has phosphodiester linkage in just reverse or 5' → 3' direction. Further, both polynucleotides strands remain separated by 20 Å distance. The coiling of double helix is right handed and a complete turn occurs every 34 Å. Since each nucleotide occupies 3.4 Å distance along the length of a polynucleotide strand, ten mononucleotides occur per complete turn. (The base pairs are rotated 36° with respect to each adjacent pair). The helix has two external grooves, a deep wide one, called major groove and a shallow narrow one, called minor groove; both of these grooves are large enough to allow protein molecules to come in contact with the bases.

Z-DNA (or left-handed-DNA)

Crystallog-raphic studies on synthetic nucleotides consisting of alternating purines and pyrimidines such as GCGCGCGC have shown that left-handed-DNA can also exist. This DNA is called Z-DNA because of its zigzag structure. Z-DNA can also be found in solutions of high ionic strength, such as 2M NaCl. In Z-form of DNA, the molecule still consists of two antiparallel chains, but otherwise, it is quite different from A or B form. The helix of Z-DNA is 18 angstorm in diameter, containing 12 base pairs per turn.

The differences and similarities of Z-DNA and B-DNA have been summarized as below:

Similarities between Z-DNA and B-DNA

(*i*) Both are double helical.

(*ii*) In both DNAs, two polynucleotide strands of double helix are antiparallel.

(*iii*) Both forms exhibit G≡C pairing.

Differences Between Z-DNA and B-DNA

(*i*) Z-DNA has left handed helical sense, while B-DNA has right handed helical sense.

(*ii*) The phosphate backbone of Z-DNA follows a zigzag course, while in B-DNA this backbone is regular.

(*iii*) In Z-DNA, the adjacent sugar residues have opposite orientation, while in B-DNA they have same orientation. Due to this, repeating unit in a dinucleotide in Z-DNA as against a mononucleotide unit in B-DNA (Fig. 5.11).

(*iv*) In Z-DNA, one complete helix (*i.e.*, a twist through 360°) has twelve base pairs of six repeating dinucleotide units, while in B-DNA one complete helix has only 10 base pairs or 10 repeating units.

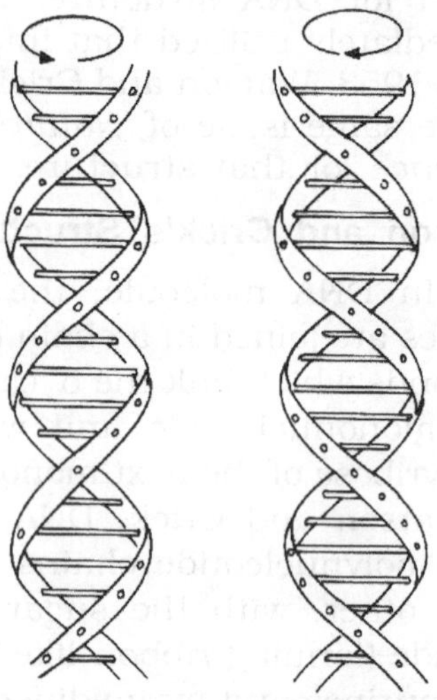

Right handed B-DNA molecule
A

Left handed Z-DNA molecule right left
B

Fig. 5.11. A. The normal right handed helical sense of B-DNA and B. rare left handed helical sense of Z-DNA.

(*v*) The angle of twist (rotation) per repeating unit (dinucleotide) in Z-DNA is 60° than the 36° of mononucleotide in B DNA.

(*vi*) In Z-DNA, one complete turn of helix is 45 Å long, while in B-DNA it is 34 Å long.

(*vii*) Since bases get more length spread out in Z-DNA so that angle of tilt is 60°, as a result they are more closer to the axis. Due to this fact, the diameter of Z-DNA molecule is 18 Å as compared to 20 Å diameter of B-DNA.

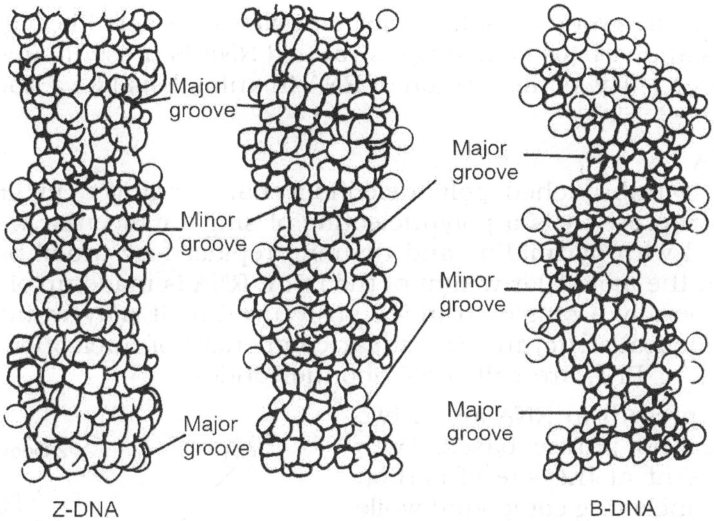

Fig. 5.12. Mode of orientation of adjacent sugar residues. Z-DNA has opposite orientation resulting in dinucleotide unit, while B-DNA has same orientation resulting in mononucleotide units.

Region of Z-DNA may be involved in gene regulation in the cells of "higher" organisations. Short regions of such drastically altered helical geometry could be specifically recognized by gene regulatory proteins, and hereby have important biological roles (see *Alberts et al.* 1989). Lastly, the presence of any alternate configuration (*e.g.*, Z-DNA) suggests that DNA is a more flexible molecule than it was previously thought and that it can adopt in the genome of a variety of forms (*Rich*, 1980).

RIBONUCLEIC ACID OR RNA

The second type of nucleic acid is ribonucleic acid.

Occurrence

Ribonucleic acid is generally found in the cytoplasm and the nucleolus. In the cytoplasm, it is found freely but also found associated with ribosomes. *RNA* is also found in mitochondria,

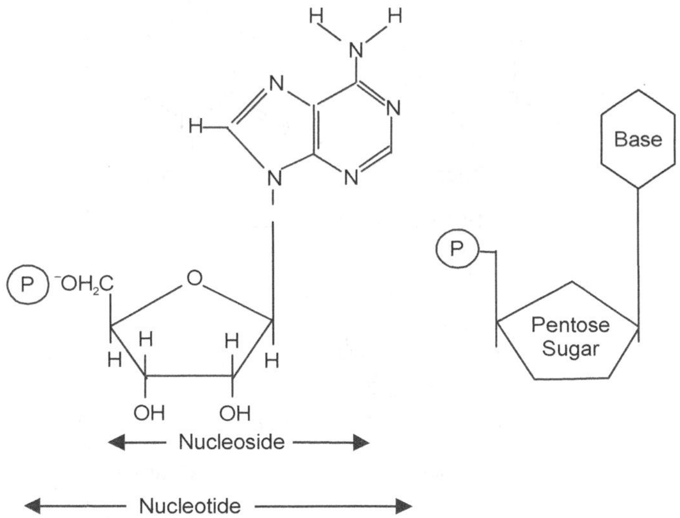

Fig. 5.13. Diagrammatic representation of molecular structure of nucleotide and nucleoside.

chloroplast, and in eukaryotic cells. In some plant viruses is acts like a genetic material. In most of the plants and viruses etc., single stranded RNA is present such as TMV, wound, and tumour virus etc. The configuration of RNA in animals is like a double helix such as Reo viruses etc.

Structure of RNA

Generally it is unbranched polynucleotide chain arranged in linear sequence or single stranded structure. It is a polynucleotide of single molecular weight. In it, ribose sugar is associated with pyrimidine and thymine replaced by uracil base. Its molecular weight is less than the molecular weight of the DNA. RNA is made up of many nucleotides like DNA. It is generally a single stranded structure but it may wound back itself and may form double helical structure. The sugar component of nucleotides of RNA is ribose sugar. Nucleotides of RNA are called as ribonucleotides.

Bases. The bases present in RNA are of two types—pyrimidine and purine bases. Both the bases are present at the site of carbon 1. Pyrimidine is a single ring compound while purine is a double ring compound. Cytosine and uracil—pyrimidines and adenine and guanine—purines are present in RNA.

Nucleoside. Composite form of one molecule of base and one molecule of a sugar is known as nucleoside. Following nucleosides are present in the RNA: (*i*) Cytidine, (*ii*) Uridine, (*iii*) Adenoside, and (*iv*) Guanoside. C_1 position of sugar of a nucleoside joined with N_3 position of the Nitrogen of pyrimidine or N_9 position of purine nitrogen.

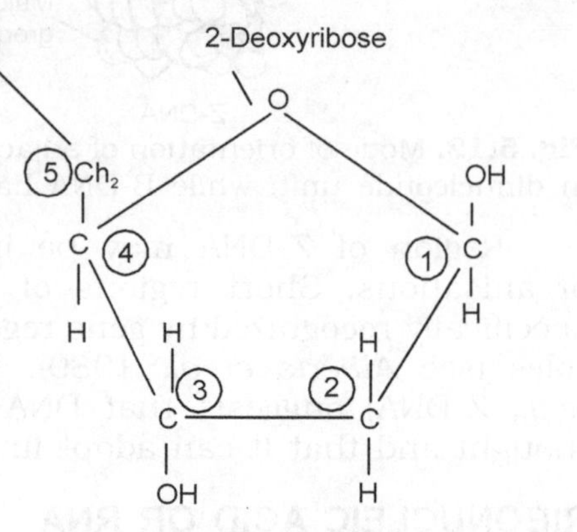

Fig. 5.14. Structural formula of ribose sugar.

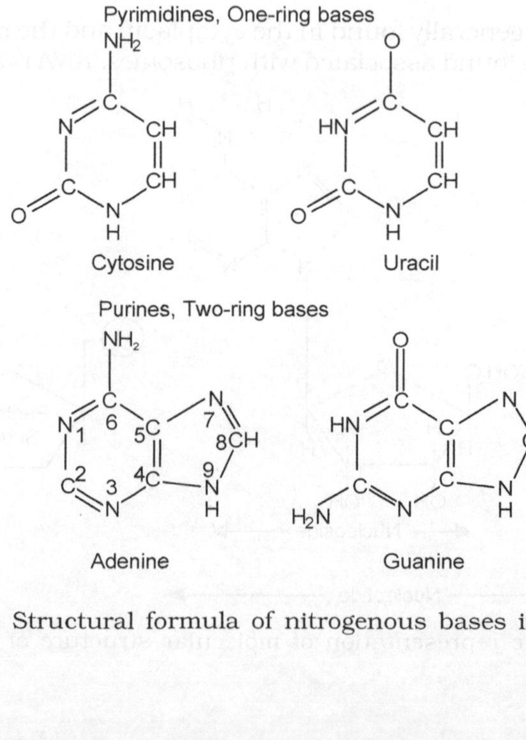

Fig. 5.15. Structural formula of nitrogenous bases in RNA.

Nucleotide. By the combination of one molecule of phosphoric acid with one nucleoside forms a nucleotide. Phosphate group linked with C_5 position of carbon of sugar molecule or C_3 position of carbon group. According to the linkage of phosphate group with sugar, nucleotides are called as 5'P^{3}OH nucleotide or 3'P^{5}OH nucleotides. In ribose sugar, phosphate group can linked at the position of C_2, because at position 2 hydroxyl group is present. But in the living beings phosphate group is linked generally at the position of 5 or at 3rd position in sugar. So, nucleotides present in the RNA are: (*i*) Cytidylic acid, (*ii*) Uridylic acid, (*iii*) Adenylic acid, (*iv*) Guanylic acid.

Types of RNA

Generally DNA is always genetic in nature, while RNA is sometime genetic in nature. So in some organisms RNA transmit hereditary information, but the most of RNAs found in the organisms take part in the reactions taking place in them. So RNAs are of two types—Genetic and non-genetic.

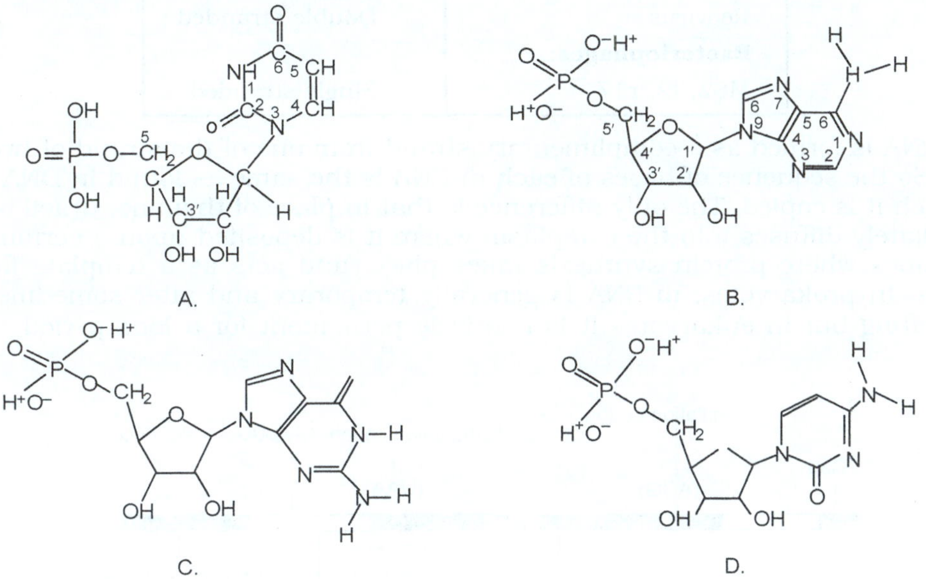

Fig. 5.16. The four nucleotide in RNA. A. Uridylic acid. B. Adenylic acid. C. Guanylic and. D. Cytidylic and.

Genetic RNA

In most of the plant viruses and in some bacterial viruses the genetic material is RNA. RNA contains all those information, which are found in DNA of higher grade of animals. RNA found in viruses might be single stranded or double stranded. When it is double stranded then the base pairing will be in the same way as found in RNA.

Non-genetic RNA

In most of the animals in which DNA is present in the form of genetic material, RNA is found in excessive quantity, but it is of non-genetic nature. Its synthesis is always on DNA template and is of the following three types:

1. Messenger RNA or m-RNA 2. Ribosomal RNA or r-RNA 3. Transfer RNA or t-RNA.

1. **Messenger RNA or m-RNA.** m-RNA is that RNA which carries genetic information from chromosomal DNA to Ribosome for protein synthesis, so it functions as a

messenger, therefore *Jacob* and *Monod* in 1961 named this RNA as m-RNA. It is 5% to 10% of the total quantity of RNA present in the cytoplasm. The molecular weight of an average sized m-RNA is 500000 dalton.

Table 5.3. Different RNA Viruses and the Nature of RNA Associated with them

Virus	*Types of RNA*
Plant Viruses:	
TMV	Single stranded
Wound tumour	Double stranded
Animal Viruses:	
Influenza virus	Single stranded
Rous sarcoma	Single stranded
Poliomyelitis	Single stranded
Reovirus	Double stranded
Bacteriophages:	
MS2, f2, r17	Single stranded

m-RNA is formed as a complimentary strand from one of the strand of two strands of DNA. So the sequence of bases of each m-RNA is the same as found in DNA molecule from which it is copied. The only difference is that in place of thymine, uracil is present. It immediately diffuses into the cytoplasm where it is deposited upon a certain number of ribosomes where protein synthesis takes place; and acts as a template for protein synthesis. In prokaryotes, m-RNA is generally temporary and after sometime it starts disintegrating but in eukaryotes it is relatively permanent for a long period.

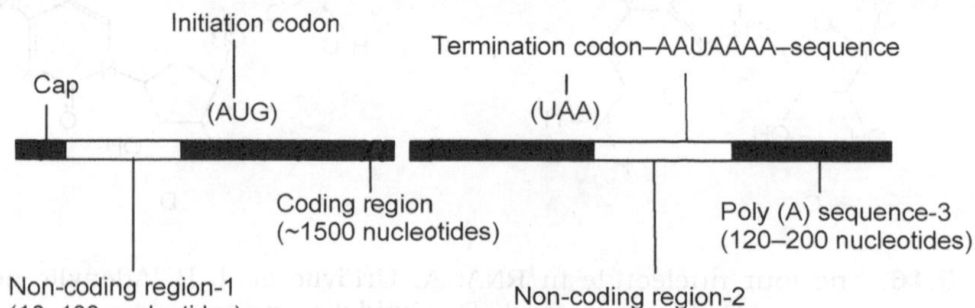

Fig. 5.17. General features of eukaryotic m-RNA.

(i) *Heterogeneity of m-RNA.* The molecules of m-RNA are heterogenous, because they occur in different sizes and have different molecular weight. Their heterogeneity depends on two factors:

1. Size and number of cistron
2. Size of the protein molecule

On the basis of number of cistron, m-RNA is of two types:

(*a*) ***Monocistronic m-RNA.*** This type of m-RNA contains the codon of a single cistron. It can synthesize only one molecule of protein.

(*b*) ***Polycistronic m-RNA.*** This type of m-RNA contains codons of more than one cistron. It can synthesize more than one protein molecules.

(*ii*) ***General Structural features of m-RNA.*** In eukaryotic m-RNA following special structural features are present:

(*a*) *Cap.* In most of the eukaryotic cells and in some animal viruses a 'cap' is found at the 5' end of the m-RNA in which methylated guanine is present. The rate of protein synthesis depends on the presence of a cap. In the absence of cap, m-RNA molecules bind very poorly with ribosomes.

(*b*) *Non-Coding region-1 = NCI.* Immediately after the cap a region of 10 to 100 nucleotides is present. In this region A and U are found in excess quantity and does not translate protein.

(*c*) *Initiation Codon.* In both prokaryotes and eukaryotes primary codon AUG is found.

(*d*) *Coding Region.* This region consists of about 1500 nucleotides. Coding, region also translates protein.

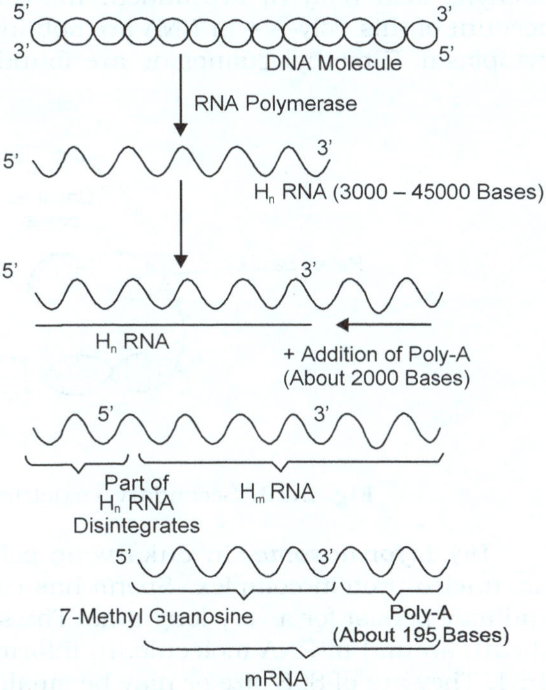

Fig. 5.18. Formation of mRNA from h_n-RNA in eukaryotic cell.

(*e*) *Termination Codon.* The termination of the translation of m-RNA is brought about by termination codon. The termination codon in eukaryotic cells are UAA, UAG and UGA.

(*f*) *Non-Coding Region 2 = NC2.* In this region 50–150 nucleotides are found and it does not translate protein. This region consists of AAUAAA sequence.

(*g*) *Poly-A Sequence.* At the end of 3', polyadenylate or Poly-A sequence is found, which initially consists of 200–250 nucleotides, which reduces gradually with the age. Poly-A sequence is added in the nucleus before the m-RNA.

(*iii*) ***Biosynthesis of m-RNA.*** Synthesis of m-RNA occurs in only one strand amongst the two strands of DNA. Synthesis occurs in 5'-3' direction. RNA polymerase enzyme unites at the initiation site of the DNA cistron, and opens the smallest part of DNA. Now m-RNA is synthesized by DNA. Synthesis of m-RNA is similar to replication of DNA or synthesis of DNA except some differences, such as: (*i*) Instead of deoxyribose nucleotides, ribose nucleotides are used, (*ii*) Adenine combines with uracil only, (*iii*) Only one strand of DNA transcribes the m-RNA.

Before the formation of m-RNA, the transcribed strand undergoes processing. At the time of processing the elements which are no longer of any use are removed and these elements are added which make the m-RNA functionally more effective. Prokaryotic mRNA generally undergoes very small processing. Methylation and Polyadenylation occur in eukaryotic m-RNA.

In eukaryotes the transcribed RNA molecules are larger than mature m-RNA by 20%. They are called as heterogenous nuclear-RNA or hnRNA. By the process of polyadenylation at the 3' end of hnRNA, approximately 200 nucleotide chains of poly-adenylic acid (Poly-A) are added. Immediately after this hnRNA starts breaking. On account of this Poly-A + m-RNA are obtained. Free Poly-A + m-RNA then diffused into the cytoplasm. 7 methyl guanosine are found at the 5' end of m-RNA.

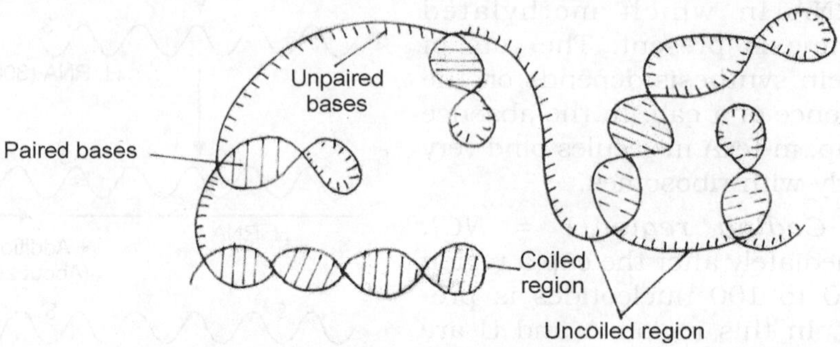

Fig. 5.19. Secondary structure of r-RNA: as an extended strand.

(*iv*) ***Informosome.*** In eukaryotic cells, m-RNA in association with protein forms ribonucleo-protein complex, *Sperin* has called it as Informosome. They are more stable and may persist for a very long time. The stability is attributed by the presence of protein sheath around m-RNA molecule. In informosomes protein and m-RNA are in the ratio of 4 : 1. They are of 80S size or may be small. When there is delay in translation then these are used in protein synthesis.

2. **Ribosomal RNA or r-RNA.** As the name indicates, they are found in ribosomes. It comprises about 80% of the total RNA of the cell. Ribosomes are made up of protein and RNA. In prokaryotes 70S ribosome contains 30S and 50S as two subunits. 16S r-RNA are present in 30S subunit while 23S and 16S r-RNA are present in 50S subunit.

In eukaryotes 80S ribosome have 40S and 60S as two subunits. 18S r-RNA are present in 40S subunits while 28-29S, 5.8S and 5S r-RNA are present in 60S subunits of vertebrate animals. 16-18S r-RNA are present in 40S subunit while 25S and 5S, 5.8 r-RNA are present in 60S subunit of plants and invertebrate animals. In *E. coli* only 23S and 16S r-RNA are present.

Above description reveals that on the basis of sedimentation and molecular weight ribosomal RNA are of three types:

(*a*) The first type includes those r-RNA having high molecular weight of more than a million. *Example*, 21-29S RNA.

(*b*) The second type have high molecular weight r-RNA whose molecular weight is less than one million. *Example*, 12S-18S r-RNA.

(*c*) The third type includes r-RNA of low molecular weight. *Example*, 5S r-RNA.

(*i*) ***Base Composition.*** r-RNA differs in base contents from t-RNA and m-RNA. The quantity of guanine and cytosine is relatively much more. Molar ratio of base components in *E. coli* is of the following type—Adenine 21; uracil 19; guanine 36 and cytosine 23. In many different organisms the base ratio of r-RNA is very similar, which suggests a general structural similarity. On the basis of ionic strength, temperature and pH, r-RNA molecules may be in shapes of short compact rod, a compact coil, or an extended strand. Ribosomal RNA consists of single and the same strand twists upon itself in some parts and forms helix. Helix shows negative or positive interactions. In the helix region, most of the base pairs are complementary and are joined with hydrogen bonds. The rest part is unfolded or single stranded. Complementary bases are absent, so r-RNA are not having purine and pyrimidine bases equally. r-RNA remains stable at least upto two generations.

(*ii*) ***Biosynthesis of r-RNA.*** In prokaryotes 70S ribosome consists of 30S and 50S as two subunits. In 30S subunits 16S r-RNA and in 50S subunit 23S and 5S r-RNA are present. DNA by which r-RNA is transcribed is known as ribosomal DNA. In bacterial gene subunits 16S, 23S and sometimes 5S are arranged in a series. The RNA which is formed from DNA is in the form of 30S subunit is called as P30S. Practical evidences suggest that in 30S subunit 16S part is present at 5' end and 23S at 5' and in between them a spacer unit is present. In some prokaryotes 5S r-RNA is transcribed at or near the 3' end. During processing P30S transcription unit breaks up by RNase III enzymes into 25S and 18S segments. These are then transformed into P23S and P16S by RNase III enzymes. After this, with the help of RNase IV enzyme or maturase enzyme, these are then trimmed into 23S and 16S respectively. In *E. coli* by trimming P16S about 200 nucleotides are removed. Similar changes are observed in nucleoside too. Methylation of nucleoside in 23S r-RNA occur at the P30S stage, whereas modifications found in 16S r-RNA take place at the final stage of processing. In primary r-RNA of 5S r-RNA are only a few nucleotides more than mature 5S r-RNA. In *Bacillus*, two types of large r-RNA precursors are present—(1) P^5 A—is 180 nucleotides long; (2) P^5 B—which is 140–150 nucleotides long. Both these types are transcribed on different 5S gene. P5A, by enzyme RNase M5 breaks up into three parts (*i*) S' terminal pieces of 20 nucleotides, (*ii*) mature 5S r-RNA having 118 nucleotides and (*iii*) a 3' terminal piece of about 40 nucleotides. 5'-3' terminal pieces are broken down to their mononucleotides by an exonuclease enzyme. No action of this enzyme on mature 5S r-RNA. Precursor P^5B splits by RNase M5 into 22 nucleotides 5' terminal pieces and also into 5S r-RNA and some smaller 3' terminal pieces.

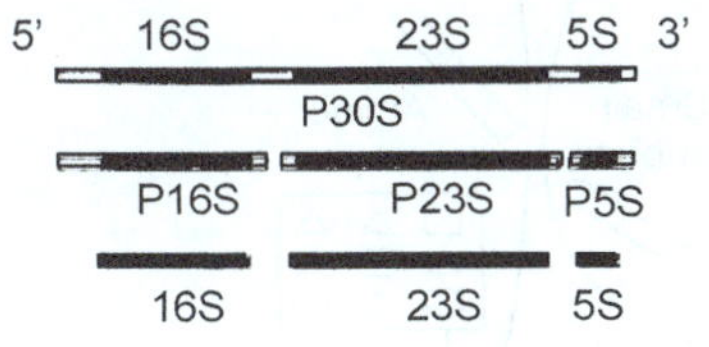

Fig. 5.20. Processing of prokaryote 30S RNA.

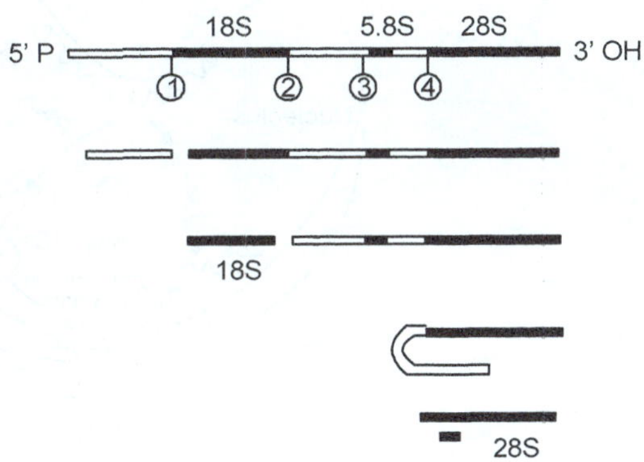

Fig. 5.21. Processing of eukaryotic 43S r-RNA in yield 28S, 18S and 5.8S r-RNA.

In eukaryotes 80S ribosome consists of two subunits 48S and 60S. The 40S subunit consists of 16–18S r-RNA and the 60S subunit consists of 25–28S r-RNA, 5.8S r-RNA, and 5S r-RNA. In eukaryotes the transcription of 28S and 18S takes place in that part of DNA which is present in nucleolus. The transcribed spacer sequence is situated at the 5'P end of 45S r-RNA and 28S component at the 3' OH end. 45S transcriptional unit r-RNA undergoes cleavage at four places. place-I removes the transcribed spacer sequence at 5'P end. Place-II separates 18S and a segment of 140 nucleotides with a sedimentation coefficient of 5.8S, situated in between III and IV place, remains linked to the segment of place-III. Place-IV results in final termination of r-RNA.

5S RNA is transcribed by extra nucleolar organizer and this is attached with larger subunit of ribosome, function of 5S is not known.

18S RNA in combination with alkaline protein forms a small subunit of ribosome which becomes free in cytoplasm. 28S RNA in combination with protein forms a large subunit, which enters into cytoplasm. Along with its 5S r-RNA is attached afterwards. Methylation of 45S takes place in nucleus.

3. **Transfer RNA or t-RNA.** t-RNA molecules are also called as soluble RNA or supernatant-RNA. t-RNA is about 10–15% of the total weight of RNA of the cell. Its molecular weight is 25000 to 30000 dalton. The sedimentation co-efficient of t-RNA is 3.8S. It is made up of 73–93 nucleotides. Each bacterial cell probably consists of about 100 or more different types of t-RNA. The function of t-RNA is to identify amino acids and carry them to m-RNA. It was thought in the beginning that, 20 amino acids are present in protein synthesis, so t-RNA are also 20 in number — for each amino acids one t-RNA is present. But in many events for one amino acids two t-RNA are also present. Therefore, the number of t-RNA is much more than the number of amino acids. t-RNA recognises the codon present on ribosome and m-RNA.

The synthesis of t-RNA is by DNA template in nucleus only 0.025% of DNA codes for t-RNA.

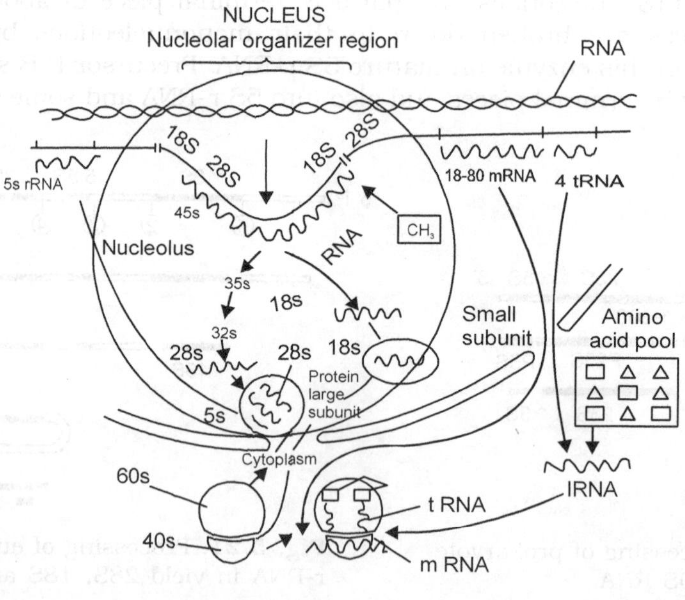

Fig. 5.22. Major events in the transcription of different types.

Structure. On account of its small size its structure has been studied in detail in last few years.

(a) *Two Dimensional Structure of t-RNA or Clover Leaf Model. Robert Holley* in 1965, first of all described the structure of t-RNA. The name of the t-RNA was Yeast alanine t-RNA. After this, many t-RNA were studied. But the basic structural plan of all was studied by Holley and was similar to that of t-RNA. This basic structural plans of the t-RNA was similar to clove leaf (Fig. 5.23), hence it is called as Clover Leaf Model.

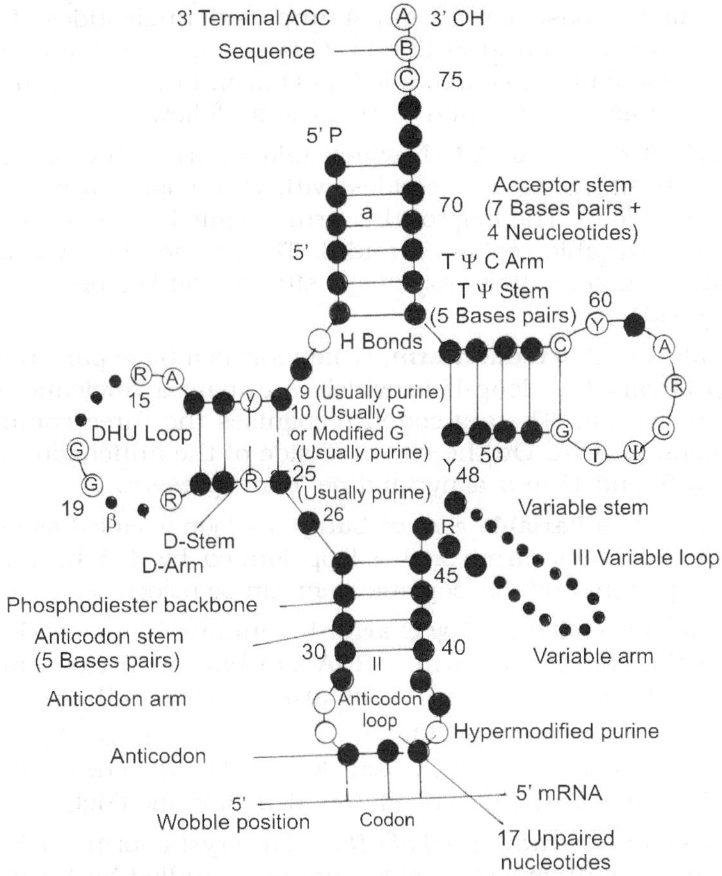

Fig. 5.23. General features of Yeast t-RNA numbering system has been followed.

On the basis of detailed structure of different species of t-RNA, following structural characteristics can be given:

- In all types of t-RNA a large number of irregular nucleosides are present. Some of the important ones are as follows—pseudouridine, Ionosine I, Dihydrouridine (DHU). These bases can be obtained by the methylation of common bases.
- A : U and G : C ratio is about in units; by this it can be concluded that t-RNA also have double helical segments similar to DNA.
- G : C base pairs arc prcscnt in excess.
- Tertiary structure of all t-RNA are present for the stability of their structure, the concentration of Mg ions is necessary.

- In each t-RNA 4 loops—I, II, III IV and V and stems a, b, c, d and e are present.
- According to clover leaf model a single polynucleotide chain of t-RNA is folded back on itself, by which 3' end and 5' end come together. In this way, 5 arms are formed. Each arm contains one stem and one loop. In loop, base pairs are present. Only stem is present in one arm and loop is absent. This arm is called as acceptor arm. CCA sequence is present at 3' end, on which a special amino acid is attached with the help of Amino acyl synthetase enzyme. This arm contains 7 base pairs, and 4 unpaired nucleotides. In four unpaired nucleotides CCA is found at 3' end. Guanine or cytosine is present at 5' end of t-RNA. Rest four arms are called as O arm, anticodon arm, variable, T ψ C arm. The detailed description of them is as follows:

Second arm is called as **D or DHU,** which takes part in the binding of aminoacyl synthetase. It consists of 15–18 nucleotides with 3–4 base pairs in stem and 7–11 unpaired nucleotides in loop. The loop of this arm is called as Loop-I or Dihydrouridine loop and contains two variable regions α and β. The amino acyl synthetase site which recognises the amino acid activating enzymes is situated on D loop and some part of the acceptor stem on 5' side.

Third arm is called as **Anticodon arm,** which forms a base pair from specific codon of m-RNA. Its loop is called as loop-II in which 7 unpaired nucleotides are present of which three forms anticodon. The anticodon recognises the three complimentary bases which forms the codon m-RNA. On the site of 3' side of the anticodon is Hyper modified Purine (H), while on 5' end U and a pyrimidine Y are present.

Fourth arm is called as **Variable arm of Lump.** Its loop is called as loop-III. Sometime stem is not present in variable arm, only a loop formed by 4–5 bases. In other t-RNA 13–21 nucleotides are present and the loop and stem can be recognised easily and separately.

Fifth and final arm is called as **T ψ C arm.** It contains 7 nucleotides and stem with 5 bases pairs. With the help of this arm, t-RNA can bind with ribosome. Therefore, its loop is called as Ribosomal binding loop. This arm recognises the ribosome.

Beside usual bases A, U, G and C, t-RNA consists of some unusal bases-Methyl guanine (GMe), Dimethyl guanine (GMe2), Methyl Cytosine (CMe), Ribothymine (T), Pseudo uridine (ψ), Dihydro uridine (DHU, H@U, UH2), Inosine (I) and Methylinosine (MeI).

(*b*) *Three Dimensional Structure (TDS) of t-RNA.* The crystal form of t-RNA was available in 1968 after that its three dimensional structure was studied by X-ray cyrstallography method. For its study various models were proposed and the model which was proposed by *S.H. Kim* in 1973 was acepted unimously. According to this TDS of t-RNA is somewhat like alphabet of english vernacular 'L' shaped having its thickness 20 Å.

Classification. Study of various t-RNA shows that the structure of acceptor stem, anticodon arm and T ψ C arm are constant. But differences are present in between D and variable arms. On the basis of these differences, t-RNA is classified into following three classes:

Class-I = D stem with 4 base pairs and in variable loop 4–5 bases.
Class-II = D stem with 3 base pairs and in variable loop 4–5 bases
Class-III = D stem with 3 base pairs and large variable arm.

On the basis of variable arm t-RNA is of two types:

Class-I = The variable loop of it have 4–5 base pairs.
Class-II = The arm is large with 13–21 bases.

Its origin from two dimensional clover model can be easily done. In the conformation of TDS, its arm can be extended without any deformity in it. Its CCA stem is projected out and can be arranged separately.

Biosynthesis of t-RNA. DNA of the nucleus form precursor t-RNA through RNA Polymerase. It consists of 120 –130 bases. After processing it is transformed into mature t-RNA. Processing of precursor t-RNA takes place by many processes, in which some nucleotides are added and some are transformed. The cleavage of precursor t-RNA is by RNase P and RNase P_2 enzymes.

Table 5.4. Comparison between m-RNA, t-RNA and r-RNA

S. No.	*Characters*	*m-RNA*	*t-RNA*	*r-RNA*
1.	Base of relationship with DNA	m-RNA shows relationship to DNA. It is formed from all regions of DNA.	No relationship with DNA. t-RNA is formed from only small part of the DNA.	Similar as in t-RNA
2.	Number of nucleotides	900 to 1500 nucleotides- *E. coli*	73 to 93 nucleotides.	5S RNA = 120 nucleotides, 16–18S RNA = 1600–2500 nucleotides, 23–28S RNA = 3200–5200 nucleotides.
3.	Molecular weight	500000	20000-30000	1.1×10^4 = 23S RNA, 0.5×10^6 = 30S RNA
4.	Unusual bases	Small	High	Small
5.	Percentage of total RNA of cell	3-5%	10–20%	80%
6.	Beginning of synthesis	m-RNA are present in ovum. New m-RNA is synthesized during cleavage.	At the end of cleavage t-RNA synthesis occur.	At Gastrulation.
7.	Site of synthesis	In nucleous on DNA template.	Similar to m-RNA	From nucleolar DNA
8.	Sedimentation coefficient	8S	3.8S	28S, 18S, 5.8S, 5S, 23S, 16S, 5S
9.	Function	Transmit genetic information from DNA to ribosome where it takes part in Protein Synthesis.	Helps in the attachment of amino acids to m-RNA template	Basis of m-RNA and t-RNA binding to ribosome.

Difference between DNA and RNA

Deoxyribose Nucleic Acid (DNA)	*Ribonucleic Acid (RNA)*
Generally present in chromosomes.	Generally present in cytoplasm and some quantity of it also present in nucleolus and nucleoplasm.
In its one molecule of double strands (helix) of polynucleotides are present which are spirally coiled and twisted like a rope.	Normally a single strand of Polyribotides.
It contains deoxyribose sugar.	It contains ribose sugar only.
It contains pyrimidines and purines in equal quantity.	It is not necessary that these may be present in equal quantity.
Base pairing—adenine with thymine and guanine with cytosine.	In base pairing thymine base replaced by uracil base.
In DNA the base composition is A/T = G/C = 1	It is not the same in RNA.

6 STRUCTURE OF DNA

DEOXYRIBOSE NUCLEIC ACID: AN INTRODUCTION

DNA is formed in the cells of all animals, plants, prokaryotes and in most viruses. In eukaryotes, it is present in combination with proteins forming nucleoproteins. In prokaryotes (*e.g., E. coli*), the genetic material consists of a single giant molecule of DNA about 1000 microns in length, without any associated proteins. DNA is present mainly in the chromosomes. It has also been reported in cytoplasmic organelles like mitochondria and chloroplasts. The DNA of all plants and animals and many viruses (polyoma virus, smallpox virus, bacteriophages T2, T4 and T6) is double stranded. In the bacteriophage ϕ X 174, however, it is single stranded. In some viruses, the genetic material is RNA as in the tobacco mosaic virus (TMV). In tobacco virus, influenza virus, poliomyelitis virus and the bacterial viruses F2 and R17, the RNA is single stranded. In the reovirus and the wound tumour virus (in bacteria) and in higher plants and animals both DNA and RNA are present. Viruses usually contain either DNA or RNA.

Table 6.1. Genetic Material of Organisms.

Double stranded DNA (dsDNA)	Higher animals and plants, bacteria Polyoma virus and smallpox virus The T-even bacteriophages (T2, T4, T6)
Single Stranded DNA (ssDNA)	The bacteriophage ϕX 174 and several bacterial viruses
Double Stranded RNA (dsRNA)	Reo group of viruses Wound tumour virus
Single Stranded RNA (ssRNA)	Tobacco mosaic virus A tobacco virus Poliomyelitis virus Bacterial viruses F2 and F17.

STRUCTURE OF DNA

The most accepted molecular model of DNA is the double helix structure proposed by *Watson* and *Crick* (1953). The DNA molecule consists of two helically twisted strands connected together by 'steps' (Figs. 6.1 and 6.2). Each strand consists of alternating molecules of deoxyribose (a pentose sugar) and phosphate groups. Each step is made up of a double ring purine base and a single ring pyrimidine base. The purine and pyrimidine bases are connected to deoxyribose sugar molecules. The two strands are intertwined in a clockwise direction, *i.e.*, in a right hand helix, and run in opposite directions. The strand completes a turn each 34Å. Each nucleotide occupies 3.4Å. Thus, there are 10 nucleotides per turn. Each successive nucleotide turns 36 degrees in the horizontal plane. The width of the DNA molecule is 20Å. The twisting of the strands results in the formation of *deep* and *shallow spiral grooves.*

The DNA molecule is a polymer consisting of several thousand pairs of nucleotide monomers. Each nucleotide consists of the pentose sugar deoxyribose, a phosphate group and a nitrogenous base which may be either a purine or a pyrimidine. Deoxyribose and a nitrogenous base together form a nucleoside. A nucleoside and a phosphate together form a nucleotide.

Nucleoside = Deoxyribose + Nitrogenous base

Nucleotide = Deoxyribose + Nitrogenous base + Phosphate.

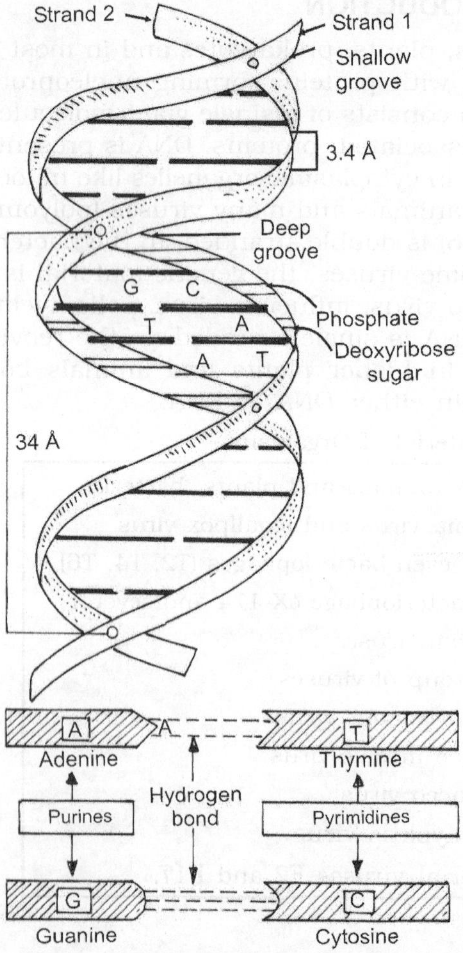

Fig. 6.1. Diagramatic representation of the DNA molecule.

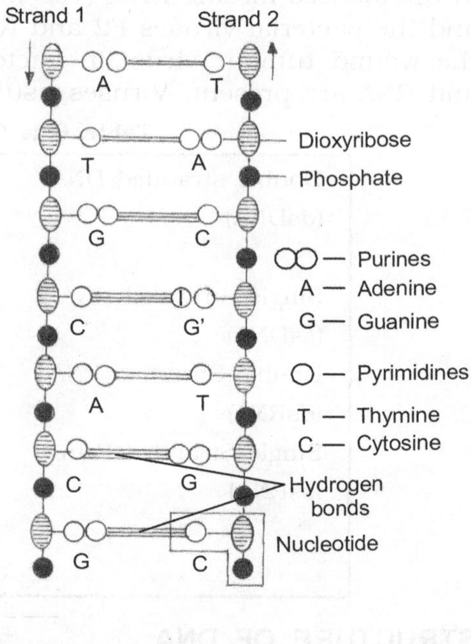

Fig. 6.2. The DNA double helix.

Deoxyribose is a Pentose Sugar with Five Carbon Atoms

Four of the five carbon atoms plus a single atom of oxygen form a five membered ring. The fifth carbon atom is outside the ring and forms a part of a CH_2 group. The four atoms of the ring are numbered 1' 2' 3' and 4'. The carbon atom of CH_2 is numbered 5'. There are three OH groups in positions 1', 2', 3' and 4'.

Ribose, the pentose sugar of RNA, has an identical structure except that there is an—OH group instead of H on carbon atom 2'.

All the sugars in one strand are directed to one end, *i.e.*, the strand has polarity. The sugars of the two strands are directed in opposite directions.

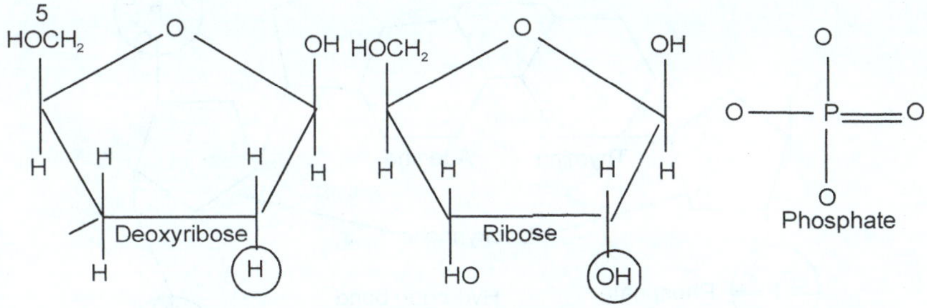

Fig. 6.3. Deoxyribose, ribose and phosphate.

Nitrogenous Bases

There are two types of nitrogenous bases, pyrimidines and purines. The pyrimidines are single ring compounds, with nitrogen in positions 1' and 3' of a 6 membered benzene ring. The two most common pyrimidines of DNA are cytosine (C) and thymine (T). The purines are double ring compounds. A purine molecule consists of a 5 membered imidazole ring joined to a pyrimidine ring at positions 4' and 5'. The two most common purines of DNA are adenine (A) and guanine (G).

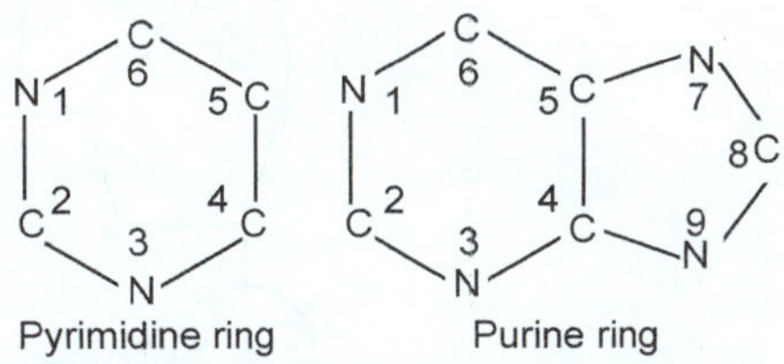

Fig. 6.4. Pyrimidine and purine rings.

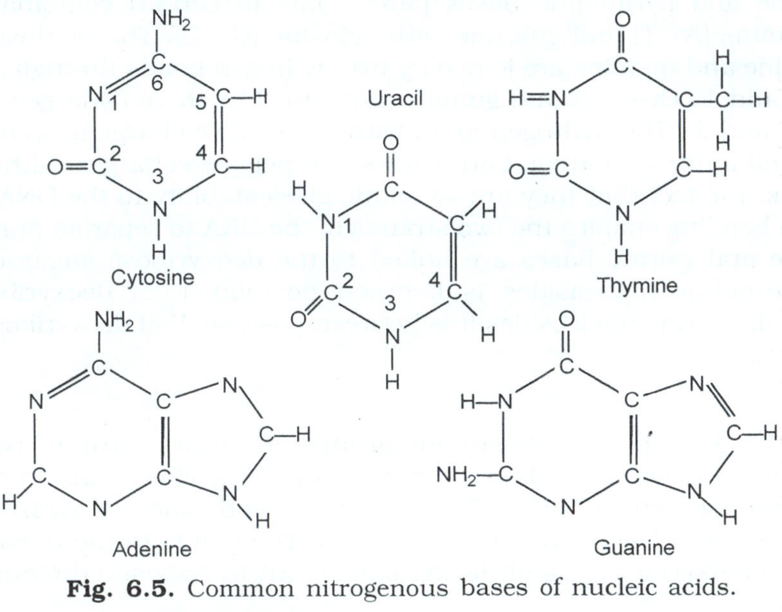

Fig. 6.5. Common nitrogenous bases of nucleic acids.

Base Pairing

Each 'step' of the DNA ladder is made up of a purine and a pyrimidine pair, *i.e.*, of a double ring and a single ring compound. Two purines would occupy too much space, while two pyrimidines would occupy too little. Because of the purine-pyrimidine pairing, the total number of purines in a double stranded DNA molecule is equal to the total number of pyrimidines. Thus A/T = 1 and G/C = 1 or A + G = C + T. The ratio A + T/ G+C, however, rarely equals 1, and varies with different species from 0.4 to 1.9. This ratio is commonly low in microorganisms and high in higher animals.

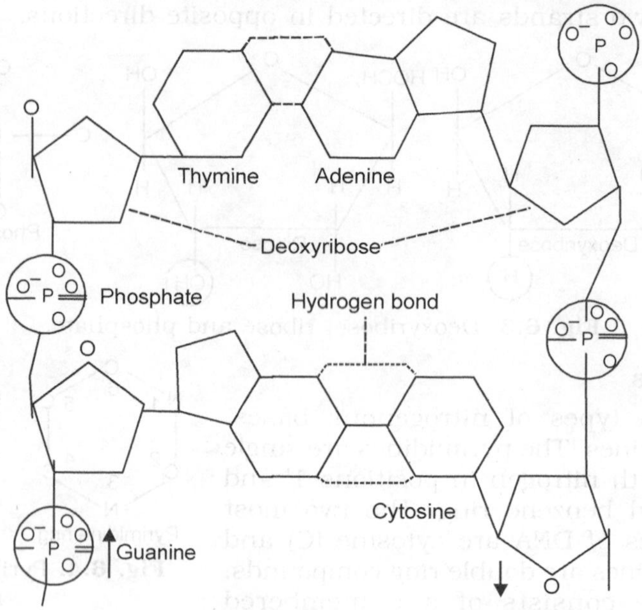

Fig. 6.6. Segment of DNA molecule showing pairing between two nucleotides.

The purine and pyrimidine bases paired only in certain combinations. Adenine pairs with thymine (A : T) and guanine with cytosine (G : C). The total width of the pair is 10.7Å. Adenine and thymine are joined by two hydrogen bonds through atoms attached to positions 6' and 1'. Cytosine and guanine are joined by three hydrogen bonds through positions 6', 1' and 2'. The hydrogen atom with its positive charge is shared between an oxygen atom and a nitrogen atom, both with slight negative charges. Although hydrogen bonds are weak, the fact that they are so many, gives stability to the DNA molecule. The weak hydrogen bonding enables the two strands of the DNA to separate during replication. The pyrimidine and purine bases are linked to the deoxyribose sugar molecules. The linkage in pyrimidine nucleosides is between position 1' of deoxyribose and 3' of the pyrimidine. In purine nucleosides it is between position 1' of deoxyribose and position 9' of the purine.

Phosphate

In the DNA strand the phosphate groups alternate with deoxyribose. Each phosphate group is joined to carbon atom 3' of one deoxyribose and to carbon atom 5' of another. Thus each strand has a 3' end and a 5' end. The two strands are oriented in opposite directions. The 3' end of one strand corresponds to the 5' end of the other. Consequently the oxygen atoms of deoxyribose point in opposite directions in the two strands.

Nucleosides and Nucleotides

A sugar molecule and a nitrogenous base form a nucleoside and a nucleoside plus a phosphate group form a nucleotide.

In other words a nucleoside is a base sugar combination and a nucleotide is a nucleoside phosphate. The nucleotides of RNA are called ribonucleotides, and those of DNA deoxyribonucleotides. Ribonucleotides contain the sugar ribose, and deoxyribonucleotides the sugar deoxyribose.

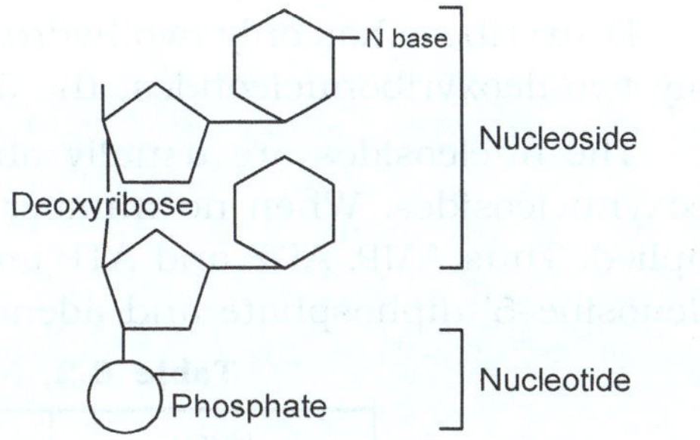

Fig. 6.7. A nucleoside and a nucleotide.

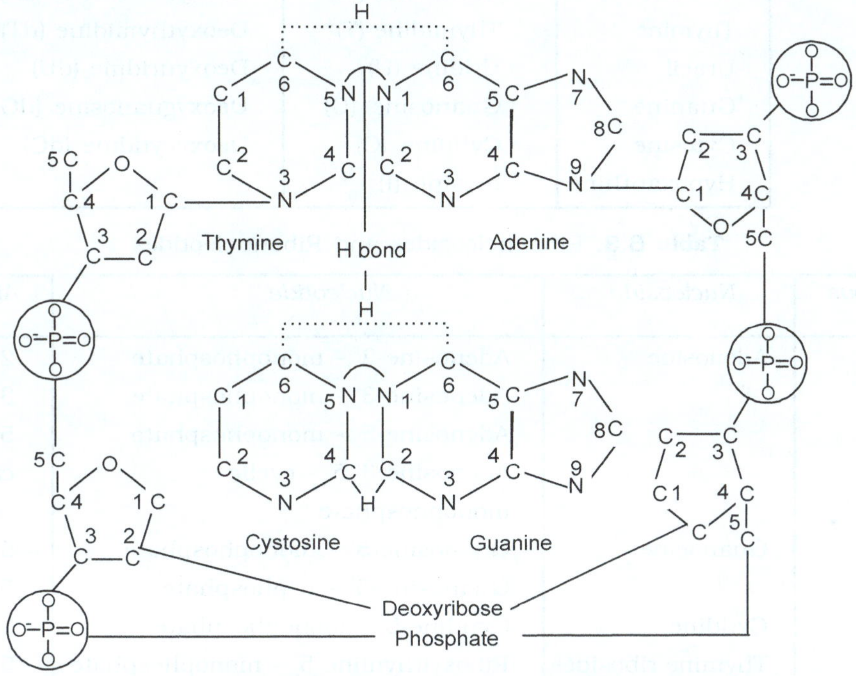

Fig. 6.8. Two pairs of molecules showing bonding between phosphate, deoxyribose and nitrogenous bases.

The ribose in a nucleoside has free hydroxyl groups on carbon 2' 3' and 5'. The phosphate can be attached at any of these positions. Although all the three nucleoside, monophosphate have been found, the most commonly occurring ones are 5' phosphates and 3' phosphates.

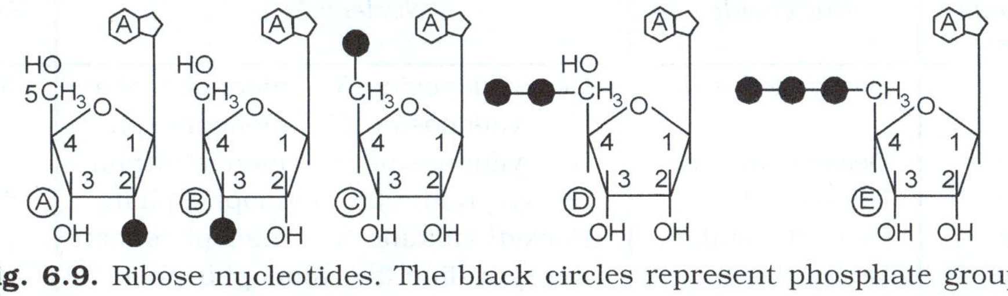

Fig. 6.9. Ribose nucleotides. The black circles represent phosphate groups.

A. Adenosine – 2' — monophosphate B. Adenosine – 3' — monophosphate
C. Adenosine – 5' — monophosphate (AMP) D. Adenosine – 5' — diphosphate (ADP)
E. Adenosine – 5' — triphosphate (ATP)

Deoxyribose has only two hydroxyl groups (on carbons 3' and 5') and can thus form only two deoxyribonucleotides, the 3' and the 5' phosphate derivatives.

The nucleosides are usually abbreviated as A, T, G or C, with d as a prefix for deoxynucleosides. When no number appears before the abbreviation the 5' — ester is implied. Thus AMP, ADP and ATP are abbreviations for adenosine – 5' monophosphate, adenosine–5' diphosphate and adenosine–5, — triphosphate, respectively.

Table 6.2. Nucleosides of RNA and DNA

Base	*Ribonucleosides (RNA)*	*Deoxyribonucleosides (DNA)*
Adenine	Adenosine (A)	Deoxyadenosine (dA)
Thymine	Thymidine (T)*	Deoxythymidine (dT)
Uracil	Uridine (U)	Deoxyuridine (dU)
Guanine	Guanosine (G)	Deoxyguanosine (dG)
Cytosine	Cytidine (C)	Deoxycytidine (dC)
Hypoxanthine	Inosine (I)	

Table 6.3. Ribonucleosides and Ribonucleotides

Nitrogenous base	*Nucleoside*	*Nucleotide*	*Abbreviation*
Adenine	Adenosine	Adenosine-2' - monophosphate	2'-AMP
		Adenosine-3' - monophosphate	3'-AMP
		Adenosine-5' - monophosphate	5'-AMP
		Adenosine-3',5' - cyclic monophosphate	cAMP
Guanine	Guanosine	Guanosine-5' - monophosphate	5'-GMP
		Guanosine-5' - triphosphate	5'-GTP
Cytosine	Cytidine	Cytidine-5' - monophosphate	5'-CMP
Thymine	Thymine ribosides (Ribosylthymine)	Ribosylthymine-5' - monophosphate	5'-rTMP
Uracil	Uridine	Uridine-5' - monophosphate	5'-UMP

Table 6.4. Deoxyribonucleosides and Deoxyribonucleotides

Nitrogenous base	*Nucleoside*	*Nucleotide*	*Abbreviation*
Adenine	Deoxyadenosine	Deoxyadenosine 3' – monophosphate	3'-dAMP
		Deoxyadenosine 5' – monophosphate	5'-dAMP
Guanine	Deoxyguanosine	Deoxyguanosine 3' – monophosphate	5'-dGMP
Cytosine	Deoxycytidine	Deoxycytidine 3' – monophosphate	5'-dCMP
Thymine	Deoxythymidine	Deoxythymidine 3' – monophosphate	3'-dTMP
Uracil	Deoxyuridine	Deoxyuridine 3' – monophosphate	3'-dUMP

* Formerly it was believed that thymine occurred only in DNA, and so the term thymidine was used without the deoxy-prefix to describe the nucleoside of DNA.

Polynucleotides are Polymers of Nucleotides: They are designated by prefixing 'poly' to the repeating unit. *Examples*:

Poly (A) — polyadenylate or polyadenylic acid.
Poly (T) — polythymidylate or polythymidylic acid.
Poly (G) — polyguanidylate or polyguanidylic acid.
Poly (C) — polycytidylate or polycytidylic acid.
Poly (U) — polyuridylate or polyuridylic acid.
Poly (A-U) — alternating copolymer of A and U.
Poly (A, U) — random copolymer of A and U.
Poly (A_2, U) — random copolymer of A and U with average composition 2 : 1.
Poly (A_2, U) 150—as above, with average length of chain, 150 residues.

Polynucleotides can form double-stranded, helical, Watson-Crick complexes. They can also from helical structures with two or more strands which may or may not have Watson-Crick base pairs, *e.g.*, Poly (A) and Poly (I).

Polynucleotides containing the same repeating unit are known as homopolynucleotides, *e.g.*, poly (A) and poly (T).

Unusual Bases in DNA

Although the bases most commonly present in DNA are adenine, guanine, cytosine and thymine, other bases have also been found. In some viruses (*e.g.*, PBS1 and PBS2) uracil occurs in place of thymine in DNA. Also, in some bacterial viruses (bacteriophages) cytosine is replaced by 5-hydroxymethyl-cytosine (HMC). The variations of C, A and G in DNA can be considered to be the result of methylation of these bases.

Single Stranded DNA

Although DNA of most organisms consists of two strands, single stranded DNA is present in the bacteriophage virus ϕX174 and in several other bacterial viruses. Single stranded DNA differs from double stranded DNA in the following respects:

1. Ultra-violet absorption of double stranded DNA remains constant from 0°C to 80°C, and then rises rapidly (80°C, is the critical melting point). In single stranded DNA, ultra-violet absorption increases steadily from 20°C to 90°C.
2. Double stranded DNA is resistant to the action of formaldehyde. In single stranded DNA the reactive sites are exposed, and therefore it is not resistant.
3. In double stranded DNA A = T and G = C. In single stranded DNA, ϕ × 174 of the proportion of A : T : G : C is 1 : 1. 33 : 0. 98 : 0.75.
4. Double stranded DNA is linear while single stranded DNA is circular. During replication, single stranded DNA becomes double stranded (replicative form).

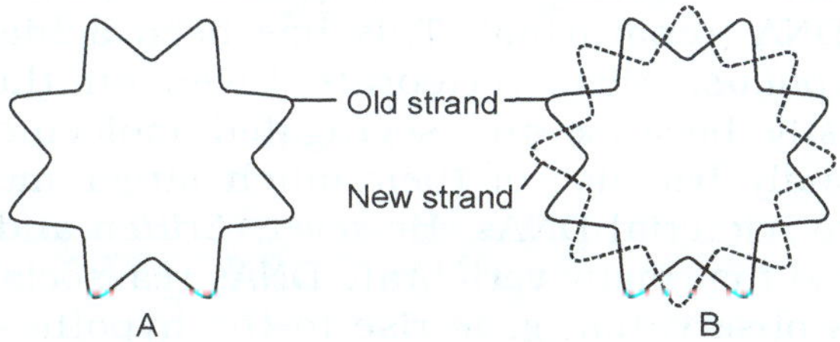

Fig. 6.10. A. Single stranded DNA B. Single stranded DNA During replication.

Forms of DNA

There are six different morphological forms of DNA double helix have been described. Such as A, B, C, D and Z forms. Except B and Z the other forms occur in rigidly controlled experimental conditions. Some forms of DNA are interconvertible. The difference in these DNA forms are associated with:

(*i*) The number of base pairs present in each turn of DNA helix.

(*ii*) The angle between each base pair.

(*iii*) The helical diameter of DNA molecule.

(*iv*) The handedness of double helix.

Sugar puckering is the most important characteristic for distinguishing the DNA forms. The A form has 3'-endo puckering while the B and C form have 3'-exopuckering, and 2'-endo puckering respectively.

The B form (B-DNA) is the structure proposed by Watson and Crick, and is the native conformation of DNA in solution. The B-form X-ray diffraction in pattern is obtained at high humidity (>66%) and in the presence of excess salt. It consists of a right handed antiparallel double helix of sugar phosphate backbone, with purine pyrimidity.

(λ) to the normal of the helix is 6.3°. One turn of the helix consists of 10 base pairs (10 fold helix). The rise of the helix per base pair (*h*) is 3.37Å.

The A form (A-DNA) has 11 base pairs (11 fold helix). The base pairs are considerable tilted for the axis of the helix (γ = 20.2°). Because of this displacement, the depth of the deep groove is increased and that of the shallow groove decreased. The axial rise (*h*) is only 2.56 Å. The B form is observed at lesser hydration and salt content than the A form.

Reduction of hydration of the B form below 66%, with excess of salt still present, results in the C form (C-DNA). The helix symmetry of the C form is 28/3, or 9½ *i.e.*, there are fewer residues per turn than in the B form. The axis of the helix is located in or near the minor groove, and there is a pronounced negative tilt of the base pair plane (γ = 7.8°). The helix rise per base pair is 3.32 Å.

The D form is the metabolically stable configuration. It undergoes transition to the A, C or D conformations at relative humidity lower than 92%, depending upon the nature of the counter ion, the nucleotide sequence and the concentration of excess salt. D→A transition has been observed at low relative humidities. The A form is, however, metastable and rapidly returns to the D form. RNA always occurs in the A form only, RNA-DNA hybrids are also in the A form leading to the belief that DNA must undergo a B→A transition during transcription.

REPETITIVE DNA (SATELLITE DNA)

If DNA is subjected to heat it becomes denatured, *i.e.*, the two strands of double helical DNA separate. Under certain conditions, the two strands reassociate to form active double helical DNA (renaturing). This has been achieved experimentally in the bacterium *Haemophilus*. Most vertebrate DNAs on the other hand do not reassociate easily; firstly because the segragated molecules form heterogenous populations, and secondly because of their much larger molecular weights (upto X 1000 as compared to bacterial DNAs. However, *Britten* and *his coworkers* (1966, 1968) have demonstrated that many vertebrate DNAs reassociate, especially if broken into small pieces. This observation gave rise to the hypothesis that, 'certain short sequences of bases are repeated hundreds of times. Such DNA has been called repetitive DNA or satellite DNA'.

Repetitive DNA consists of short identical genes which are repeated in tandem several hundred or thousand times. Such DNA is found in the region of the chromosome adjacent to the centromere. In many cases the base competitions of the repeating sequences are unlike that of the rest of the DNA. It is therefore easy to separate repetitive DNA by ultracentrifugation.

In *Drosophila virtis* DNA, there are three other highly repetitive satellite DNAs. These satellite DNA can be isolated by density gradient centrifugation in neutral caesium chloride (CsCl) as they have distinctive bouyant densities. Each satellite DNA consist of a repeating sequence of seven nucleotide pairs.

Table 6.5. Satellite DNAs of Drosophila virtis (Gall *et. al.* 1974)

Repetitive DNA	*Bouyant Density*	*Repeat Sequence*
Satellite DNA I	1692	5'-ACAAACT-3' 3'-TGTTTGA-5'
Satellite DNA II	1688	5'-ATAAACT-3' 3'-TGTTTGA-5'
Satellite DNA III	1671	5'-ACAAATT-3' 3'-TGTTTT

All eukaryotes, except perhaps yeast, contain repetitive DNA. In *Drosophila*, about 25% of the DNA is of the repetitive type. In humans, 30% of the DNA contains sequences which are repeated at least 20 times.

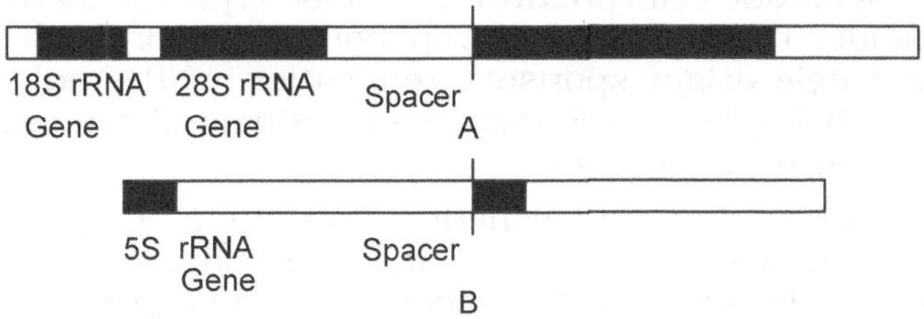

Fig. 6.11. Repetitive genes in *Xenopus laevis*. A. Genes for 18S and 28S rRNA separated by spacer region. B. Genes for 5S rRNA separated by spacer region.

The function of highly repetitive DNA is not known. This DNA can replicate but cannot transcribe RNA for protein synthesis. This is probably because the abort sequences lack promoter sites on which RNA chains can be initiated by RNA polymerase. Repetitive DNA is therefore inert and is partly dispensable.

In the African elawed toad *Xenopus laevis*, the genes for 40S precursor RNA which gives rise to 28S and 18S RNA, are repeated about 450 times (Fig. 6.11). The genes are tandemly arranged and are separated by spacer regions of about 5000 base pairs. Genes for 5S rRNA, also separated by spacer regions, are arranged in clusters of 100 to 1000 repetitive units at the ends of most of the 18 chromosomes.

7 DNA Replication in Prokaryotes and Eukaryotes

INTRODUCTION

Cell has only one chromosome in prokaryotes or has many chromosomes as in eukaryotes. The whole genome must be replicated. Replication has started, it continues until the entire genome has been duplicated.

The unit of DNA in which an individual act of replication occurs is called a replicons.

In prokaryotes the initiation of replication is a single event involving a unique site on the bacterial chromosomes but in eukaryotic cells, initiation of replication is identified by the start of S-phase a protracted period, during which DNA synthesis occurs.

A genome in a prokaryotic cell constitutes a single replicon, so the units of replication and segregation coincide. The largest such replicon is that of the bacterial chromosome itself. Initiation at a single origin sponsors replication of the entire genome, once for every cell division. Each haploid bacterium has a single chromosome, so this type of replication control is called single copy.

Bacteria may contain additional genetic information in the form of plasmids. *A plasmid is an autonomous circular DNA genome that constitutes a separate replicon.* A plasmid replicon may replicate with the bacterial genome *pari passu* (showing single-copy control) or may be under a different type of control. When the number of copies of a plasmid is greater than that of the bacterial chromosome, it is said to replicate under multicopy control.

REPLICATION OF DNA

The most important feature of DNA is that it can make same copies of itself. It is called replication which is the basis of life. Two strands of DNA double helix are united by hydrogen bonds between the purine and pyrimidine. When the hydrogen bonds break the two strands, are separated and unwinded. Adenine, Guanine, cytosine and thymine are present in nucleus of free nucleotide. These free nucleotides unites with two separated strands of DNA by hydrogen bonds. The free adenine pairs with thymine and guanine pair with cytosine. In this way, new strand is form which is an identical copy of the old strand of double helix DNA. The result of this double helix so formed have one parental strand and the other one is called as offspring. This process is called semiconservative.

EXPERIMENTAL EVIDENCE FOR SEMICONSERVATIVE DNA REPLICATION

Meselson and Stahl's Experiment

M. Meselson and *F. W. Stahl* in 1958 had publised the results of their experiments, it was done to observe that DNA can replicate by semi-conservative methods. For this experiments, Meselson and Stahl have cultured a species of bacteria *E. Coli.* They have cultured the cells of *E. coli* in a culture medium containing N^{15} isotope of nitrogen for 14 generations. So that each nitrogen of the nitrogen bases of DNA instead of N^{14} may be depicted as N^{15}. After this, these cells were suddenly transferred into a cultured medium of N^{14}. If the replication is completed by semi-conservative methods then in the next generation of DNA one strand will be N^{15} and second one as N^{14}. After this by replication method the future generation may have N^{14} respectively or density may be reduced gradually.

After this to separate heavy DNA from light DNA, the desnity gradient equillibrium centrifugation mehtod may be used. DNA is put into concentrated solution of 8.8M CsCl in water and then putting the solution into a transparent quarts tube and centrifuged at high speed for 5 to 8 hours approximately. When centrifugal force starts functioning over CsCl it starts forming sediments because CsCl is much more dense than water. The nature of this sedimentation is opposed by the diffusion but after some hours equillibrium position arises. The distribution of CsCl inside the tube is such that more concentrated density is towards the bottom of tube and the lower density is towards the upper part of the tube, and all the medium density is found in between these two. The DNA molecule of CsCl present in the tube assemble or accumulate at that density gradient point where the density gradient is similar or equal to that of DNA molecule. The definite position of DNA is established by the ultraviolet rays. In it the photo of tube is taken by ultraviolet rays because these rays are absorbed only by DNA band and not by CsCl. If density gradient of CsCl is seen or observed in the form of DNA dark band, according to the position of this DNA band, the density of DNA present in it can be known.

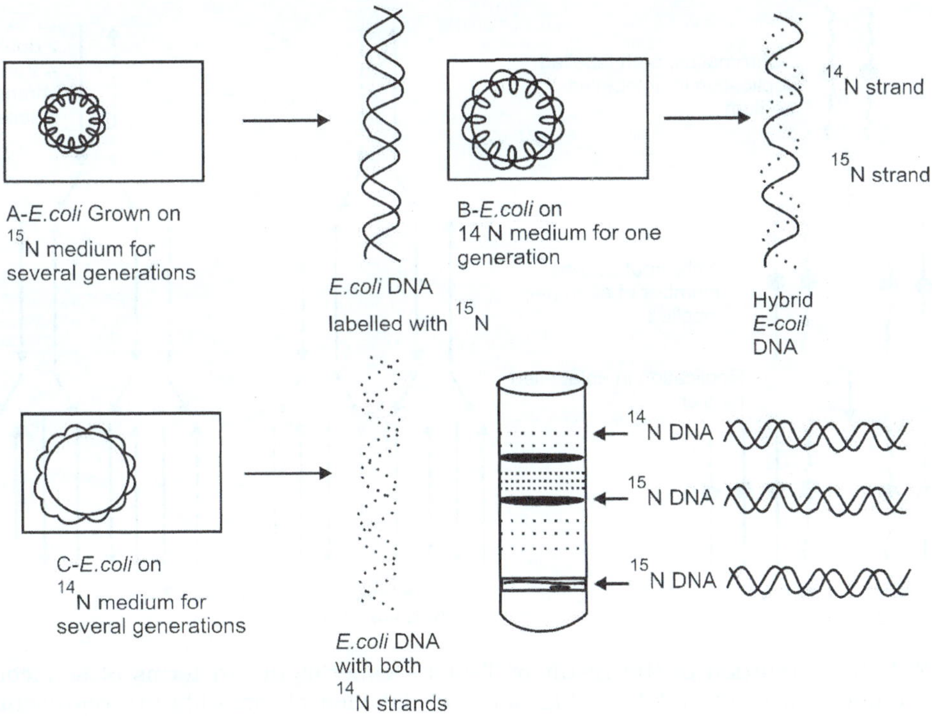

Fig. 7.1. Experiment of Meselson and Stahl.

Semiconservative Replication of Chromosomes

J.H. Taylor and his *co-workers* for the study of chromosomes in the root tips of *Vicia faba* have used the method of autoradiography. The result of these experiments were published in 1857. When roots after developing in titrated thymidine medium were transferred into unlabelled thymidine, then in the first generation of duplication, both the half chromosomes were labelled, or in each half chromosome DNA was double helical and amongst both the strands of DNA one was labelled. After two rounds of replication in each chromosomes two pairs of half chromosomes were found (tetraploid) out of which in each chromosome one half chromosome is labelled while the second one is unlabelled. In the third round of replication in each chromosomes, four pair of half chromosome were present (octoploid) out of which in two pairs, one was labelled and the second was unlabelled chromatid. While in other two pairs both chromatids were unlabelled. This indicates that cytological experiments confirms the semiconservative replication of DNA.

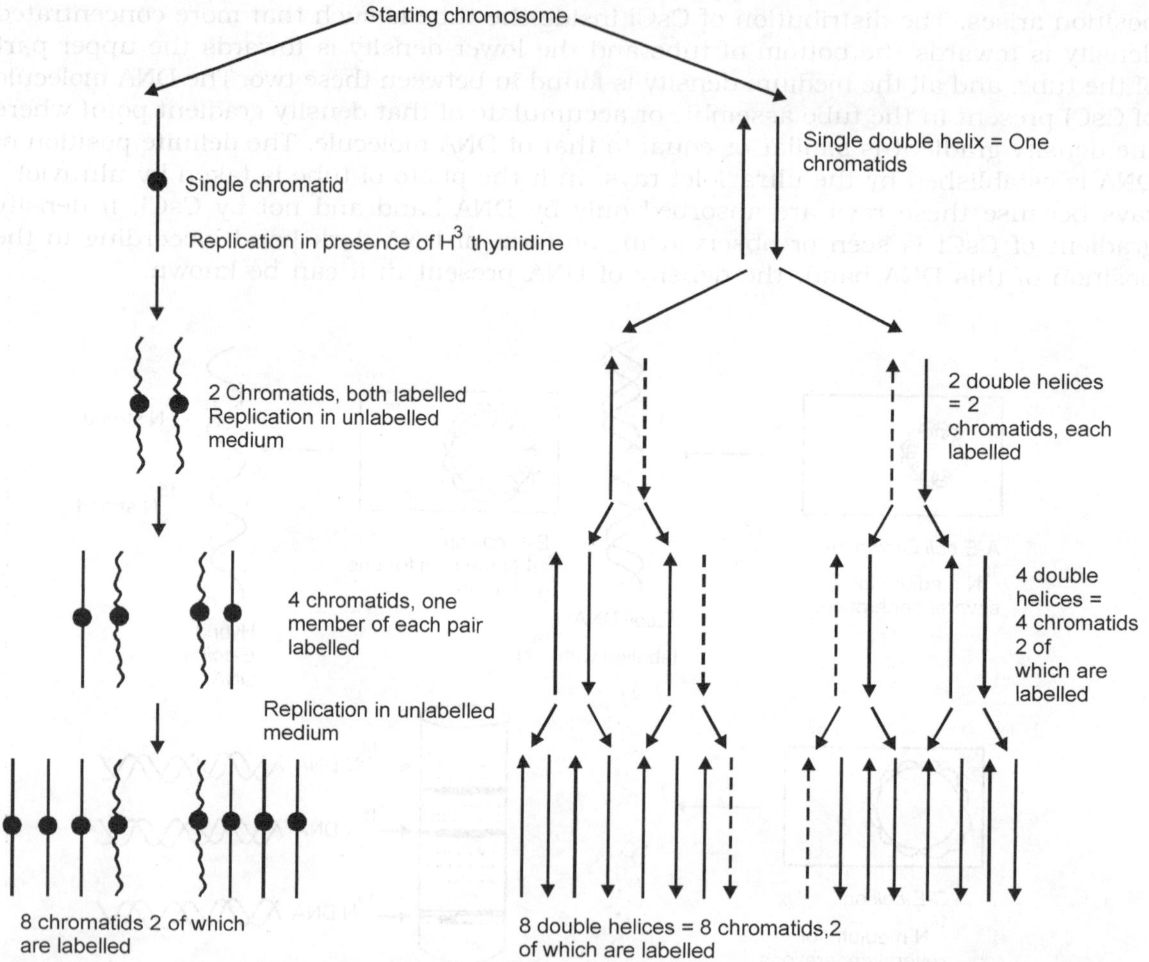

Fig. 7.3. Interpretation of the result of Taylor's experiments in terms of semiconservative replication of the DNA double helix, H^3 labelled chromatids are represented by wavy or dash lines

Continuous and Discontinuous Replication of DNA

As already described that at the time of DNA replication, the chains which are formed are antiparallel to template. So on 3'–5' strand a complimentary strand 5'–3' is formed. In the same way on 5'–3' strand a complimentary strand 3'–5' is formed. In 5'–3' direction whatsoever replication take place in it, new nucleotide joined at 3–OH end, while replication takes place in 3'–5' direction, the new nucleotide must join at 5'–P end. So for the synthesis of new chain separate enzymes should be in both directions. But the enzyme synthesize in the direction 5'–3'. Therefore, the chain which is formed in the direction of 5'–3' is always formed continuously while the replication in the second strand is in discontinuous form and small fragments are formed, each fragment is formed in 5'–3' direction. These fragments are called as **Okazaki fragments** or **segments** or **pieces.** These fragments are then found together by Ligase enzyme. New chain which is continuously formed are called as **Leading strand** and the chain formed in a discontinuous way is called as **Lagging strand.**

Cairn's concluded that the synthesis of DNA starts from a definite point of chromosomes and moves only in one direction. After this it was presumed that, the results of the Cairn's can be defined or explained in the forms of **bi-directional replication.**

Semidiscontinuous DNA Replication

Various experimental evidences have suggested that DNA synthesis is continuous on one strand (3' to 5' strand), called **leading strand** and discontinuous on the other strand (5' to 3' strand), called **lagging strand.** Since DNA synthesis always proceeds in 5' to 3' direction, so, on the lagging strand, synthesis takes place discontinuously in pieces, called **okazaki fragments** (after the name of discoverer *R. Okazaki*, 1968). Later on, these pieces are fused with the help of ligase enzyme to form an intact lagging strand. Such a DNA replication, where the leading strand is synthesized continuously and the lagging strand is synthesized discontinuously, is called **semidiscontinuous replication.**

Unidirectional and Bidirectional DNA Replication

A molecule of DNA engaged in replication has two types of regions. When replicating DNA is viewed by electron microscopy, the replication region appears as an eye within the non-replicated DNA. The non-replicated region consist of the parental duplex, this opens into the replicated region where the two daughter duplexs have formed. A replication fork moves sequentially along the DNA, from its starting point at the origin. Replication may be unidirectional or bidirection.

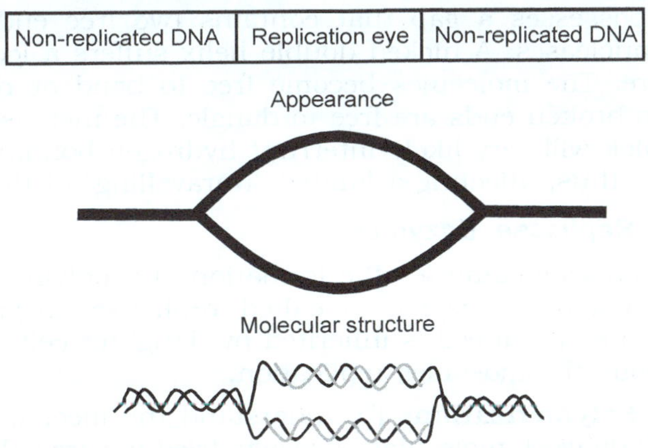

Fig. 7.3. Replicated DNA is seen as a replication eye flanked by non-replicated DNA.

The appearance of a replication eye does not distinguish between unidirectional and bidirectional replication. As depicted in Fig. 7.4, the eye can represent either of two structures. If generated by unidirectional replication the eye represents one fixed origin and one moving replication fork and if generated by bidirectional replication, the eye represents a pair of replication forks. In either case, the progress of replication expands the eye until ultimately it encompasses the whole replicon.

When a replicon is circular, the presence of an eye forms, the θ-structure. The successive stages of replication of the circular DNA of polyoma virus are visualized by electron microscopy.

Whether a replicating eye has one or two replication forks can be determined in two ways. The choice of method depends on whether the DNA is an indentifed molecule or an unidentified region of a cellular genome.

Enzymes of DNA Metabolism

In eukaryotes and prokaryotes cells, three type of nuclease enzymes present to perform enzymatic activities that act on DNA, namely nuclease, polymerase and ligase.

1. Nuclease Enzymes

Polynucleotide chain of nuclease enzymes breakdown or hydrolyze into its component nucleotides. A polynucleotide is held together by 3', 5' phosphodiester bonds and a nuclease enzyme will attack either the 3' or 5' end of this linkage. Nuclease enzymes are of two types.

(*i*) ***Exonuclease Enzyme.*** A nuclease enzyme which begins its attack from a free end of a polynucleotide is called exonuclease. Therefore, depending on the specificity of the enzyme, an exonuclease will either begin at a free 3'-OH end of a polynucleotide and progressively cleave the bonds on the 3'-OH side of the phosphodiester backbone or it will begin at a free 5'-P end and digest the polynucleotide in a 5' → 3' direction. In both cases the enzyme travels along the chain in a stepwise manner, liberating single nucleoside monophosphate molecules and eventually digesting the entire polymer.

(*ii*) ***Endonuclease Enzyme.*** Endonuclease enzyme also attacks one of the two sides of phosphodiester linkages, but they react only with those bonds that occur within the interior of a polynucleotide chain. If the polynucleotide chain is single-stranded (*e.g.*, viral DNA), such an attack will obviously cut the chain into two pieces. If, however the polynucleotide strand is a member of a DNA helix (*e.g.*, prokaryotic and eukaryotic DNA), a single endonucleolytic cut will create a nick in the helix. The helix remains in one piece but it now possesses a gap that contains two free ends, which can serve as substrates for exonucleases. A nicked double helix suffers a localized disruption of its secondary structure. The molecules become free to bend or rotate around its intact strand and the two broken ends are free to dangle. The increased molecular motion in the region of the nick will very likely interrupt hydrogen bonding between bases in the vicinity of the nick, thus, affecting a limited “unravelling” of the helix.

2. Polymerase or Replicase Enzymes

A polymerase enzyme catalyses the formation of a polymer or cellular polymerase enzymes. This polymerase enzyme is called replicase enzyme when the copy of polynucleotide chain so produced is inherited by daughter cells or virus, *i.e.*, when the enzymes brings about chromosomes replication.

***In vitro* DNA Polymerization.** To understand the mechanism of DNA replication inside the living cell (*in vivo*), molecular biologists tried *in vitro* polymerization of DNA and found that in addition to DNA polymerase enzyme, three classes of organic molecules are essential for an *in vitro* reaction. The first are deoxynucleoside triphosphates. These are the familiar deoxynucleotide monophosphates (*i.e.*, dAMP, dCMP, dGMP, and dTMP) with two additional phosphate groups attached to the initial or α-phosphate group (*e.g.*, dATP, dCTP, dGTP and dTTP). The second class of essential molecules for many DNA polymerase enzymes are polynucleotide chains with free 3'-OH ends (often called primer strands), meaning that they cannot initiate the *de novo* synthesis of a new strand. The third class of essential molecules for DNA polymerase enzymes are template strands. All biologically important DNA polymerases possess a critical property. They will add nucleotides to a primer strand only in response to the base sequence found on a second template strand. Just as Watson and Crick suggested in their original model of DNA replication, the polymerase enzymes observe the rule of complementary base pairing. Thus, if a free 3'-OH group on a primer strand lies opposite a thymine on a template strand, a polymerase enzyme will add only an adenine group to the primer, even when dCTP, dTTP, dGTP are also present in the reaction mixture.

Once the appropriate molecules are present and the appropriate ionic conditions are maintained *in vitro* polymerization reaction will occur. The phosphate of a nucleoside triphosphate molecule forms a 3', 5', phosphodiester bond with a free 3'-OH in the growing polynucleotide chain, and molecule of pyrophosphate (P~P) is simultaneously released. Pyrophosphate contains a "high-energy" or "~" bond, meaning that when the released pyrophosphate is hydrolyzed into two phosphate molecules, energy is liberated which drives the polymerization process forward. The resultant polymerization will always proceed in a net 5' → 3' direction.

1. Prokaryotic DNA polymerase. Three different DNA polymerase are found in *E.coli* are:

(*i*) ***DNA polymerase-I.*** Enzyme was isolated by *Arthur Kornberg* in 1960 and also suggested this enzyme to be involed in DNA replication. It is called *Kornberg enzyme.* This enzyme have five active sites—template site, primer site, 5'→3' cleavege or exonuclease site, nucleoside triphosphate site and 3'→5' cleavege site.

(*ii*) ***DNA polymerase-II.*** This enzyme resembles DNA polymerase-I activily, but a DNA repair enzyme. It brings about the growth in 5'→3' direction, with free 3'-OH group.

(*iii*) ***DNA polymerase-III.*** This polymerase enzyme plays an important role in DNA replication. It is multimeric enzyme having ten subunits such as alpha (α), beta (β), epsilon (ε), theta (θ) tau (τ), gamma (γ), delta (δ), dash (δ′), chi (*x*) and psi (ψ).

Table 7.1. Comparison of Different Characters of Three Types of DNA Polymerases of *E. coli.*

	Character		*DNA polymerases*	
		I	II	III
1.	Polymerization 5' → 3'	Yes	Yes	Yes
2.	Exonuclease 3' → 5'	Yes	Yes	Yes
3.	Exonuclease 5' → 3'	Yes	No	No
4.	Use of primer single strands	Yes	No	No
5.	Use of nicked duplex or helix	Yes	No	No
6.	Molecular weight	109000 (Single chain)	120000 (Single chain)	> 250000 (hetero-multimeric chain)

7.	Molecules per cell	400	Not known	10–20
8.	Gene	pol A	pol B	pol C (DNA E), DNA N, DNA ZX, DNA Q, DNA T
9.	Nucleotides polymerized at 37°C, molecules/minute	up to 1000	up to 50	up to 15000
10.	Affinity for nucleoside triphosphate (TPs) precursors.	low	low	high

The core enzyme comprises three subunits α, β and θ.

2. Eukaryotic DNA polymerase. Eukaryotic cells like yeast, rat liver, human tumour are found to contain five types of DNA polymerase.

(*i*) ***DNA polymerase alpha* (α).** The relatively high molecular weight enzyme is also called cytoplasmic polymerase. It is found in both nucleus and cytoplasm.

(*ii*) ***DNA polymerase beta* (β).** This is called nuclear polymerase or small polymerase and is found only in vertebrates.

(*iii*) ***DNA polymerase gama* (γ).** This enzyme is called mitochondrial polymerase and is encoded in the nucleus.

(*iv*) ***DNA polymerase delta* (δ).** This enzyme is found in mammalian cells and is PCNA Proliferating Cell Nuclear Antigen dependent and occurs in synthesis processivity.

(*v*) ***DNA polymerase epsilon* (ε).** It was previously known as DNA polymerase **δ**11. This enzyme is PCNA independent and occurs in mammalian. For example Hela cells and budding yeast.

3. DNA ligase

These enzymes are capable for catalyzing phosphodiester bond formation between 3'←OH and free 5'–P groups of nick of DNA which is created by endonuclease enzyme. Many DNA ligases have already been discovered. The ligase enzyme from *E. coli* requires the presence of oxidized nikotinamide adenine dinucleotide (NAD^+) a cofactor, whereas the ligase enzyme is specified by T_4 bacteriophage requires ATP to bring about the joining reactions.

Roles of RNA Primers in DNA Replication

No known DNA polymerase can initiate synthesis of DNA without the availability of a primer RNA strand. So, before actual DNA replication starts, short RNA oligonucleotide segments, called RNA primers or simply the primers, have to be synthesized by DNA primase enzyme, utilizing ribonucleoside triphosphates. This RNA primer is synthesized by copying a particular base sequence for one DNA strand and differs from a typical RNA molecule in that, after the synthesis the primer remains hydrogen bonded to the DNA template (*Freifelder*, 1985). The primers are about 10 nucleotides long in eukaryotes and they are made at intervals on the lagging strand where they are elongated by the DNA polymerase enzyme to begin each okazaki fragment. These RNA primers are later excised and filled with DNA with the help of DNA repair system in eukaryotes (or DNA polymerase I in *E. coli*).

In bacteria, two different enzymes are known to synthesize primer RNA oligonucleotides–RNA polymerase (on the leading strand) and DNA primase (on the lagging strand).

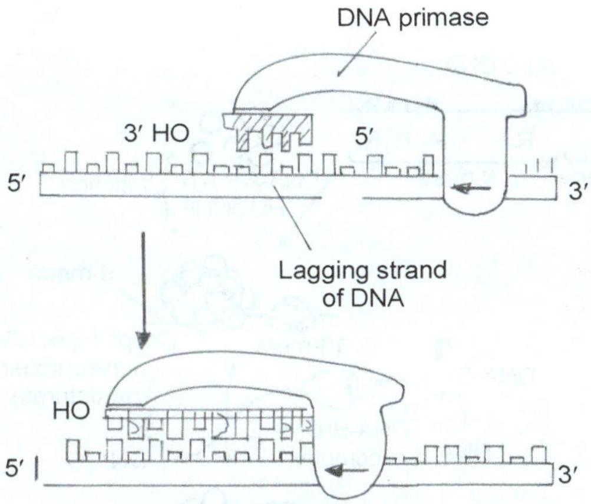

Fig. 7.4. Formation of RNA primer on lagging strand by DNA primase enzyme. Unlike DNA polymerase, this enzyme can start a new polynucleotide chain by joining two nucleotide triphosphates together (after *Alberts et al.*, 1989).

Replicons

DNA replication in prokaryotes and eukaryotes is attained in discrete units, called replicons. The number of replicons may vary in a genome from one in bacteria (*E. coli*) and 500 in yeast to several thousands in plants and animals (Table 7.2). For example in the *E. coli* there is a single replicon with the origin, identified as a genetic locus ori C (245 bp). The origin is A : T *rich*, a feature that is related to unwinding of DNA to initiate replication. In *E. coli*, there are also termination sites (ter A-F), each consisting of 23 bp. The process of termination of DNA replication requires the product of *tus* gene (*Tus* protein or TBP, *i.e.*, *ter* binding protein) which recognizes *ter* or termination sites.

Table 7.2. Prokaryotic and Eukaryotic Replicons

	Organism	*Number of Replicons*	*Average Length (kb)*	*Fork Movement (bp/min)*
1.	Bacteria (*E. coli*)	1	4200	50000
2.	Yeast (*S. cervisiae*)	500	40	3600
3.	Fruit fly (*D. melanogaster*)	3500	40	2600
4.	Toad (*Xenopus laevis*)	15000	200	500
5.	Mouse (*Mus musculus*)	25000	150	2200
6.	Bean (*Vicia faba*)	35000	300	Not known

MECHANISM OF DNA REPLICATION IN PROKARYOTES

In *E. coli*, the process of DNA replication involves the following steps are:

1. Initiation of DNA replication
2. Elongation of DNA chain.

1. Initiation of DNA Replication. This process complete in three steps.

(*i*) Recognition of the origin(O)

(*ii*) Opening of DNA duplex to generate a region of single strand DNA.

(*iii*) Capture of DNA-B protein.

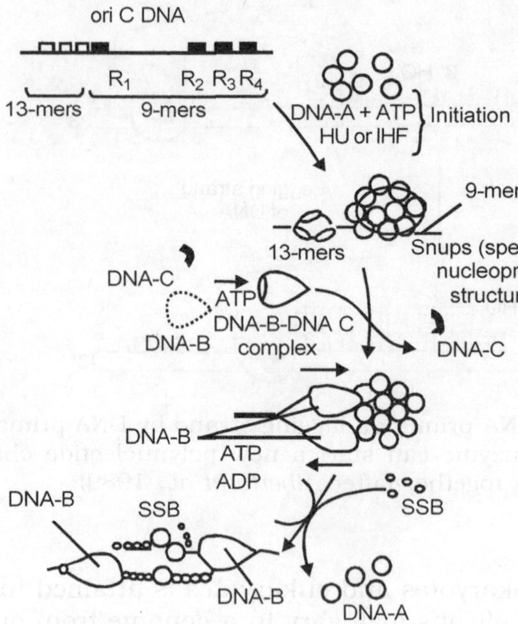

Fig. 7.5. A mechanism of initiation of DNA replication in *E.coli.*

In the DNA-A, –ATP complex bind with 9-bp inverted repeat region are R_1, R_2, R_3, R_4 of DNA-C of *E.coli* and promotes opening of the DNA duplex in a region of three direct repeats of 13-bp sequence. The opening occurs for right 13 mer leftwards and required negetively supercoiled DNA and HU or HIF is initiator proteins. DNA is transferred to exposed single strand DNA and causes unwinding of DNA in the presence of ATP, SSB protein, and DNA gyrase. DNA duplex and the replication from ori C proceeds in both directions such as, SSB binds with single stranded regions and DNA-B complexed are loaded one on each strand.

2. Elongation of DNA Chain. The next step requires the presence of the following enzymes and factors are;

(*i*) DNA-B or helicase

(*ii*) Primase

(*iii*) DNA polymerase holoenzyme

(*iv*) SSB protein

(*v*) RNase H which removes RNA primers'

(*vi*) DNA polymerase I which is used for filling the gap created due to RNA primers

(v*ii*) DNA ligase.

In initiation to elongation transition, the followings events occur:

1. As helicase (travels in 5' → 3') direction, it generates a replication fork by opening the DNA duplex.
2. DNA strand having helicase becomes the lagging strand. DNA primase associates with DNA-B helicase and formed primosome which synthesizes the

multiple primers for lagging strand and single RNA primers for the leading strand.

3. In the synthesis of lagging strand, the DNA pol III HE has to work on the same strand to which DNA-B helicase is bound, while it travels in directions.
4. DNA-B helicase, DNA-G primase and DNA pol III HE work together in strand elongation.

DNA REPLICATION IN EUKARYOTES

Two different type of enzymes namely DNA polymerase-α and DNA polymerase-δ are required for eukaryotic DNA replication. Beside these two enzyme six more factors are involved in Eukaryotic DNA replication.

(*a*) T antigen

(*b*) Replication factor A or RF-A

(*c*) Topoisomerase I

(*d*) Topoisomerase II

(*e*) Proliferating–cell nuclear antigen

(*f*) Replication factor C or RF–C

EXERCISE

Based on Unit–1

Q.1 Write in detail about chemical composition of chromosomes.

Q.2 What is the structural organization of chromatids, centromere, telomere and chromatin?

Q.3 Write notes on:

(*a*) Nucleosome organization

(*b*) Eu and heterochromatin

(*c*) Chromatids

Q.4 How many types of special chromosomes exist?

Q.5 Write a detailed account of:

(*a*) Polytene chromosomes.

(*b*) Lampbrush chromosomes.

Q.6 What are chromosomal aberrations?

Q.7 Distinguish between structural and nuismerical aberration with example.

Q.8 What is Genetic material and how it is chemically composed?

Q.9 Write a brief account on the nature of nucleic acids.

Q.10 Write in detail about the structure of DNA.

Q.11 What do you understand by the term DNA replication?

Q.12 Write short notes on:

(*a*) DNA replication in prokaryotes.

(*b*) DNA replication in eukaryotes.

UNIT-2

8 Molecular Basis of Life, Mendelian's Laws of Inheritance and Gene Interaction

INTRODUCTION

The living substance, which forms the cells and is responsible for all the properties of living beings is called protoplasm (Gr. *Proto*: Primitive or first + plasm). It is a viscous semifluid, jelly like substance with granules and globules of various sizes, suspended in it.

History

1772	*Corti* observed protoplasm
1835	*Dujardin* named it sarcode
1835	*Von Mohl* emphasized the importance of protoplasm
1840	*Purkinje* used the term "protoplasm"
1861	*Man Scholtze* formulated cell theory
1868	*J. Huxley* called it the *"Physical Basis of Life"*.

PHYSICAL NATURE OF PROTOPLASM

The "living state" of protoplasm depends, not only upon its highly complex, self-regulating and self-reproducing physico-chemical organization, but also upon its specific physical properties which are as follows:

1. Physically, the protoplasm is a thick and greyish, jelly-like, translucent and odourless viscid fluid mixture. Water, in it, acts as the solvent, and numerous other compounds and elements as solutes.

2. Some protoplasmic solutes, such as sugars, salts, simple acids, bases, etc., form a true or crystalline solution, but most solutes remain as electrically charged molecular aggregates or colloid particles (0.001 µ to 0.1 µ in diameter). Thus, protoplasm is a complex crystallo-colloidal mixture, yet mainly colloidal.

3. During the 19th century, divergent theories about the physical nature and appearance of protoplasm under the light microscope claimed it to be alveolar, granular, fibrillar, and reticular. According to the *"Colloidal Theory"* of *Fischer* (1894) and *Hardy* (1899), however, the *protoplasm is a heterogeneous polyphasic colloidal system which may appear differently in different phases.*

4. Like many others, protoplasm is a reversible colloidal system, *i.e.*, it can change from a thicker, highly viscous "gel" state to a comparatively more fluid and less viscous "sol" state and vice-versa. In gel state, the colloidal particles form a continuous phase

(network) entrapping the basal aqueous solution of protoplasm between the meshes of the network. In sol state, the colloid particles are dispersed, so that the basal aqueous solution is the continuous phase. Gelation imparts contractility, tensility and viscosity to the protoplasm. Solvation, on the other hand, imparts fluidity, causing streaming or "*cylosis*" of the protoplasm and active motion (Brownian motion–*R. Brown*, 1828) of the colloid particles in it, facilitating chemical exchange from the surrounding medium, and increasing reactivity of the solutes.

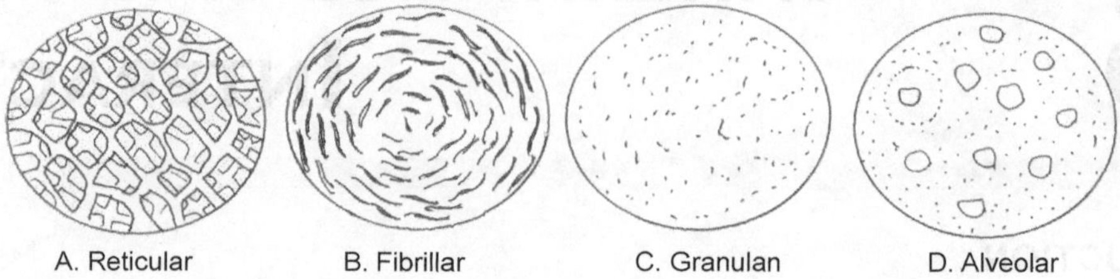

Fig. 8.1. Physical appearance of protoplasm.

5. Being a liquid mixture, the protoplasm has a surface tension. That is why, in any given mass or lump of protoplasm, some solutes (proteins and lipids), having less surface tension, form a delimiting plasma membrane and other membranelles. Being semipermeable, this membrane regulates the chemical exchange between the protoplasm and its surrounding medium.

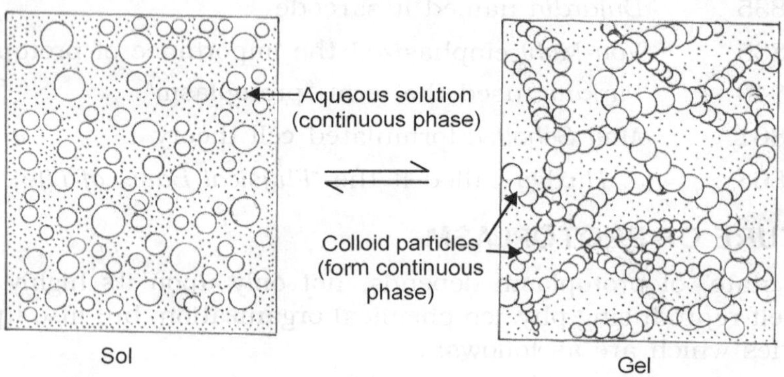

Fig. 8.2. Reversible 'sol-gel' conversion of protoplasm.

6. Heating, electrification, contact with harmful chemicals, etc., bring about immediate irreversible coagulation of protoplasm like that of egg-albumin.

ELEMENTS OF PROTOPLASM

All the 92 naturally occurring elements of the earth, in various combinations, form the non-living matter, but only about 27 elements are essential for life and hence, participate in the composition of protoplasm.

COMPOUNDS OF PROTOPLASM

Some elements occur in protoplasm as free ions, but mostly two or more elements are differently combined to form different kinds of compounds, numbering about 120. These constituents compounds of protoplasm are of two main categories—organic and inorganic.

Average quantities of the main compounds are as given in the following table.

Table 8.1. Average Quantities of Main Compounds in Protoplasm

Organic Compounds		*Inorganic Compounds*	
Name	*Average Percentage (%) Quantity*	*Name*	*Average Percentage (%) Quantity*
Proteins	14	Water	80
Lipids	3	Salts, acids, bases, gases and electrolytes	1
Carbohydrates	1		
Nucleic acids, enzymes and regulatory compounds	1		

ORGANIC COMPOUNDS OF PROTOPLASM

There are carbon containing compounds (biomolecules) which constitute about 95% solid part of protoplasm. These exhibit different degrees of complexity and are classified into micro and macromolecules.

1. Micromolecules are small and simple organic molecules of low molecular weight. These are amino acids, mono and disaccharides (simple sugar), lipids and nucleotides.
2. Macromolecules are large sized molecules with larger molecular weights and complex molecular structure. These are formed by the polymerization of micromolecules. These are proteins, (polymers of amino acids), polysaccharides, (polymers of monosaccharides), fats and the nucleic acids (polymers).

Protein

Proteins (Gr. *proteios* = first or foremost) form the major part of living protoplasm. About 75% of the dry weight of protoplasm, is proteins. *Berzelius* (1837) and *Mular* (1938) coined the term protein. These are essentially compounds of carbon, hydrogen, oxygen and nitrogen but may also contain traces of sulphur, phosphorus, iodine and iron. Usually, the protein molecules of protoplasm are large and complex macromolecules of high molecular weight. *Emil Fischer* and *Hofmeister* (1902) discovered that on complete hydrolysis, a pure protein molecule yields only amino acid molecules. This proves that the protein molecules are long-chain polymers of amino acid units of monomers.

Although about 300 amino acids occur in nature, only 20 of these enter into the composition of protoplasmic proteins. Each amino acid molecule is tetrahedral structure with a central, or alpha carbon atom covalently bonded to hydrogen atom, a positively charge (basic) amino group (NH_2), a negatively charged (acidic) carboxyl group (COOH) and a side chain carbon compound called an R-group. It is only the R-group, which differs in different amino acids. Thus, all amino acids can be indicated by a typical structural formula.

All amino acids are colourless, soluble in water, ionized in aqueous solution and have both acidic (–) and alkaline (+) properties (amphoteric compounds of zwitterions) due to their carboxyl and amino groups, respectively. When the carboxyl group of an amino acid molecule covalently links with the amino group of the other, a molecule of water is lost (Fig. 8.4). This linkage is called a peptide (amide) bond. It is, obviously, a dehydration-condensation reaction and the resultant molecule of this reaction is called a dipeptide. By linking in this way into chains, amino acid molecules polymerize, forming tripeptides, oligopeptides (short chains of a few amino acid molecules) and polypeptides (chains of several amino acid molecules). The latter, they link with each other, forming first the peptones, then proteoses and ultimately the variously folded and highly complex macromolecules of proteins.

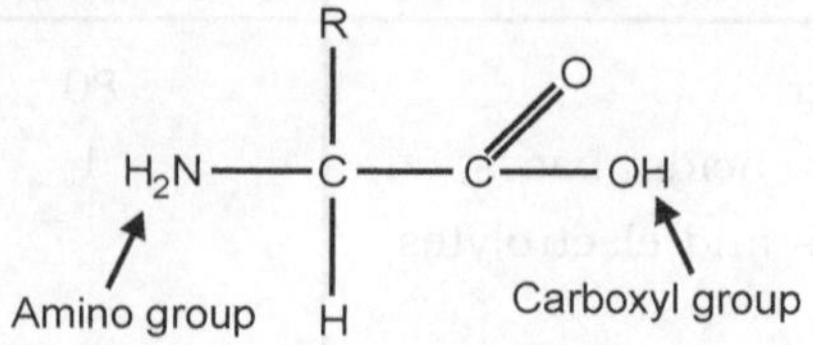

Fig. 8.3. Typical structure of amino acid molecules.

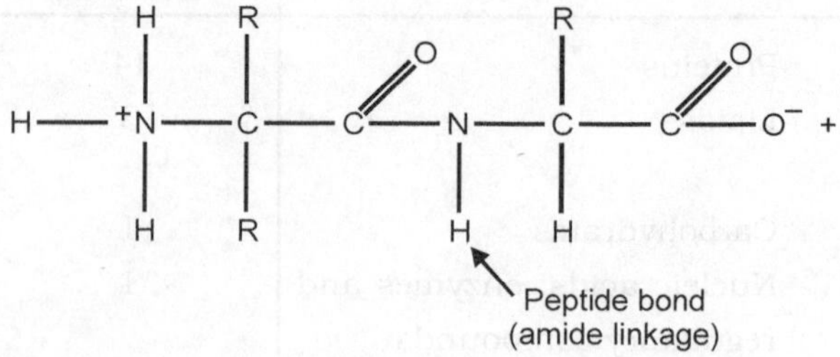

Fig. 8.4. Formation of a peptide bond.

Hundreds or thousands of molecules of 2 to all the 20 kinds of amino acids may participate in the synthesis of various proteins in various combinations and sequences as if they are the 20 "alphabets" of the "bioprotein language". That is why, proteins are not only the most abundant but also the most variable biomolecules of protoplasm. Each protein has its own unique amino acid sequence which endows it with a specific shape, size and function. One category of proteins, called structural proteins, are responsible for the colloidal nature and structural framework of protoplasm and for the 'sol-gel' and other physical changes in it. The other category of proteins, called globular proteins, act as biological catalysts or enzymes and, hence, are responsible for all chemical changes which occur in protoplasm incessantly to keep it in the living state. Obviously, these are the proteins that are responsible for all the diversity of protoplasmic composition, structure and function found amongst organisms.

Carbohydrates

These form about 1% of protoplasm. These are comparatively simpler compounds of carbon, hydrogen and oxygen in the ratio of 1 : 2 : 1. The ratio of hydrogen to oxygen is thus the same as in water (H_2O). Some carbohydrates may also have traces of nitrogen, sulphur, phosphorus etc.

Carbohydrates include all sugars and starches. Sugars are mostly sweet and simpler, water soluble (crystalline) compounds. Starches are tasteless and mostly insoluble complex compounds. The simplest carbohydrates are monosaccharide sugars. These have the empirical formula $(CH_2O)_n$ where n is equal to 3 or more, but only up to 7 (trioses, tetroses, pentoses, hexoses and heptoses). Protoplasm usually contains pentoses (5-carbon sugars) and hexoses (6-carbon sugars). The most common hexose sugars are mainly four isomers, named glucose, fructose, galactose and mannose. These have a common empirical formula $C_6H_{12}O_6$. The difference between them is only the arrangement of OH groups in their molecules. Of these, glucose (D-glucose) is the most common and most abundant monosaccharide found naturally. It is the chief source of chemical energy, *i.e.*, the main "fuel compound" in protoplasm.

The common pentose sugars of protoplasm are ribose ($C_5H_{10}O_5$) and deoxyribose ($C_5H_{10}O_4$). These respectively participate in the synthesis of RNA and DNA, the two nucleic acids of vital genetic significance.

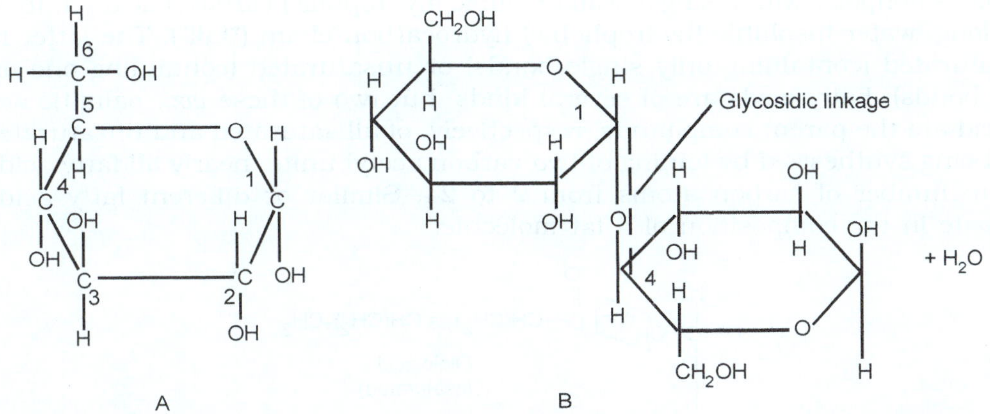

Fig. 8.5. A. Molecular structure of glucose. B. Formation of glycosidic linkage.

When two monosaccharide molecules link by means of a "glycosidic bond (α-1, 4)", a molecule of water is released, and a disaccharide molecule is formed (Fig. 8.5B). Obviously, this is also a dehydration-condensation reaction, and the monosaccharides are monomers of carbohydrates:

$$C_6H_{12}O_6 + C_6H_{12}O_6 \longrightarrow C_6H_{22}O_{11} + H_2O$$

Disaccharides, trisaccharides and in fact, all saccharides having a few (up to 10) linked monosaccharide molecules are generally called oligosaccharides. These all are soluble sugars.

Linkages of several (more than 10) monosaccharide molecules result in polymerization of complex carbohydrates, called polysaccharides (glycans), indicated by the general formula $(C_6H_{10}O_5)_n$. More than 200 different naturally occurring monosaccharides have been identified, but only a few of these occur as regular constituent monomers of the polysaccharides. The latter are insoluble and form the bulk of the carbohydrates of protoplasm. Cellulose, starch and insulin are the most common polysaccharides in plants, and chitin and glycogen, in animals. Cellulose of plants and chitin of animals are common *structural polysaccharides*, while starch and insulin of plants and glycogen of animals principally serve as '*glucose reserves*' (nutrient polysaccharides), *i.e.*, "Reserve fuel".

Lipids (Fats)

Lipids are oily or greasy organic compounds forming about 3% of protoplasm. These are also compounds of carbon, hydrogen and oxygen but, unlike carbohydrates, the hydrogen and oxygen, in them, are never in 2 : 1 ratio (H_2O); the amount of oxygen is considerably less. Being insoluble in water and soluble in organic solvents like acetone, chloroform, benzene, hot alcohols, ether, kerosene, petroleum, etc., lipids occur in protoplasm as minute globules.

The simplest and most abundant lipids are neutral or true fats, called simple lipids. These are esters (alcohols + acids) of long-chain fatty acids and glycerol. Each fat molecule is formed by combination of usually three fatty acids (aliphatic carboxylic acid) molecules, linked with one molecule of glycerol and, hence it is called a triglyceride (Fig. 8.6). This linkage is called "ester bond". In this type of reaction three molecules of water are released hence, this is also a condensation-dehydration synthesis. A fatty acid molecule is nonpolar with a single, water soluble (hydrophilic) carboxyl acid group (COOH) and a long, water insoluble (hydrophobic) hydrocarbon chain ("tail"). The latter may be fully saturated (containing only single bonds) or unsaturated (containing one or more double bonds). Fatty acids are of several kinds, but two of these *viz.*, palmitic acid and oleic acid are the parent compounds, respectively, of all saturated and unsaturated fatty acids. Being synthesized by joining of two-carbon acetyl units, nearly all fatty acids have an even number of carbon atoms from 2 to 24. Similar or different fatty acids may participate in the composition of a fat molecule.

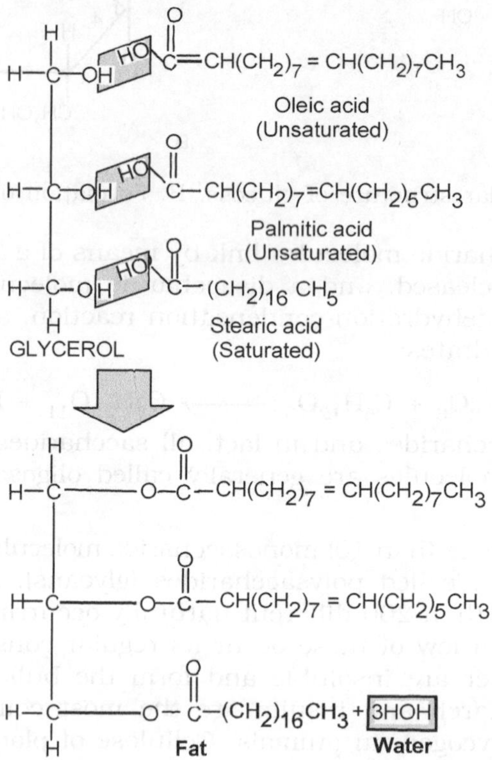

Fig. 8.6. Structural representation of the formation of a triglyceride molecule.

Like glucose, fats may also be oxidized for energy. These release more than twice as much usuable energy in the form of ATP, as does glucose. Thus, these serve as the most economic reserve of spare chemical energy, *i.e.*, as "stored food".

Protoplasm also contains certain complex, conjugated lipids like, phospholipids and glycolipids. In a phospholipid molecule, the glycerol is linked only by two fatty acids molecules. Its third linkage site is linked to a phosphate group which is, in turn, attached to a small hydrophilic compound (ethanolamine, choline, or serine). The two fatty acids form the hydrophobic "tail" and the phosphate group with associated compounds forms a highly polar hydrophilic "head". Due to their peculiar structure, phospholipid molecules are fitted to form the "lipid bilayer" of all the cell membranes, including plasma membrane. In glycolipids, glycerol is replaced by another alcohol (sphingosine) which is also linked to sugars in addition to fatty acids. These particularly occur in nervous tissue and upon plasma membrane of all cells. The third category of fats is of derived lipids which include all products of lipid metabolism like fatty acids, glycerol and other alcohols, fatty aldehydes, ketone bodies, steroids (*e.g.*, cholesterol), etc.

Nucleic Acids

Friendrich Miescher (1868) discovered the presence of these compounds in protoplasm, but *Altmann* (1889) introduced the term "Nucleic acid". These acids are the largest and most complex compounds of carbon, hydrogen, oxygen, nitrogen and phosphorus, and of utmost biological importance. There are two types of nucleicacids, occur in biological systems, *viz.*

1. Deoxyribonucleic acid (DNA), and
2. Ribonucleic acid (RNA).

Just as proteins and carbohydrates are polymers, respectively, of amino acid and monosaccharide units or monomers, nucleic acids are also a long-chain polymers (= macromolecules or biopolymers) of nucleotide monomers (polynucleotides). This polymerization is also an example of dehydration-condensation process, becasue in which, nucleotide monomers link together by elimination of water.

Each nucleotide molecule is itself a complex compound of a nitrogenous organic base, a pentose sugar and a phosphoric acid group linked together by dehydration-condensation in two steps. First, a base and sugar molecule linked by a N-glycosidic bond to form a nucleoside. Then, the C-5 hydroxyl of the sugar molecule of nucleoside joins with a phosphate group to form a nucleotide molecule. Five nitrogenous bases enter the composition of various nucleotides. These are of two kinds—pyrimidine derivatives with a single ring and purine derivatives with a double ring structure. Pyrimidine derivatives are Cytosine, Uracil and Thymine. Purine derivatives are Adenine and Guanine. The pentose sugars of nucleotides are also of two types—deoxyribose and ribose. Thus, the nitrogenous bases, combining with pentose sugars and a phosphate group, form eight types of nucleotide molecules as follows:

1.	Cytosine	+	Ribose sugar	+	Phosphate	=	*Cytidylic acid*
2.	Cytosine	+	Deoxyribose sugar	+	Phosphate	=	*Deoxycytidylic acid*
3.	Uracil	+	Ribose sugar	+	Phosphate	=	*Uridylic acid*
4.	Thymine	+	Deoxyribose sugar	+	Phosphate	=	*Thymidylic acid*
5.	Adenine	+	Ribose sugar	+	Phosphate	=	*Adenylic acid*
6.	Adenine	+	Deoxyribose sugar	+	Phosphate	=	*Deoxyadenylic acid*
7.	Guanine	+	Ribose sugar	+	Phosphate	=	*Guanylic acid*
8.	Guanine	+	Deoxyribose sugar	+	Phosphate	=	*Deoxyguanylic acid.*

As it is clear from the above, thymine combines only with deoxyribose sugar and uracil, only with ribose sugar. Other bases can combine with either of the two sugars.

Each nucleic acid molecule (polynucleotide) is a linear polymer, in which the phosphate group of one nucleotide molecule links with 3' phosphodiester linkage between the two sugar molecules. However, an RNA molecule differs from a DNA molecule in three ways—(*i*) the sugar phosphate backbone contains ribose sugar instead of deoxyribose; (*ii*) the uracil base is present, but never thymine; and (*iii*) it is usually a single-stranded molecule instead of a double-stranded helix.

A DNA molecule (*i*) is a long and highly complex, spirally twisted, right handed double-helix, ladder-like structure, formed of two polynucleotide strands; (*ii*) thc sugar phosphate backbones of both strands have deoxyribose sugar instead of ribose; and (*iii*) the base thymine is present, but never uracil. *Wilkins* and *Franklin* (1951–53) discovered some primary clues about the structure of DNA molecule by X-ray diffraction technique. On the basis of these clues, *Watson* and *Crick* (1953) could propose a model of the molecular structure of DNA. Wilkins, Watson and Crick shared the 1962 Nobel prize for this work. The two polynucleotide chains or strands of the DNA molecule are linked by means of weak hydrogen bonds between the nitrogenous base molecules of their complementary nucleotide monomers. Adenine (purine) always links with thymine (pyrimidine) by two hydrogen bonds and guanine (purine) with cytosine (pyrimidine) by three hydrogen bonds. Hence, the two strands are linearly opposite (antiparallel) and complementary. Obviously, the ratio of adenine to thymine, of guanine to cytosine, and of purines to pyrimidines is always 1 : 1. The double helix is 20 Å (2 nm = 2 mμ = 0.000002 nm) thick. It completes a spiral at every 10 nucleotide pairs at a length of 34 Å (= 3.4 nm), which means that each pair of nucleotide molecules occupies 3.4 Å (= 0.34 nm) length of the molecule.

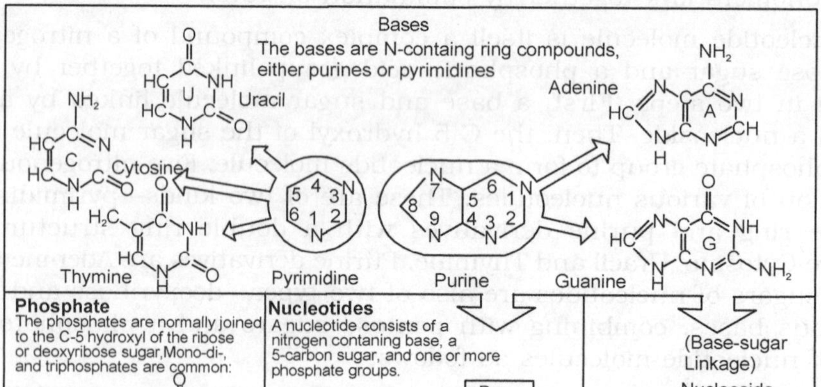

Fig. 8.7. Component molecules of a nucleotide and their linkages.

A single molecule of DNA may have millions of pairs (3,400,000 pairs in *E. coli*) of nucleotide monomers. Sequences of these monomers may present innumerable variations. Evidently, innumerable varieties of DNA exist in the organism. A DNA is always linear with free terminals.

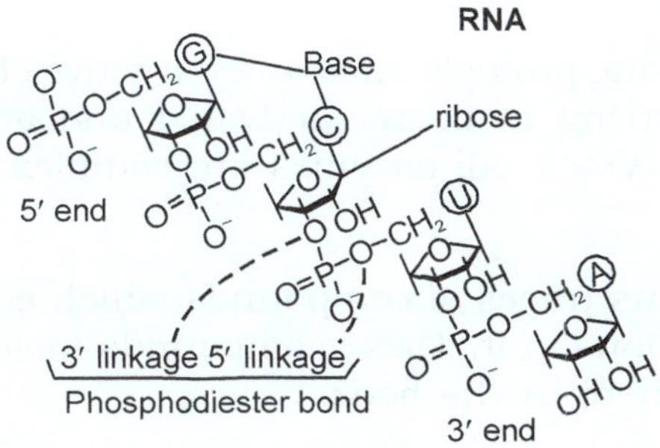

Fig. 8.8. A four-nucleotide section of an RNA molecule: G, C, U and A are abbreviations used for the bases guanine, cytosine, uracil and adenine.

DNA contains the "blueprint" of all organization and activities in protoplasm, with the help of RNA, it directs the synthesis of all structural and functional proteins of protoplasm. Thus, it alone is responsible for transmission of hereditary characteristics. Derivatives of some free nucleotides of protoplasm, such as ADP and ATP, which are derivatives of adenylic acid (AMP), serve as energy-carriers and for other functions. Some are used in specific signalling molecules in the cells (*e.g.*, cyclic AMP).

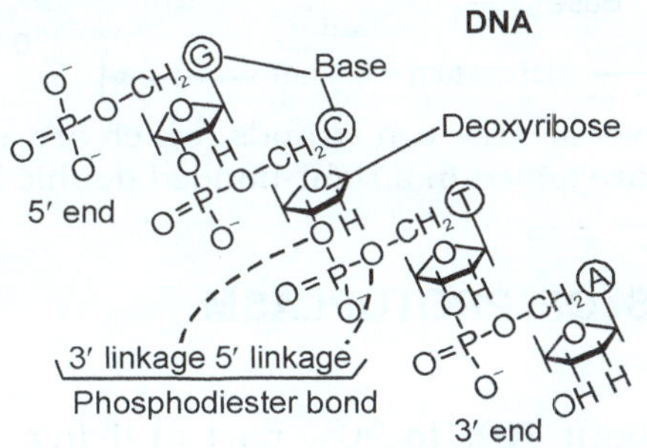

Fig. 8.9. A four-nucleotide section of one of the two strands of DNA molecule: G, C, T and A are abbreviations for the bases guanine, cytosine, thymine and adenine.

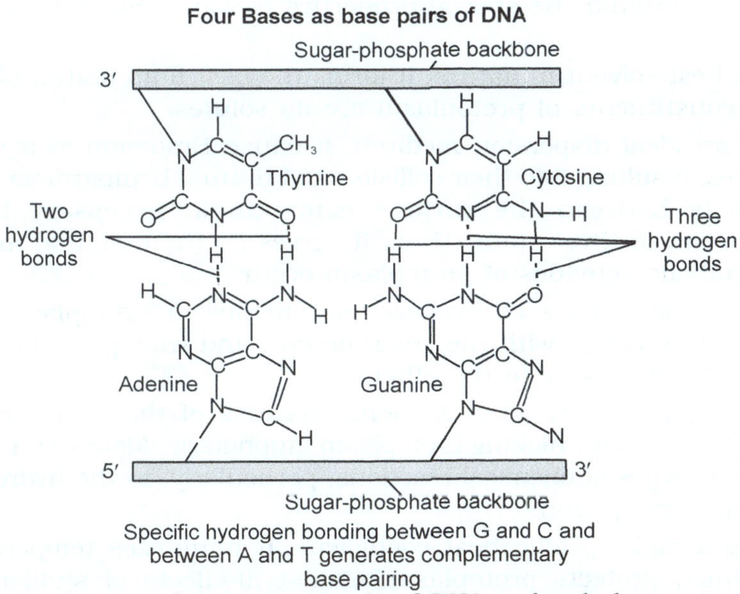

Fig. 8.10. Joining of the two strands of DNA molecule by pairing of specific opposite base pairs by means of 2 or 3 hydrogen bonds.

Enzymes

As mentioned before, living protoplasm is an ever active chemical system. All chemical reactions, incessantly occurring in it, are activated and accelerated by thousands of organic catalysts, called enzymes. All enzymes are complex proteins.

Regulatory Compounds

Protoplasm also contains traces of compounds which control and regulate physical changes and chemical reactions in it. These compounds include vitamins and hormones which are dealt with elsewhere in the book.

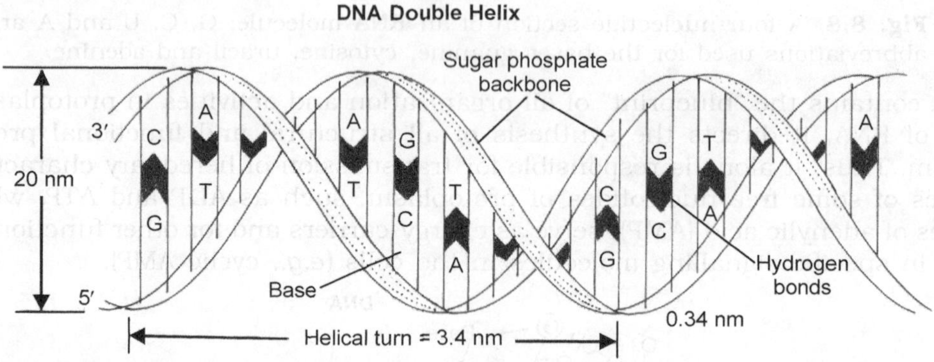

Fig. 8.11. A section of DNA molecule; the two strands, which are antiparallel and complementary in their nucleotide sequences, are joined in a right-handed double helix with about ten nucleotide pairs per helical turn.

INORGANIC COMPOUNDS OF PROTOPLASM

Water

Water usually forms about 70% to 90% part of living protoplasm. Life upon earth originated in the ocean and the conditions of that aqueous primeval environment have been retained by organisms as the basic integral part of their chemistry. All organisms have been designed around the special properties of water. Some important properties of water are:

1. As the best solvent in nature, it forms the best fluid matrix of protoplasm. All other constituents of protoplasm are its solutes.
2. Being an ideal dispersion medium, it cause *Brownian movement* of colloidal particles, resulting into their collision and mutual bombardment. This facilitates reactivity between the various compounds, necessary for maintaining protoplasm in "live" state. Thus, it serves as the essential medium in which all metabolic activities of protoplasm occur.
3. It causes streaming or cyclosis movements in protoplasm, facilitating its chemical exchange with the environment, and transportation of solutes from one part of the body to the others.
4. It dissociates into H^+ and OH^- ions, because of this, it becomes a very good electrolytic solvent, making protoplasm amphoteric. Moreover, it itself participate in certain types of chemical reactions, particularly, in the hydrolytic breakdown of complex compounds.
5. Having a high specific heat coefficient, it minimizes temperature variations and, thus, protects protoplasm against ill-effects of sudden rise or fall of temperature in the environment.

Salts, Acids and Bases

Inorganic salts of protoplasm are mostly chlorides, phosphates, carbonates, bicarbonates and sulphates of calcium, sodium, potassium, magnesium, copper, and iron, etc. Inorganic acids of protoplasm are mostly hydrochloric and phosphoric acids. Although, these serve important functions as follows:

1. These ions are responsible for electric conductivity, rendering protoplasm, and also irritable and responsive to environmental changes.
2. These provides linkage or chemical bonds in many chemical reactions. Some metallic and other ions, such as magnesium (chlorophyll), iron (haemoglobin) act as cofactors in enzymatic activities. Thus, these ions are vital to biological functions.
3. These prevent pH variation by maintaining a carbonic acid-bicarbonate buffer system (acid-base equilibrium).
4. These regulate the osmotic pressure and chemical exchange of protoplasm from its environment.

Gases

Gases of oxygen, carbon dioxide and nitrogen may, sometimes, occur in the protoplasm.

MENDELIAN'S LAWS OF INHERITANCE

Organisms regularly transmit their characteristics from one to the next generation. This phenomenon is called "heredity" or "inheritance" and the characteristics which transmitted are called hereditary or genetic. "Heredity" is also sometimes defined as "Sum total of an individuals inherited characteristics" or "the genetic continuity between successive generations."

The term Genetics (Gr. *Gen* = to generate or to become, or grow into) was introduced by *W. Bateson* (1905) for the "*Branch of biology dealing with Heredity and Variations.*" or say "*Similarities and dissimilarities between organisms related by descent*" or simply "*The study of heredity and hereditary variations.*"

The basic concept of genetics is that heredity is controlled by a vast number of submicroscopic particles in each organism called *gene.*

EARLY HISTORY

Since ancient times, antagonists of "Spontaneous Generation (A biogenesis)" *i.e.*, the protagonists of "Biogenesis" have been trying to explain how "Organisms form from pre-existing organisms" or "How life is continued generation after generation." Following theories emerge from early speculations:

1. **Vapour Theory.** Greek philosopher Pythagoras (500 B.C.) proposed that, during coitus, moist vapours from all parts of a male's body give rise to a similar body in female's womb.
2. **Fluid Theory.** *Aristotle* postulated that, during coitus, the female releases an "*inert*" fluid to which "life" is given by a "vitalizing fluid (semen)" released by male and considered as "highly purified blood."
3. **Particulate Theory.** After the invention of microscopes in 17th century, *Graaf*, Malpighi and *Swammerdam* discovered eggs. While *Leeuwenhock* and *Hamm* discovered spermatozoa. These discoveries took human knowledge from

the era of speculation into the era of reality. It was soon realized that gametes as minute particles formed by male and female parents unite in sexual reproduction to form new individuals.

Darwin (1868) in his "Theory of pangenesis", postulated that all parts of a living body release in blood their miniature copies called pan genes or gemmule which collect the gonads to form gametes.

Austrian priest and scholar, *Gregor Johann Mendel* (1866), ultimately proved the truth in "Particulate Theory of Continuity of Life" by discovering the "Basic Laws of Heredity". Mendel laid the actual "Foundation of Genetics" and became popular as "*The father of Genetics or Modern Genetics.*" He hypothesized laws of genetics in 1865 and published them in 1866 in "Annual Proceedings of the Natural History of Society of Brunn." Unfortunately, his research remained unknown to scientists for about 34 years, till 1900 when it was rediscovered independently by *Carl Correns* in Germany, *Hugo de Vries* in Holland and *Eric von Tschermak*, Seysenegg in Austria. These scientists incidentally came across Mendel's paper and cited his hypothesis in their own publications.

MENDEL'S HYBRIDIZATION EXPERIMENTS

Selection of Garden Pea Plant. Starting in 1856, Mendel conducted his hybridization experiments for about eight years upon the garden pea *(Pisum sativum)* plant, in his monastery garden. Mendel made the selection because of the following attributes of this plant as an experimental material:

1. Several distinct types or varieties of this plant, differing in sharply defined contrasting characteristics, grew in the monastery garden.
2. Mendel found that all varieties of pea plants growing in his monastery garden were true breeding pure lines, passing their respective special characteristics to their offsprings, generation after generation.
3. Pea plants are monoecious with bisexual (= hermaphrodite) flowers and the flower structure is such that normally only self pollination (autogamy) can occur. These two factors were responsible for the establishment of pure line varieties.

Mendel favoured pea plants also because these were small and easy to handle and grow, their flowers ripened in a shorter period, and these produced a large number of seeds in each generation. In fact, the selection of pea plant for his experiments itself revealed Mendel's remarkable insight into the methodology, necessary for good experimental biology.

Selection of Hereditary Traits. For his experiments, Mendel selected, with meticulous care, seven hereditary traits each with two (a pair) sharply distinguishable contrasting expressions (display) in different pea plants as follows:

1. Stem length	Tall or short (dwarf)
2. Flower position	Axial or terminal (apical)
3. Flower colour (or colour of seed coat)	Purple or white (gray-brown or white)
4. Colour of unripe pod	Green or yellow
5. Shape of ripe pod	Inflated or constricted between seeds
6. Shape of seeds	Round or wrinkled
7. Colour of cotyledons	Yellow or green

MENDEL'S MONOHYBRID CROSSES

Mendel began his experiments with monohybrid crosses, *i.e.*, by crossing pure line, pea plants displaying contrasting expressions of single genetic traits. For example, he crossed (allogamy) a plant of tall stem variety (stem 6 to 7 feet high) with a plant of short stem variety (stem only 9 to 18 inches high) by transferring pollen grains from the former on to the stigma of the pistils of the latter (Fig. 8.12). To prevent self-pollination (selfing or autogamy), he earlier removed all stamens from the flowers of the dwarf plant. Mendel marked these plants as P_1 (plants of parental generations). He carefully collected all seeds formed in the flowers of the dwarf P_1 and sowed these in his garden. All plants that grew from these seeds had tall stems. Mendel marked these as plants of F_1 (first filial–*filius* = son or offspring) generation. He allowed these to reproduce by self-pollination

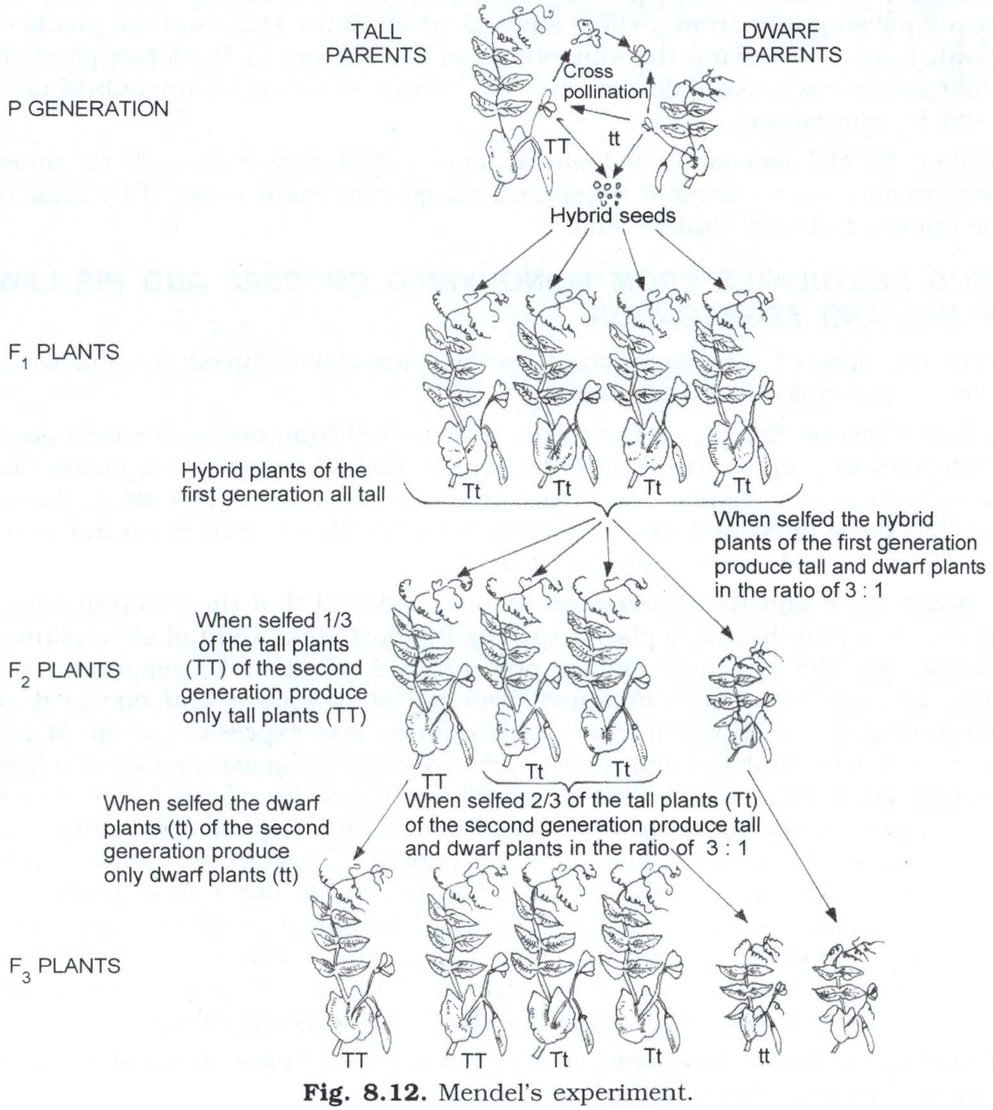

Fig. 8.12. Mendel's experiment.

and collected their seeds numbering 1064. Next, when he sowed these seeds, 787 of these grew into tall stem plants and 277 into short stem plants of F_2 (second filial) generation, giving nearly a 3 : 1 ratio of tall to dwarf plants. The most significant feature of this experiment was that dwarfness disappeared in F_1 generation, but reappeared roughly in 25% plants of the F_2 generation.

Mendel allowed all F_2 generation plants also to reproduce by selfing (autogamy), raising the F_3 generation. He observed that all short stem plants of F_2 generation were pure breeding because these produced only short stem progenies, but of the tall stem plants of F_2 generation only about one third were pure breeding for tallness, while the remaining two-third produced both tall stem and short stem plants in 3 : 1 ratio exactly like the hybrid plants of F_1 generation.

RECIPROCAL CROSS

Mendel also experimented with reciprocal crosses by reversing the sources of male and female gametes between P_1 plants. For example, he conducted the above cross by transferring pollen grains from a short-stem plant on to the stigma of the pistils of a tall stem plant. Earlier, removing the stamens from the flowers of the latter plant, Mendel found that reciprocal crosses do not alter the results of normal monohybrid crosses in F_1, F_2 and F_3 generations.

Further, Mendel also conducted similar monohybrid crosses to study the inheritance of the contrasting expressions of remaining six genetic traits selected by him. In each case, he obtained almost similar data.

MENDEL'S POSTULATES FROM MONOHYBRID CROSSES AND HIS LAWS OF DOMINANCE AND SEGREGATION

From the data of his monohybrid crosses, Mendel deduced the following basic postulates or principles of inheritance:

1. Unit Factors. Every genetic trait is transmitted from one to the next generation by an independent potential unit partially, which Mendel called a *unit factor*. From the 3 : 1 ratio of contrasting expressions of each trait in F_2 generation, Mendel mathematically deduced that there must be two unit factors for a hereditary trait in an individual, but only one factor in each gamete produced by it.

2. Dominance and Recessiveness. Mendel deduced that the two unit factors of a trait present in a pure breeding plant (*e.g.*, the P_1 short stem and tall stem plants) must be identical, but the two unit factors present in a plant of F_1 generation must be dissimilar, *i.e.*, one of tallness obtained from tall stem parent and one of dwarfness obtained from short stem parent. Since only tallness was expressed in the plants of F_1 generation, Mendel postulated that, *when factors of contrasting expressions of a hereditary trait combine in an individual (called a hybrid), the character of one factor is expressed while that of the other is suppressed.* He called the former factor as *dominant* (expressive) and the latter as *recessive* (suppressive or hidden). Thus, in the above experiment tallness proved dominant to dwarfness. Similarly, in his other monohybrid crosses, Mendel found axial position of flowers dominant to apical position, purple colour of flowers dominant to white colour, green colour of unripe pod dominant to yellow colour, inflated shape of ripe pod dominant to constricted shape, round shape of seeds dominant to wrinkled and yellow colour of cotyledons dominant to green colour.

Mendel's postulate of dominance was probably raised to the status of "Mendel's Law of Dominance" (Second Law) by *Carl Correns* (1901).

3. Segregation. When together, the dominant factor suppresses the expression of recessive from the dominant factor in 50% gametes (both pollens and ovules) produced by F_1 plants due to random combinations of pollens and ovules of F_1 plants therefore, 25% plants of F_2 generation receive both factors of dominance, 25% receive both factors of recessiveness and 50% receive one factor of dominance and one of recessiveness. Thus, 25% of F_2 plants resemble dominant P_1, 25% resemble recessive P_1 and 50% resemble the F_1 hybrids, displaying the dominant character. In other words, the F_2 plants are segregates of F_1 plants (hybrids) or say plants formed by 'splitting' of F_1 plants. This postulate of Mendel formed the basis for his First Basic Law, "Mendel's Law of Segregation" formulated by Carl Correns. Since this law postulates that factors of contrasting expressions of a trait may remain together for any length of time, yet they do not affect each other's purity, a gamete always receives a pure factor (dominant or recessive). That is why this law is also known as "The Law of Purity of Gametes", or "The Law of Splitting of Hybrids."

CHROMOSOMAL THEORY OF INHERITANCE

As early as 1883, *W. Roux* guessed that chromosomes were somehow connected with inheritance. Just after rediscovery of Mendelism at the turn of present century, *Montgomery* discovered homologous pairing of maternal and paternal chromosomes in zygotes, and *Boveri* (1902) and *Sutton* (1902), independently discovered that segregation of chromosomes during gamete formation and their reunion in fertilization, follow the same pattern of behaviour as described by Mendel for his hereditary factors. This indicated that the unit hereditary factors of Mendel were either the chromosomes or smaller particles located in them. On this basis, Sutton (1902) formulated *"The Chromosomal Theory of Inheritance"*. He also discovered later that hereditary factors were smaller particles located in chromosomes.

GENE THEORY OF INHERITANCE

In 1909, *Johannsen* coined the term 'gene' for the unit hereditary factors, already proved by Sutton to be present in chromosomes. Soon after, *Morgan* (1910) undertook extensive work on the location of hereditary factors of genes in the chromosomes of fruit fly (*Drosophila melanogaster*). On the basis of his work, he formulated units in chromosomes arranged in linear order like beads in a string.

SYMBOLIZATION

Mendel used the symbol P_1 for plants of parental generation and F_1, F_2, F_3 and so on for plants of the subsequent generations in his hybridization experiments. Similarly, he symbolized the unit hereditary factors as English alphabets, using capital letters for dominant and small letters for recessive factors. For instance, he symbolized the factor for tallness as 'T' and that for dwarfness as 't'.

Mendel's method of symbolization has been retained by modern geneticists, but certain changes have been introduced. The parental generation is symbolised simply as P and not as P_1. Further, it has been visualized that recessive factors have mostly arisen by mutation of original 'wild type' dominant factors. The modern geneticists, therefore derive the symbols from the name of recessive characters. Thus, the factor for tallness is now symbolized as D and that for dwarfness as d.

CHECKER BOARD (PUNNET'S SQUARE) METHOD

The possible genotypes and phenotypes resulting from random combinations of gametes during fertilization in hybridization experiments are more conveniently determined by using checker board method, devised by *Punnet* (1906). This board is a geometrical device, a square with vertical and horizontal columns. Possible types of female gametes (according to the gene of a trait these can receive from the female parent) are shown in small spheres above the vertical columns. Similarly, the possible types of male gametes are shown in spheres on the right of horizontal columns. The possible genotype combinations and their phenotypes are then entered in the boxes of the square.

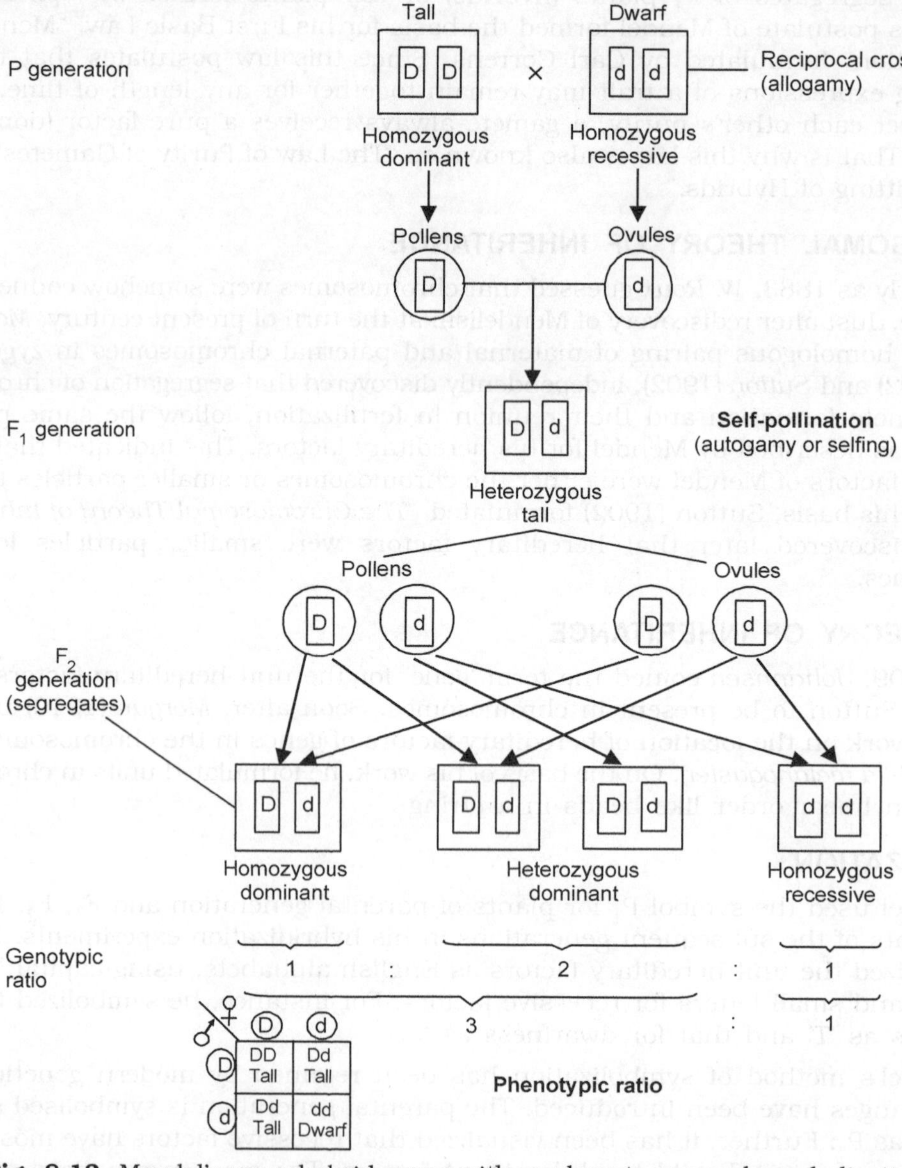

Fig. 8.13. Mendel's monohybrid cross with modern terms and symbolization.

BACK CROSSES AND TEST CROSSES

To further confirm the purity of his unit factors, Mendel sometimes crossed the F_1 hybrids with either one of the parental types of plants (Fig. 8.13). Such a cross is called a *back cross*. An individual showing the dominant expression of a trait may be genotypically homozygous (DD) or heterozygous (Dd) for this trait. To determine its genotype, it is backcrossed with the recessive parental (homozygous dd) genotype. This is called a *test cross*. If the unknown genotype of the tested individual is homozygous (DD), the *testcross* yields all hybrid dominants (Dd) as in a normal monohybrid cross, but if heterozygous (Dd), the test cross yields dominants and recessives in 1 : 1 ratio.

MENDEL'S DIHYBRID CROSSES

In the next phase of his experiments, Mendel crossed his pea plants to study inheritance of any two of the selected genetic traits together and to clarify their mutual relationship in inheritance. These crosses are called *dihybrid crosses*. For example, he crossed a pure-breeding plant that produced round seeds with yellow cotyledons (WWGG) with a pure-breeding plant that produced wrinkled seeds with green cotyledons (wwgg Fig. 8.14). In his monohybrid crosses, he had already determined that the round shape

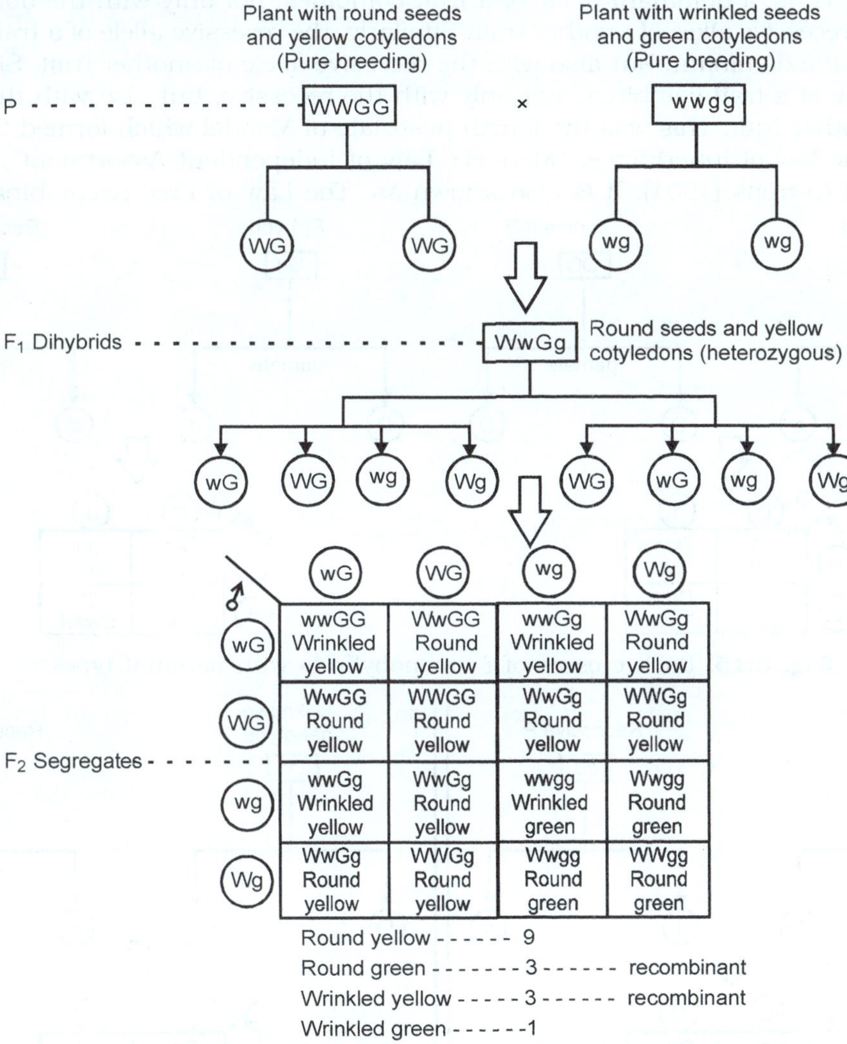

Fig. 8.14. A dihybrid cross of Mendel.

of seeds was dominant to wrinkled shape and yellow colour of cotyledons was dominant to green colour. All F_1 plants (dihybrids – WwGg) of this cross were allowed to reproduce by self-fertilization (selfing or autogamy) and as was expected, these produced round seeds with yellow cotyledons. He collected 556 seeds and sowed these to study the condition in F_2 generation. He found that, on selfing 315 of the F_2 plants produced round and yellow seeds, 108 produced round and green seeds, 101 produced wrinkled and yellow seeds and 32 produced wrinkled and green seeds.

MENDEL'S LAW OF INDEPENDENT ASSORTMENT

The F_2 generation of the above dihybrid cross obviously yielded, not only the two parental phenotypes, but also two new phenotypes (called 'recombinant types'), *viz.* round seeds with green cotyledons and wrinkled seeds with yellow cotyledons. It thus, revealed the fact that segregation of the two alleles of a given genetic trait is entirely independent of the segregation of the alleles of any other trait. In other words, an allele (dominant or recessive) of a given trait freely combines with either one (dominant or recessive) of the alleles of another trait. Thus, a dominant allele of a trait combines, not only with the dominant, but also with the recessive allele of another trait. Similarly, the recessive allele of a trait combines, not only with the dominant, but also with the recessive allele of another trait. Similarly, the recessive allele of a trait combines, not only with the recessive, but also with the dominant allele of any other trait. This was the fourth postulate of Mendel which formed the basis for his third basic law of inheritance. "Mendel's Law of Independent Assortment", formulated again by Carl Correns (1901). It is also known as "The Law of Free Recombination."

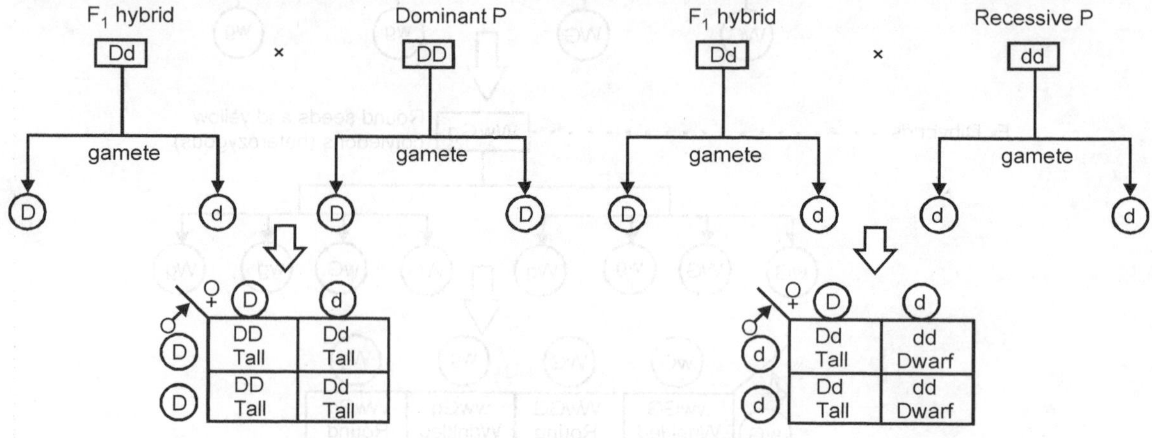

Fig. 8.15. Back crosses of F_1 monohybrids with parental types.

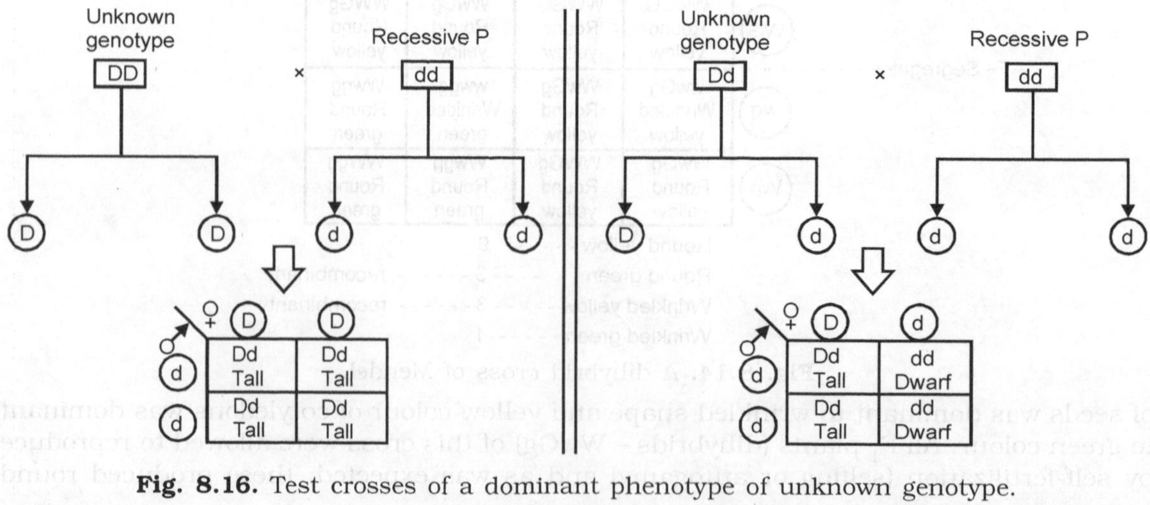

Fig. 8.16. Test crosses of a dominant phenotype of unknown genotype.

As it is clear from Mendel's dihybrid crosses and becomes more clear if we ourselves work out such a cross with checkerboard method, the number of F_2 phenotypes closely fits to a ratio of 9 : 3 : 3 : 1 as follows:

9 round seeds with yellow cotyledons
3 round seeds with green cotyledons
3 wrinkled seeds with yellow cotyledons
1 wrinkled seed with green cotyledons

If we consider the two traits of this dihybrid cross separately, their F_2 ratios in this cross are similar to those of monohybrid crosses. For example, the ratio of round (12) to wrinkled (4) or of yellow (12) to green (4) is 3 : 1 in accordance with perfect segregation. Similarly, when considered separately, the F_2 ratio for the two new phenotypes in this cross are also 3 : 1, viz. round (12) to green (4) and yellow (12) to wrinkled (4). This confirms the "*Law of Independent Assortment.*"

In dihybrid crosses, the ratios of monohybrid crosses are raised to the power of two. Thus, the monohybrid F_2 phenotypic ratio 3 : 1 becomes $(3 : 1)^2$ = 9 : 3 : 3 : 1 of dihybrid cross. Similarly, the number of F_2 genotypes (3) in a monohybrid cross become $(3)^2$ = 9 in a dihybrid cross as follows:

1 WWGG; 2 WwGG; 2 WWGg; 4 WwGg; 1WWgg
2 Wwgg; 1 wwGG; 2 wwGg; and 1 wwgg.

A dihybrid testcross of F_1 hybrid (WwGg) with the double recessive parental type (wwgg) yields the usual four phenotypes in 1 : 1 : 1 : 1 ratio.

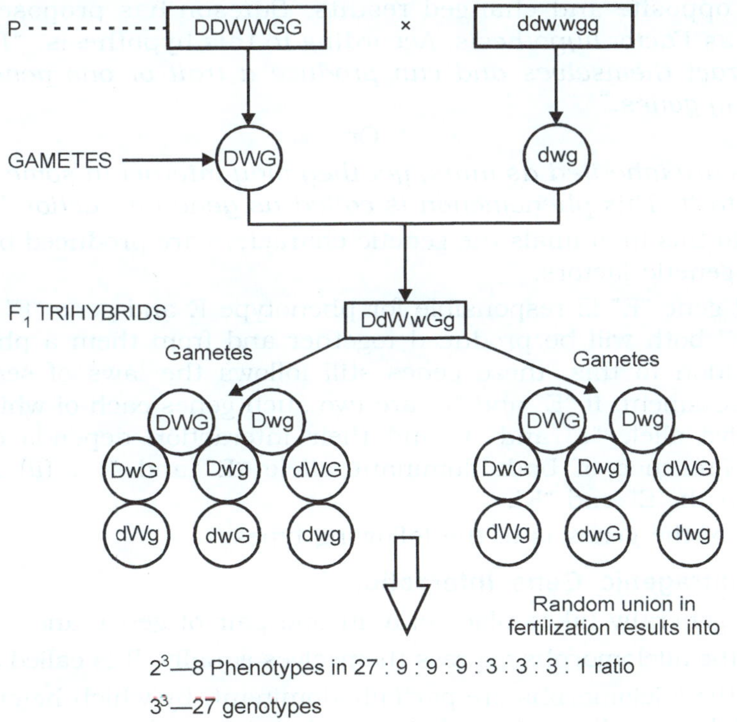

Fig. 8.17. Mendel's trihybrid cross.

MENDEL'S TRIHYBRID CROSSES

Mendel further demonstrated that identical processes of segregation and independent assortment apply to simultaneous inheritance of three or more genetic traits, each transmitted by an allelic pair of factors. If, for example, a pea plant which is pure-breeding for tallness, round seeds and yellow cotyledons (DDWWGG) is crossed with another plant which is pure breeding for dwarfness, wrinkled seeds and green cotyledons (ddwwgg), the F_1 trihybrids (DdWwGg-Fig. 8.17) will be tall stem plants producing round seeds with yellow cotyledons. Segregation and independent assortment will yield eight types of pollens in the flowers of each F_1 plant. A checkerboard of 64 boxes is, therefore to be constructed in this case to find out the number of F_2 phenotypes and genotypes combinations. These number are, as expected equal to the power of three to the number of monohybrid crosses. For example the F_2 phenotypic number in a trihybrid cross is $(2)^3 = 8$, and genotypic number is $(3)^3 = 27$.

The number of F_2 phenotypes and genotypes in cross for more than three genetic traits (multihybrid crosses) can be similarly calculated by 2" and 3".

GENE-INTERACTION

Introduction on the basis of hybrid crosses, Mendel have concluded that each trait in plants is regulated by either one gene or only one factor. Similarly, there was one more conclusion that form the pair of two alleles, one is dominant and the other one is recessive, but afterwards scientists after performing many experiments could find out that there are some exceptions of the Mendelian ratio. This is not necessary that only one trait is regulated or controlled by only one gene. On the basis of above opposite and changed results, *Bateson* has proposed a hypothesis, which is called as *Factor hypothesis*. According to this hypothesis, *"Two or more than two genes interact themselves and can produce a trait or one genetic character is effected by many genes."*

Or

"The genes are inherited as units, yet they may interact in some complex fashion to produce the trait. This phenomenon is called as gene interaction."

According to this in animals the genetic characters are produced by the interaction in between the genetic factors.

Example: If gene "E" is responsible for phenotype E and gene "F" for phenotype F than "E" and "F" both will be produced together and from them a phenotype "G" will develop. In addition to this, these genes still follows the laws of segregation and of Independent assessment. If "E" and "F" are two such genes each of which are dominant on their recessive allele "e" and "f" and their interaction depends on the following condition— (*i*) Presence of both dominant allele "E" and "F", (*ii*) Absence of "E", (*iii*) Absence of both "E" and "F".

Interaction of the genes is of the following types:

Non-allelic or Intragenic Gene Interaction

Interaction generally takes place only in one pair of genes and in it:

1. Both the allelomorphs express themselves equally. It is called as codominance.
2. Both the allelomorphs are partially dominant, by which heterozygous express or exhibits medium effect. It is called as incomplete or partial dominance.
3. Out of both the allelomorphs, one is completely dominant and another one is recessive. It is called as complete dominance.

Such types of interaction are called as non-allelic or intragenic gene interaction.

Non-allelic or Intergenic Gene Interaction

In this mechanism, interaction takes place for the expression of only one trait in between two or more pairs of allelomorphic genes, present on the same or different chromosomes or one trait/one pair of allelomorph effects one or more than one set of genes. The interaction of such non-allelic genes is called as non-allelic or intergenic interaction.

Non-allelic genes interact by the following ways:

1. Supplementary Genes 9 : 3 : 4 Ratio. Sometime two such different types of genes interact together that one of them can express itself completely while no expression of the other. These two genes interact and forms a phenotype. So, supplementary genes are two independent and dominant genes, one dominant gene produces its effect whether the other is present or not, when second dominant gene comes in contact with the first gene then it changes the effect of the first, but when this second gene is alone, it cannot express itself.

Example 1: Supplementary gene completely changes the effect of normal gene.

Castle, considering the skin colour has performed many experiments on Rats, Guinea pigs and other Rodents. Scientists have crossed gray coloured mice with albino mice. In F_1 progeny, colour of the skin of all the mice was agouti. On account of self fertilization in F_2 progeny, agouti and albino mice were produced in 9 : 3 : 4 ratio. Scientists accepted that black genes are present in the albino rats but have no gene for albino colour. So in the presence of both the genes the black colour will develop, but due to the presence of supplementary genes black colour is converted into agouti colour (See Fig. 8.18). If it is supposed that

B = for black colour gene
b = recessive allelomorph of black
C = pigment promoter gene
c = recessive allelomorph of C
G = pigment converter gene
g = recessive allelomorph of G

From the above mentioned experiment conclusion is as follows:

(*i*) Mice are albino because colour developer gene is absent in them.

(*ii*) In all the mice which may be of any colour, black colour B is always present in them.

(*iii*) If along with black colours gene colour developer gene (C) is also present then black colour will develop in mice.

(*iv*) If along with both these (B, C), colour converter gene (G) is also present then agouti colour will develop.

G is the supplementary gene of C.

Example 2: Black purple colour (P) is dominant on brown colour (P) in *Sorghum.* Sometimes it was observed during cross that black colour is converted into red colour. Gene Q for changing the colour is present on another chromosome. Q when present with recessive P, there will be no expression of it but when both dominant p^1 and Q are present together then black colour is changed or converted into red. In F_2 progeny, the genotype genes are shown in Fig. 8.19 and Table 8.2.

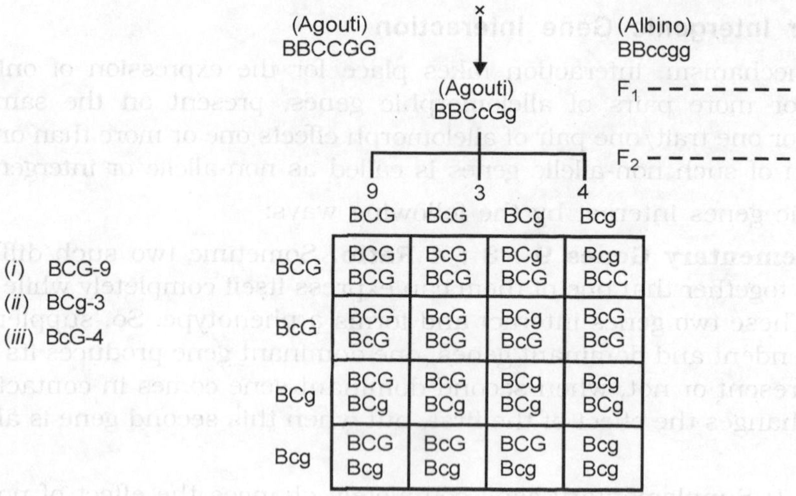

Fig. 8.18. Interaction between supplementary genes for skin colour in Mice.

Table 8.2. Abbreviated Genotype Ratio and Phenotypic Ratio Obtained in F_2 in a Cross shown in Fig. 8.18.

Abbreviated Genotype Ratio	PQ 9 :	pq 3 :	pQ 3 :	Pq 1
Phenotypic Ratio	Redish purple 9 :	Blackish purple 3 :		Brown 4

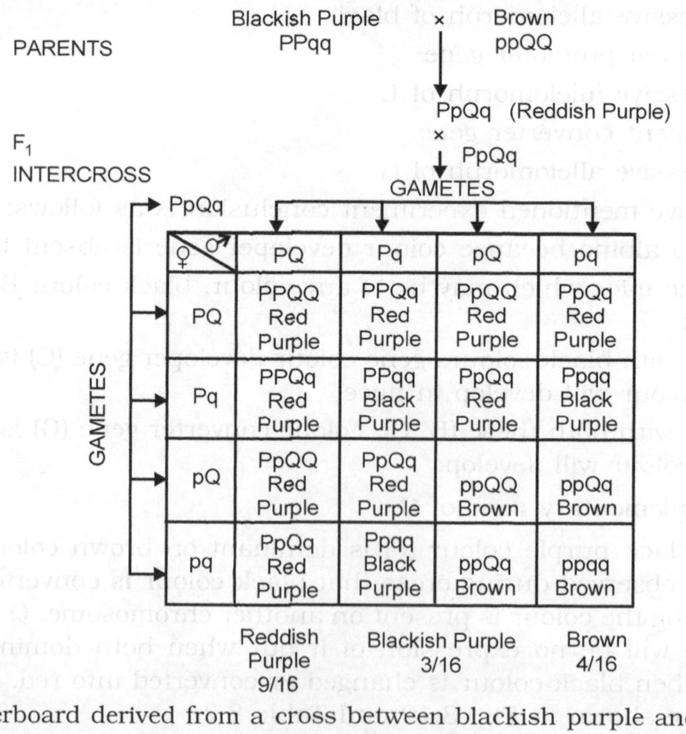

Fig. 8.19. Checkerboard derived from a cross between blackish purple and brown coloured grain character in Sorghum.

2. Complementary Genes. Even two independent genes interact itself and produces only a phenotype, but neither gene is alone sufficient to produce that gene or for one specific phenotype both the genes are essential. Such genes are called as complementary genes.

Example 1: Bateson and Punnet in their experiments have demonstrated that when two specific subspecies of white flowered sweat pea *Lathyrus odoratus* were crossed together artificially, then in F_1 progeny coloured flowers were obtained. By self fertilization in F_1 progeny, F_2 progeny was obtained, then white flowered plants were obtained in 9 : 7 ratio. According to Bateson 9 : 7 ratio is a changed/converted ratio 9 : 3 : 3 : 1 is of Mendelian ratio. According to them flower of F_1 progeny are formed by the interaction of gene C and P. When these two dominant genes are alone then white coloured flowers are formed. For red colour in the flower the presence of these two genes (C and P) is essential. Neither gene alone is sufficient to produce red colour. So C and P are complementary to each other (Table 8.3).

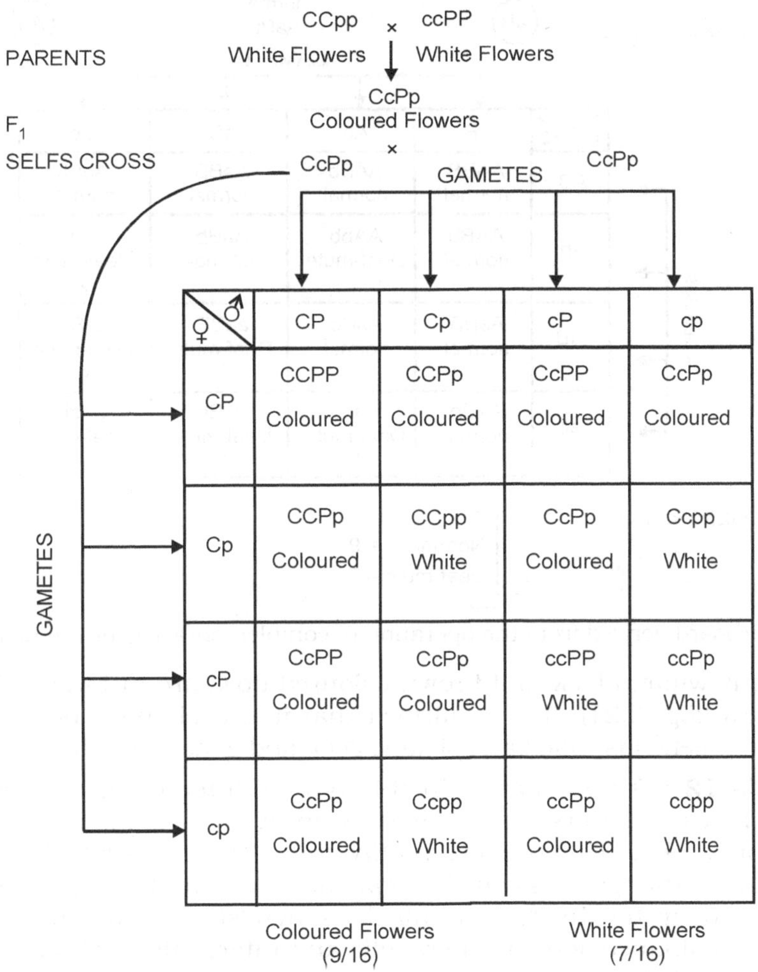

Fig. 8.20. Checkerboard derived from a cross between two white flowered varieties of sweet pea.

Table 8.3. Abbreviated Genotypic Ratio and Phenotypic Ratio Obtained in F_2 in a Cross shown in Fig. 8.18.

Abbreviated Genotype Ratio	CP 9 :	Cp 3 :	cp 3 :	cp 1
Phenotypic ratio	Coloured flowers 9 : ratio	White flowers 7		

Example 2: Deaf-mutism in Man is an example of complementary genes. In man the power of hearing and of speech is on account of interaction of dominant genes A and B. Whenever in man anyone of the gene amongst the two is recessive allelomorph homozygous then this disease or abnormality develops in man (See Fig. 8.21).

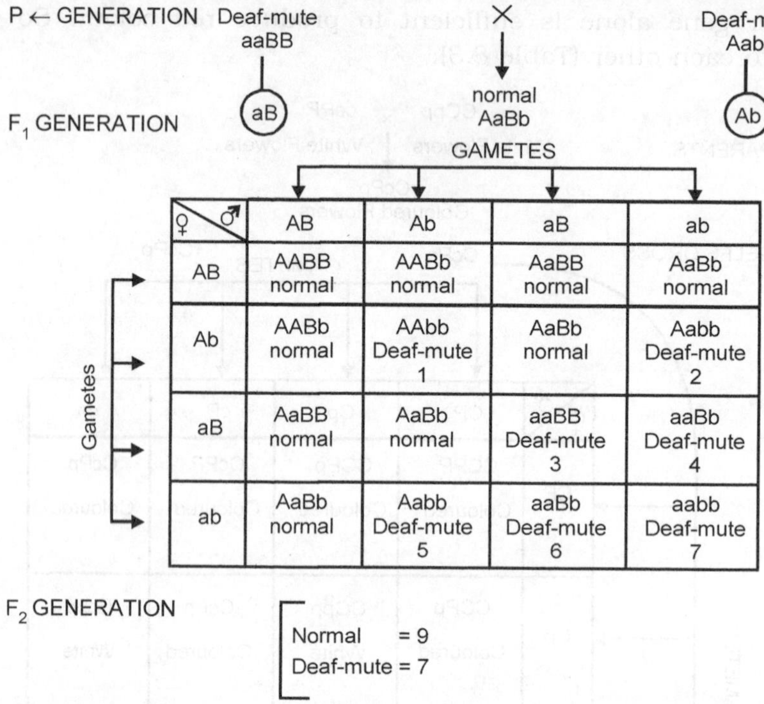

Fig. 8.21. Checkerboard derived from the operation of complementary genes in deaf-mutism in Man.

In F_2 progeny white, black and brown coloured dogs are in ratio of 12 : 3 : 1. From the checker board (Fig. 8.21), it is confirmed that it is a changed or converted dihydrid cross ratio and in it gene I is called an epistatic gene and gene B is called as hypostatic gene.

3. Epistasis 12 : 3 : 1 : Ratio. In this type of interaction the expression of one dominant gene hides or covers up the expression of other dominant gene, or prevents the expression of another. This is called as *Epistasis*. The genes which hides or prevents the expression of another gene is called as *epistatic factor* and the expression of the gene which is prevented or the hidden is called as hypostatic gene. It is just similar to dominance, but dominance acts on interallelic and intergenic level, but epistasis is only intergenic and non-allelic.

From above mentioned two genes, C is responsible for the production of a colourless chromogen. It forms such an enzyme which produces red colour pigment anthocyanin, activate for the production of a pigment coloured substance. Gene P regulates the production of the enzyme. While in pea plant, both dominant genes C and P are present resulting in the production of red flowers. Only white flowers will develop. Inspite of the presence of dominant gene P it will not able to produce raw material for anthocyanin in ccPP because in it, C gene is absent, therefore in both the conditions, absence of the gene, only white flowers will develop.

Genotype	*Enzymes*	*Phenotype*	*Number of plants*	*Ratio*
CCPP CcPP CcPp CCPp	In it for raw material an enzyme is present for the production of anthocyanin or an enzyme is present for the formation/production of anthocyanin from raw material or both enzymes are present.	Purple	1 2 4 2	9
CCPP CcPP	Enzymes of only raw material is present	White	3	
ccPP	Enzyme present for the production of	White	3	7
ccPp	anthocyanin from raw material.			
ccpP	Enzyme is absent.	White	1	

Epistasis is of two-types : (*i*) Dominant Epistasis and (*ii*) Recessive Epistasis.

(*i*) ***Dominant Epistasis.*** When amongst two genes, an allelomorph of one gene performs in the function of a dominant allelomorph inhibits the effect of another allelomorph then it is called as Dominant allelomorph.

Example 1: White colour of the skin in dog is dominant but it appears on account of the effect of dominant gene I, because the dominant gene B can inhibit the production of colour.

Gene B = Black colour
recessive allelomorph b = brown colour
Gene I = Dominant
Recessive allelomorph of gene I = i
Black coloured dogs = Bbii or Bbii
Brown coloured dogs = bbii
White coloured dogs = BBII, BbII, BBIi, BbIi, bbII, bbIi

Example 2: The grain of cholam may be pearly or chalky in appearance. The pearly grains are shining and translucent whereas chalky grains are salt white and opaque. Pearly colour of cholam grain is dominant over chalky colour. When dominant gene "Z" is present then pearly white colour develops and in the absence of it chalky colour develops. Another one gene "W" is also present. Red colour develops When either Z gene is present or absent. In the absence of "W" gene, the grains colour will be pearly or chalky which, depends on the presence of Z gene (Fig. 8.22 and Table 8.4).

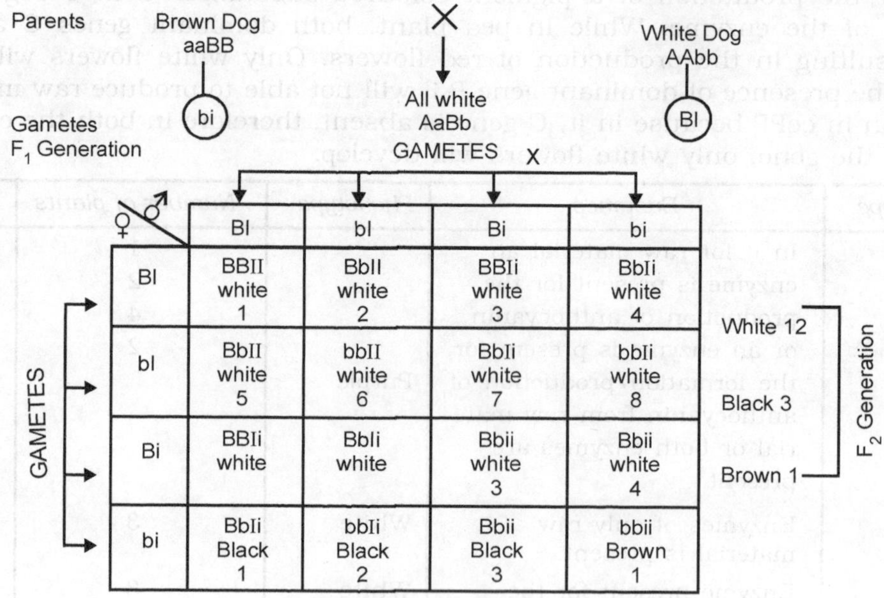

Fig. 8.22. Epistasis in dog for skin colour.

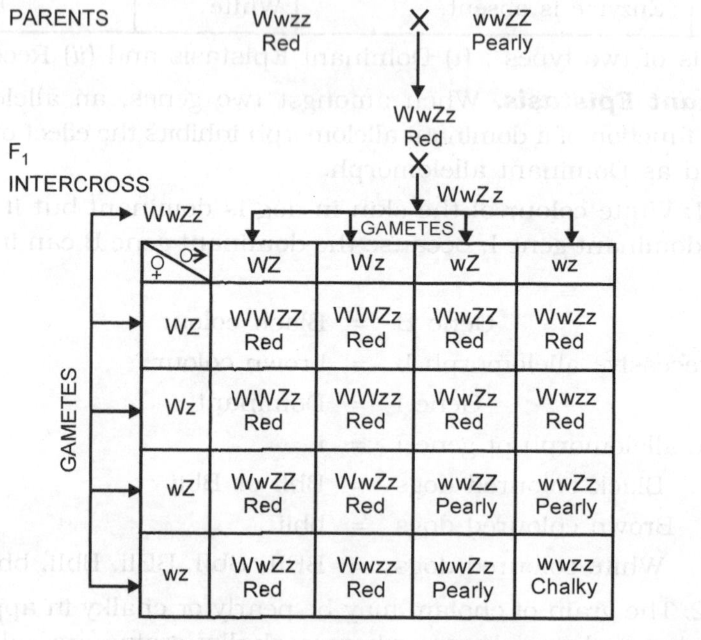

Fig. 8.23. Checkerboard derived from a cross between red grained and pearly grained varieties of cholam.

Table 8.4. Abbreviated Genotypic and Phenotypic Ratio Obtained in F_2, a Cross shown in Fig.23

Abbreviated genotype ratio	WZ 9 :	Wz 3 :	wZ 3 :	Wz 1
Phenotypie ratio	Red 12 :		Pearly 3 :	Chalky 1

(*ii*) ***Recessive Epistasis.*** In the process dominant gene is recessive and it inhibits the position of the genes. Epistasis may be recessive. If any recessive allele do not allow to be expressed by dominant allelomorph then it is called as recessive epistasis. Example— In mice wild colour is agouti. In mice when dominant gene is C then coloured mice develops, but gene C and A together develops agouti mice. But gene A could not be expressed in the presence of gene C. When coloured mice Ccaa is crossed with albino ccAA then in F_1 progeny, agouti mice will be produced as shown in Fig. 8.24.

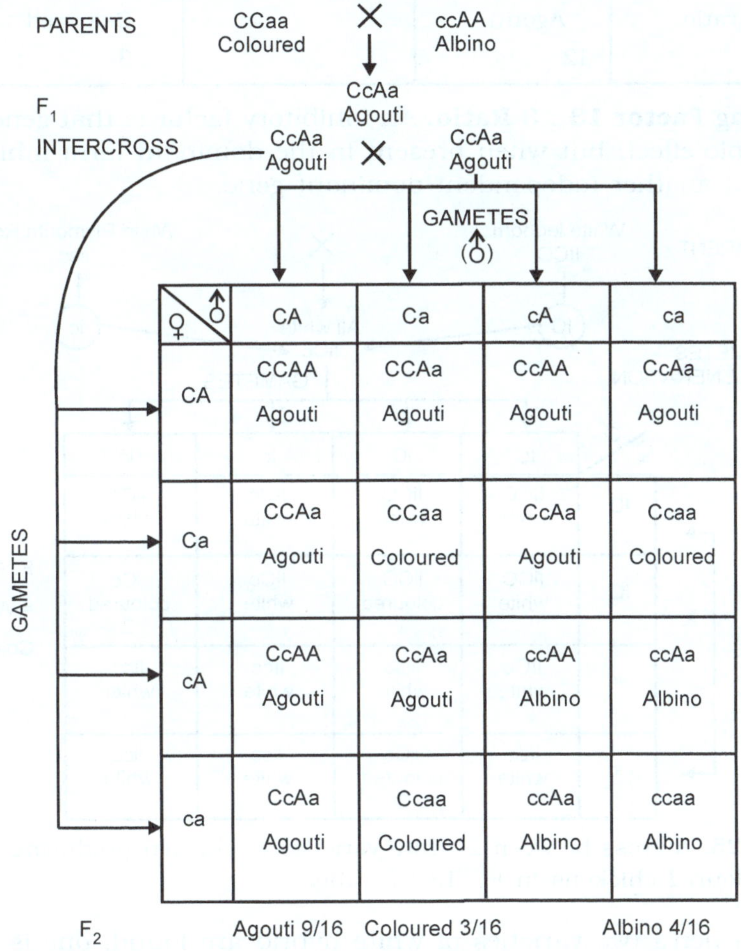

Fig. 8.24. Checkerboard derived from a cross between coloured mice.

Example: Colour of the skin in mice is regulated by recessive dominant gene/ recessive epistasis.

Gene A	=	Agouti colour
Allelomorph of A	=	a = Albino
Gene C	=	Black colour
Allelomorph C	=	c = dominant
Agouti mice	=	CA = 9
Black mice	=	Ca = 3
Albino mice	=	ccAA = 4
		CCAa

Recessive allelomorph of gene C is a dominant factor and gene A alters the expression of gene C.

Abbreviated genotype ratio	CA 9 :	Ca 3 :	cA 3 :	ca 1
Phenotypie ratio	Agouti 12 :		Coloured 3 :	Albino 47

4. Inhibiting Factor 13 : 3 Ratio. An inhibitory factor is that gene which by itself has no phenotypic effect, but when present in the dominant form inhibits or prevents the expression of another independent dominant gene.

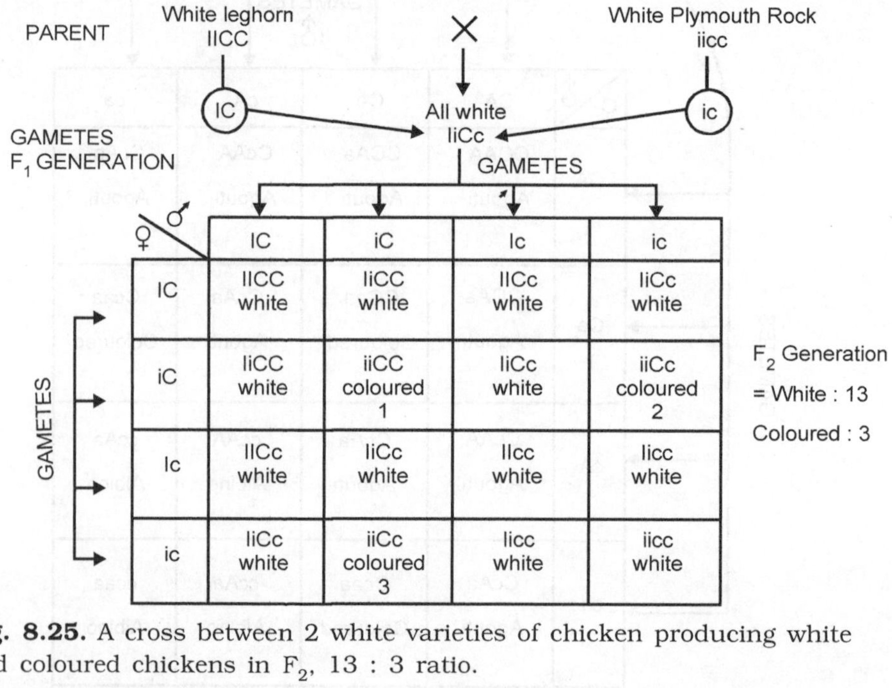

Fig. 8.25. A cross between 2 white varieties of chicken producing white and coloured chickens in F_2, 13 : 3 ratio.

Example: In hen's two varieties of white hybrid are found, one is called as White leghorn, and second white Plymouth rock. In the white leghorn, white feathers are dominant over coloured feathers, whereas in white Plymouth rock variety, the white feathers is recessive over the coloured feathers. When white leghorn is crossed with white Plymouth hen in F_1 generation, all the hen's progeny will be of white colour. But in F_2 generation, hen's of white and coloured feathers born in 13 : 3 ratio. This process is a changed ratio ro converted ratio of Mendelian dihybrid ratio. The cause of dominant factor of white leghorn is the presence of an inhibitory factor, which inhibits the effect even if the coloured dominant gene is present.

Inhibiting factor	=	white feathers develops	
recessive allelomorph of I	=	(*i*) when coloured and inhibition Both factors are present	= IC
Pigmentation factor	= C	(*ii*) when both coloured and Inhibition factors are absent	= ic
recessive allelomorph of C	= c	(*iii*) when only pigmentation factor is absent	= ccII ccIi

5. Lethal Factor. There are such genes which never effect the phenotype but effects the living capacity. The effect on the living capacity will be such that on account of which the living being will not be in a position to survive. Such types of genes which are responsible for the death of possessor gene are called as Lethal genes. Such lethal genes causes such a tremendous deviation in general development and functional nature that the possessor does not survive. Such organisms may die at any time even from fertilization to complete adulthood. Effect of such lethal gene is responsible for the death, even zygote dies immediately after the fertilization or embryo dies in the early developmental stage.

Organisms with dominant lethal genes could not survive therefore they are not inherited from one generation to another generation.

Example: In 1905, *L. Cuenot*, a French geneticist has shown the results of the heredity of coloured body in Mice; which was not according to Mendelian ratio. Scientists have shown that yellow colour of the body in mice is dominant over normal brown colour and is regulated by only one gene. Such gene is called as "Y" gene. Yellow mice are always heterozygous for dominant colour of skin on account of reproduction in yellow mice, 2 :1 ratio is present in offspring instead of 3 :1.

Steigleder in 1877, showed that on account of inter breeding, 1/4 offsprings of rats die in the embryonic condition. If the yellow colour of skin 'Y' and its allelomorphic brown colour is denoted by y then its heterozygous genetic condition will be Yy. Interbreeding in mice is shown in Fig. 8.26. From which it is clear that homozygous yellow mice (YY) dies immediately or straight away since thc gene in homozygous condition produces a lethal effect on the embryo, therefore, homozygous yellow mice (YY) never survive. Heterozygous mice develops in 2 : 1 ratio.

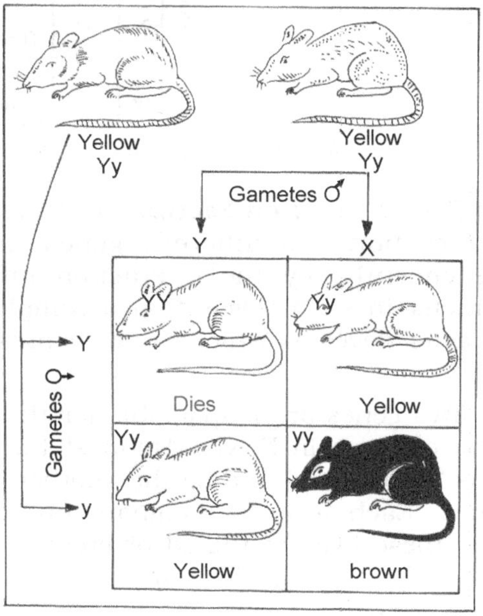

Fig. 8.26. Inheritance of Lethal genes in Mice.

6. Duplicate Factors 15:1 Ratio. When two or more genes in heterozygous chromosomes have an identical effect on a trait, it is called as an duplicate factor. Scientist *G.H. Shull* studied the *Shepherd's purse* plant of genus *Capsella*. He obtained an unusual modified ratio 15:1 from his study on *Shepherd's purse* (Normal Mendelian ratio 9 : 3 : 3 : 1). In this plant two types of seed pods are produced (*i*) Triangular (*ii*) long. On crossing these in F_2 generation, triangular and long seeds pods plants in 15 : 1 ratio were obtained. Triangular character of seed pods is regulated by 2 pairs of genes T and D. These genes are present on separate chromosomes. Both of these produces triangular traits either separately or together. In the absence of both dominant genes, elongated seed pods develops. These genes are called as duplicate genes.

From the checkerboard, it is clear that out of 16 seed pods of F_1 generation, at the most 15 have dominant gene. Therefore, triangular seed pods are present in all the plants. In one pod both the genes are recessive, hence it is elongated.

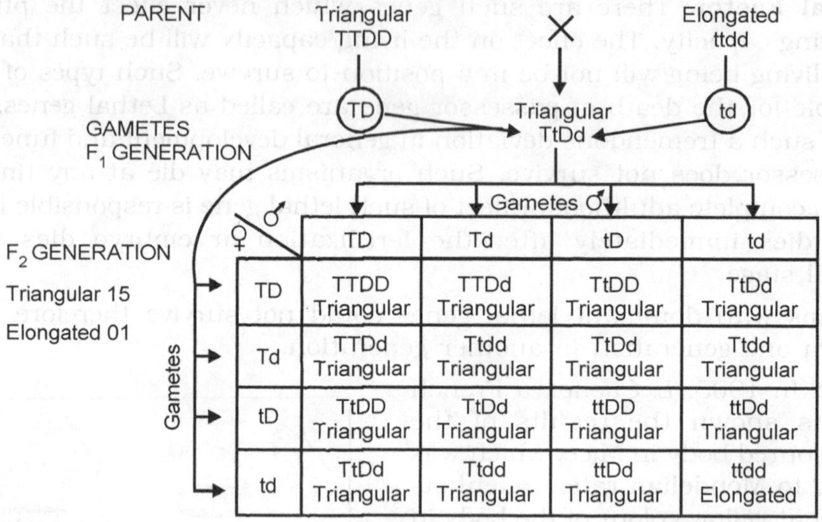

♀ \ ♂	TD	Td	tD	td
TD	TTDD Triangular	TTDd Triangular	TtDD Triangular	TtDd Triangular
Td	TTDd Triangular	TTdd Triangular	TtDd Triangular	Ttdd Triangular
tD	TtDD Triangular	TtDd Triangular	ttDD Triangular	ttDd Triangular
td	TtDd Triangular	Ttdd Triangular	ttDd Triangular	ttdd Elongated

Fig. 8.27. Interaction of duplicating genes in Shepherd's purse.

7. Collaboration Factors. 9 : 3 : 3 : 1 ratio—"In collaboration two different genes are situated independently, by the interaction influence one character in such a way that a complete different character develops which is not expressed by any gene."

Two genes are responsible for the trait of the shape of comb in Fowls. Amongst them one gene forms "Rose comb" and by the another Pea comb is formed. Each gene is dominant on single comb. When these two are united or crossed then a new comb is formed—the Walnut.

According to the experiments performed by W. Bateson and R.C. Punnet, gene R develops Rose comb and gene P develops Pea comb. By the cross of these two types in F_1 generation Walnut comb

Fig. 8.28. Different comb-shapes in poultry.

develops. It is different from the two combs. By the cross in F_1 generation four types of characters develops in F_2 generation—Walnut, Rose, Pea and Single. From these characters Walnut and Single were not in parental combination. From the cross of organisms of F_1 generation four types of characters in 9 : 3 : 3 : 1 ratio are obtained.

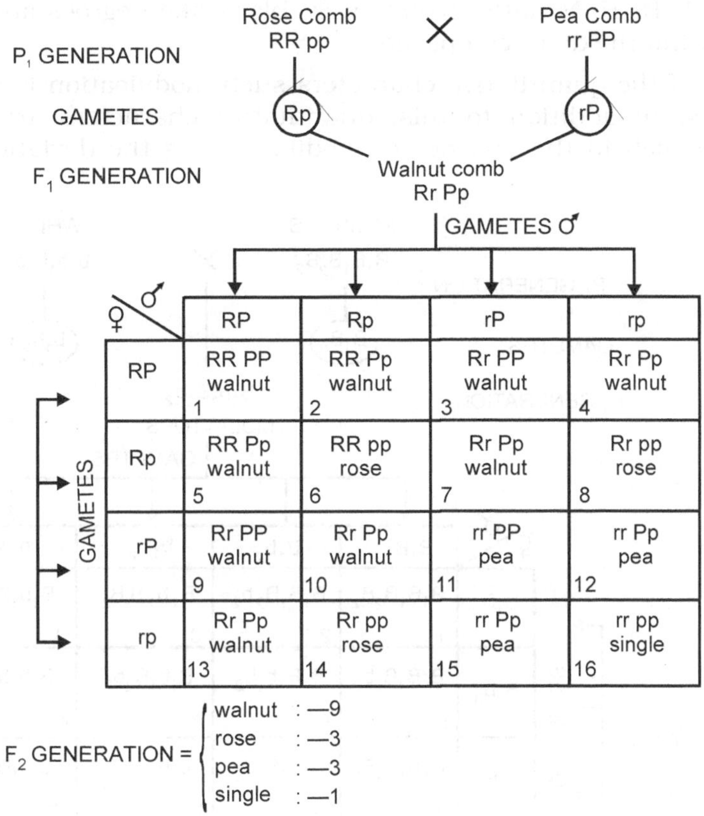

Fig. 8.29. Inheritance of Rose and Pea comb in Poultry.

8. Multiple Factors or Polygene or Cumulative Factors or Quantitative Inheritance. The cumulative genes quantitative in nature help in the multiple of the traits. This concludes that more the number of genes present more the effect of the genes increases. If more than one, two, three or four pairs of cumulative genes are present then its effect increases with that ratio.

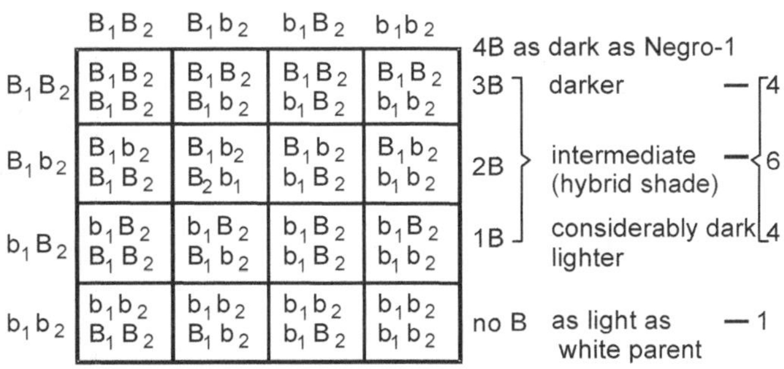

Fig. 8.30 (a)

C.B. Devenport in 1913 have studied the inheritance of skin colour in the human beings of white and Negro race of United States of America. These people of United States of America, who are born from the inter-marriage relation of Negro and white race people are known as Mulattoes. In F_2 generation, the ratio of Mulattoes and white children was 15 : 1. In 15 Mulattoes some were black like negroes and rest are usually more or less with intermediate complexion.

For majority of the quantitative characters such modification is on account of the presence of genes. In addition to this, quantitative characters are also effected by environment; by which in the absence of modified genes the deviation from real ratio may occur.

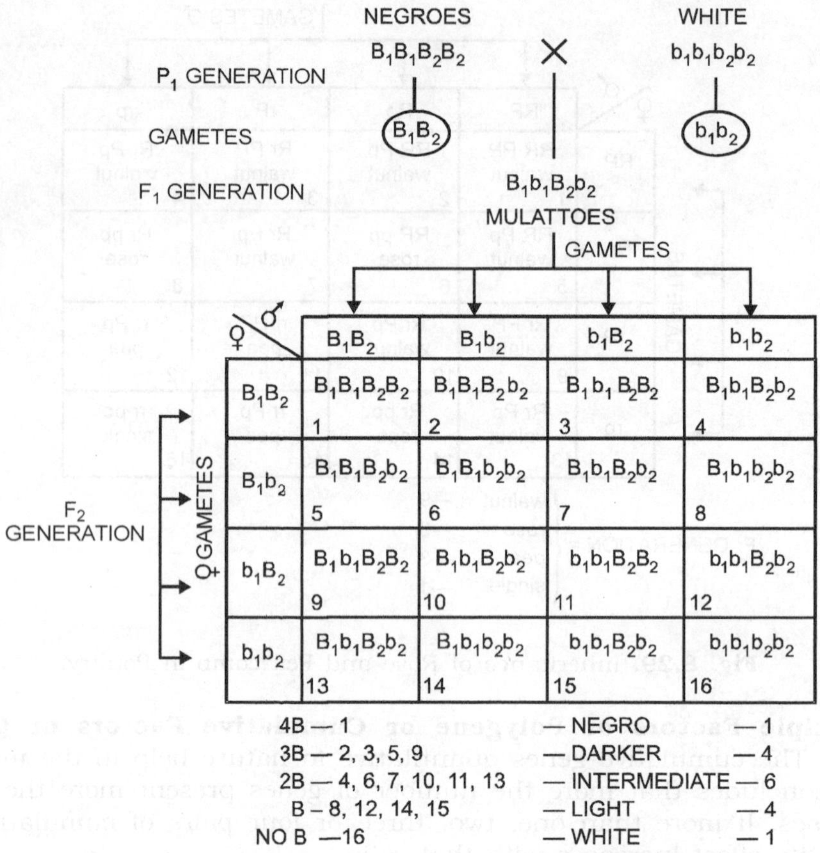

Fig. 8.30. (b) Interaction of polygenes for skin colour of man.

9 BANDING PATTERN IN HUMAN CHROMOSOMES

HISTORY

In 1912, *Winiwater* was successful attempt to count the number of human chromosomes. He was proposed that human chromosomes are 48 in women and 47 in man; in this number men having one X-chromosome and women having two X chromosome. *Painter*, in 1923, while examining the testicular material of man, observed a heteromorphic pair of sex chromosomes and proposed the XY mechanism of sex determination in man. *Tjio & Levan* (1956), cultured somatic cells from fibroblasts of human embroyos and counted the human chromosome number as 46. This chromosome number was confirmed by *Ford* and *Hamerton* while working with testicular material in the same year. *Tjio* and *Levan* provided greatly improved techniques for chromosome preparations. *Moorhead et al.*, (1960), described a simple method of culturing of lymphocytes from human blood.

Karyotyping Human Chromosomes

For karyotyping of human chromosomes, venous blood is taken and blood leucocytes are stimulated to divide (by mitosis) *in vitro* by the addition of phytohaemaglutinin. Colchicine is added to arrest cell division at metaphase stage. It is further treated with hypotonic saline solution which results in swelling of cells and dispersal and better clarity of chromosomes for counting and morphological study. Thereafter, the material is stained (*e.g.*, with Giemsa technique) to demonstrate the banding patterns of chromosomes. Finally, a suitable metaphase spread is photographed through a high power microscope. The individual chromosomes are cut out from the photograph. The chromosomes are then arranged in an orderly fashion in homologous pairs to produce a standard arrangement, the karyotype.

To charactierise a chromosome in the karyotype, the following parameters are used:

1. Shape of chromosome
2. Length of chromosome
3. Centromeric index, *i.e.*, this index is expressed in the form of ratio of the short arm length to the total chromosome length:

$$\text{Centromeric index} = \frac{\text{Short arm length}}{\text{Total chromosome length}}$$

For example, centromeric index in a metacentric chromosome is 0.5.

4. Proportion of the arms, *i.e.*, it is ratio between the long arm and short arm of the chromosome. This ratio is 1:1 in a typical metacentric chromosome.

CLASSIFICATION OF HUMAN METAPHASE CHROMOSOME

Human metaphase chromosomes were first of all classified by a conference of cytogeneticists at Denver, Colorado in 1960 and is known as the Denver classification. To follow this classification, each of the 22 pairs of autosomes has been numbered from 1 to 22 according to their decreasing size. Patau (1960), divided the human chromosomes into the following seven groups designated A to G:

1. A group : 1 to 3 pairs – Metacentric
2. B group : 4 to 5 paris – Submetacentric
3. C group : 6 to 12 pairs – Submetacentric
4. D group : 13 to 15 pairs – Acrocentric
5. E group : 16 to 18 pairs – Submetacentric (16 is metacentric)
6. F group : 19 to 20 pairs – Metacentric
7. G group : 21 to 22 pairs – Acrocentric

Group A consists of longest metacentric chromosomes. Group G consists of the shortest acrocentric chromosomes. These chromosomes have satellites that correspond to nucleolar organizers. Chromosomes of group D also contains satellites. In males, group G includes a variable Y chromosome which lacks the satellites.

The X chromosome is the member of group C and can be identified by special banding or staining methods.

BANDING TECHNIQUES

Recently banding techniques reveals structural details of chromosomes. The main banding techniques are identified by letters such as Q, G, C, R, T, F and N bands :

1. **Q Banding.** It uses fluorescent dyes (such as quinacrine mustard) and identifies the so-called Q bands.
2. **G Banding.** It uses Giemsa stain and identifies the G bands. With G banding three major types of chromatin can be recognized-euchromatin, centromeric and intercalary heterochromatin. The Q and G bands are generally similar and correspond to intercalary heterochromatin.
3. **C Banding.** It stains specifically centromeric constitutive heterochromatin.
4. **R Banding.** It gives a pattern that is the reverse of that of Q and G banding.
5. **T Banding.** It stains telomeres of chromosomes.

In some banding techniques uses the Feulgen stain (F bands) and another selective stains the nuclear organizers (N bands) which are localized in the satellite chromosomes 13, 14, 15, 21 and 22. G banding is also important tool in the analysis of mammalian, avian, reptilian and amphibians chromosomes.

Clinical Importance of Chromosome Banding

Since banding patterns are unique and constant for each normal chromosomes, in case of a large number of chromosomal abnormalities, such as loss of a very small part, insertion of an additional segment and addition of whole chromosome can be easily recognized, *e.g.*, cat-cry syndrome due to loss of small part of chromosome 5; down syndrome due to an extra chromosome 21.

10 Sex Determination in Animals, Sex Linkages

INTRODUCTION

There are physiologic and cytologic proofs that sex is determined as soon as the egg is fertilized and that it depends on the gametes. Among the physiologic evidences is the finding that, identical twins—which originate from a single zygote are always of the same sex. Furthermore, in certain species having polyembryos development (*e.g.*, armadillo), all the embryos that have develope from a single fertilized egg are of the same sex. Cytologic evidence was first obtained by *Mc Clung*, who demonstrated that the karyotype of a cell is composed of not only common chromosomes (autosomes), but also of one or more special chromosomes that are distinguished from the autosomes by their morphologic characteristics and behaviour. These were called accessory chromosomes, allosomes, heterochromosomes, or sex chromosomes.

In certain species, the gametes are not identical with respect to the sex chromosomes. One of the sexes is heterozygous, producing two types of gametes. The other is homozygous, producing only one type of gametes. Therefore only two combinations of gametes are possible in fertilization, and the result is 50 per cent males and 50 per cent females (Fig. 10.1).

The majority of organisms have a pair of sex chromosomes, which, in the course of evolution, have been specialized for sex determination. One of the sexes have a pair of identical sex chromosomes (XX), the other may have a single sex chromosome, which may be unpaired (XO) or paired with a Y chromosome (XY). The XY pair is also called *heteromorphic* because of the different morphology of the chromosomes.

In many species spermatogenesis produces two kinds of spermatozoa in similar proportion. Oogenesis on the other hand produces only one kind of gamete (ova). This type of sex determination is found in mammals, including the human and in certain insects, such as Drosophila. The male is heterogametic, whereas the female is homogametic.

In other vertebrates (birds, some reptiles, and fishes) and invertebrates (*e.g.*, insects of the order *Lepidoptera*), the female is heterogametic and the male is homogametic. In this instance there are two kinds of ova (X and Y) and only one kind of spermatozoa (X). In *orthoptera*, males are XO and females XX. In some cases sex is determined by the Y chromosome, as in the case of the axolot in which sex depends on the presence or absence of this chromosome.

In the human the Y chromosome determines the male sex. Thus, an XO individual (lacking the Y chromosome) resembles a female but lacks ovaries (Turner's syndrome).

On the other hand an XO mouse is a normal female. The Y chromosome probably determines the male sex in all mammals.

Sex chromosomes are not distinguishable in some animals. In such cases, the sex determining genes are probably confined to a short region of a pair of chromosomes.

The sex chromosomes can be thought of as composed of a homologous and a differential region. The homologous region corresponds to the pairing segment, and when recombination takes place, it is limited to this part. The differential region influences sex determination.

SEX CHROMATIN AND SEX CHROMOSOMES

Barr and *Bertram* opened an important field by their discovery in 1949 of a small chromatin body (a chromocenter) in nerve cells of the female cat—a structure which is absent in the male. These observations were then made in other tissues and animals including the human. In nuclei of the epidermis of females this chromatin body is found in much higher proportion than in males.

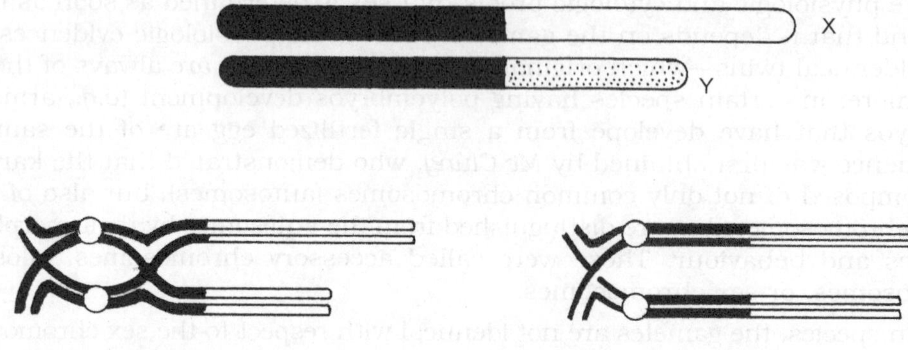

Fig. 10.1. Diagram of a pair of XY sex chromosome of a mammal. In black, the pairing or homologous segments of the X chromosomes. Stipped, the differential segment of the Y chromosomes. The configuration of the bivalent depends on the position of the chiastmata, which is produced only in the homologous segments.

The frequency with which sex chromatin can be detected in the female varies from tissue to tissue. In nervous tissue the frequency may be 85 per cent. In oral smears the frequency may vary between 20 and 50 per cent in normal females (see *Hamerton*, 1969).

Sex chromatin appears in the interphase nucleus as a small chromocenter heavily stained with basic dyes, it gives a positive Feulgen reaction. It can be found attached to the nucleus as in nerve cells of certain species, attached to the nuclear membranes, as in cells of the epidermis or of the oral mucosa; free in the nucleoplasm, as in neutrons after electric stimulation and as a nuclear expansion, the best-known example being that of the neutrophil leukocyte in which the sex chromatin appears as a small rod called the drumstick. This characteristic of the leukocyte has been utilized as a test of sex determination, along with the investigation of the basal cell of the epidermis and smears of the oral mucous (Fig. 10.2).

The study of sex chromatin has a wide field of medical applications and offers the possibility of relating the origin of certain congenital diseases to chromosome, anomalies. Among these applications, the diagnosis of sex in intersexual states in postnatal and even in fetal life (see the sections dealing with human cytogenetics).

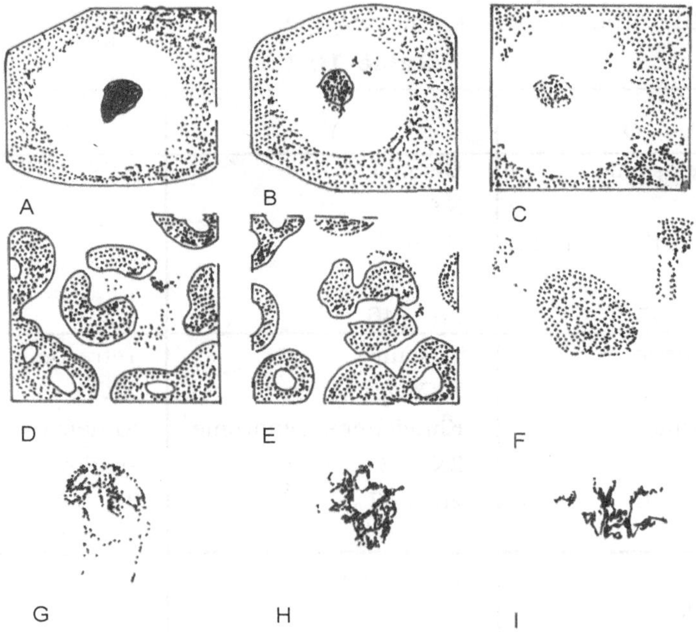

Fig. 10.2. Cellular basis of cytogenetics.

The relationship between sex chromatin and sex chromosomes has been elucidated. Sex chromatin is derived from only one of the two X chromosome, the other X is not heteropyknotic at interphase. The number of corpuscles of sex chromatin at interphase is equal to nX –1. This means that there is one Barr body fewer than the number of X chromosomes. This relationship between sex chromatin and sex chromosomes is particularly evident in some humans who have an abnormal number of sex chromosomes (See Table 10.1).

THE SINGLE X NATURE OF SEX

Chromatin

The differential behaviour of the two X chromosomes in the female led Lyon (see *Lyon*, 1966, 1972) to the so-called inactive X hypothesis by which: (*i*) only one of the X chromosomes is genetically active; (*ii*) the X undergoing heteropyknosis may be either of maternal or paternal origin, and the decision by which X becomes inactive is taken at random (Fig. 10.3); and (*iii*) the inactivation occurs early in embryonic life and remains fixed, it is now admitted that only a part of the C chromosome condenses into a Barr corpuscle; this condensed chromosome is said to contain facultative heterochromatin, as opposed to the constitutive heterochromatin found in other chromosomes.

Inactivation of the X chromosome takes place even in 3X and 4X individuals (Fig. 10.2 H and I). The inactivation starts in the human in the late blastocyst stage, on about the sixteenth day of embryonic life; the inactivated X chromosome then remains heterochromatic in the somatic cells.

The fact that X inactivation occurs at random has been demonstrated in human diseases linked to the X chromosome. The *Lesch-Nyhan syndrome*, in which a deficiency of one enzyme of the purine metabolism (*i.e.*, hypoxanthine—guanine phosphoribosyl transferase) produces mental retardation and increased uric acid levels, results from a recessive mutation in the X chromosome. When fibroblasts of heterozygous women are cultured *in vitro*, two types of cell clones are obtained. Half the clones contain the enzyme, whereas the other half lack the enzyme.

Table 10.1

X/Y	*O*	*Y*	*YY*	*Sex Chromatin*
X	Monosomic XO Turner's Syndrome 2X-1 2n = 45	Disomic XY Normal 2X 2n = 46	XYY	0
XX	Disomic XX Normal 2X 2n = 46	Trisomic XXY Klinefelter's Syndrome 2X + 2 2n = 47	Tetrasomic XXYY Klinefelter, syndrome 2X + 2 2n = 48	1
XXX	Trisomic XXX Meta female 2X + 1 2n = 47	Tetrasomic XXXY Klinefelter's Syndrome 2X + 2 2n = 48		2
XXXX	Tetrasomic XXXX Meta female 2X + 2 2X + 2 2n = 48	Penta somic XXXXY 2X + 3 2n = 49 Klinefelter		3
Phenotype	♀	♂	♂	

Heteropyknosis of the Sex Chromosomes and the Sex Vesicle

During meiosis, the XY pair is embedded in the so-called sex vesicle in most mammalian species. This vesicle is apparent during the zygonema and pachynema stages at which time the XY chromosomes are not heteropyknotic. At the end of prophase, when the vesicle disregenerates, the XY bivalent becomes heteropyknotic again.

In the mouse the sex chromosomes, and hence the sex vesicle, are concerned with the formation of the nucleolus (*i.e.*, they contain the nucleolar organizers). In this case the vesicle has two zones; (*i*) the chromatic zone, oval in shape and attached to the nuclear membrane, and (*ii*) the RNA containing or nucleolar zone, which has a crescent shape. The chromatic zone contains thin, convoluted DNA microfibrils and has some filamentous cores that represent the lateral components of the synaptonemal complex. The nucleolar zone contains RNA and is Feulgen negative. In the center of this zone there is a dense round body. The rest is a sponge like structure made of granules of about 15 nm. The human sex vesicle contains an RNA, and the nucleolar zone is absent.

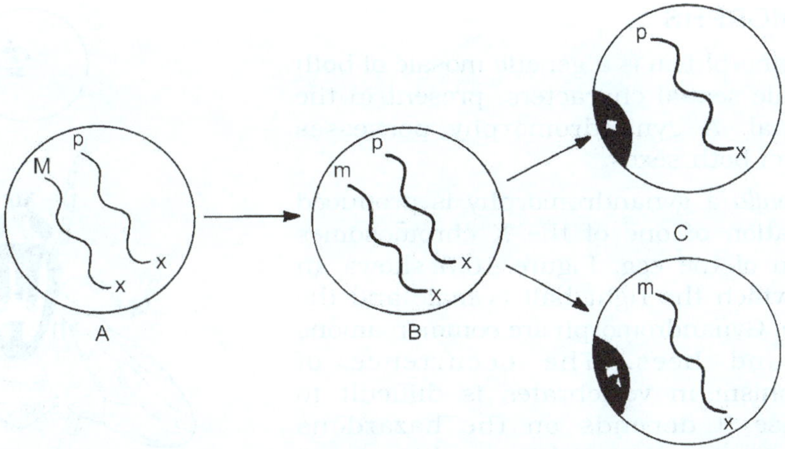

Fig. 10.3. The evolution of XX chromosomes. A. In the zygote both the paternal (P) and maternal (M)X are euchromatic. B. In the early blastocyst the same is true as in A and C. In late blastocyst 50% of the cells have a maternal heterochromatic X chromosome, the other 50% have the paternal heterochromatic X chromosome.

NEO-XY SYSTEM OF SEX DETERMINATION

In addition to the common type of sex determination based on XY chromosomes, in several species a special type of sex determination has been observed and designated the *neo*-XY system. As shown in Fig. 10.4, this arises from the fracture of the X chromosome, followed by fusion of the main fragment to one autosome. This association constitutes the *neo*-X chromosome. At meiosis, the other autosome of the pair forms the so called neo-Y chromosome and remains confined to the male sex.

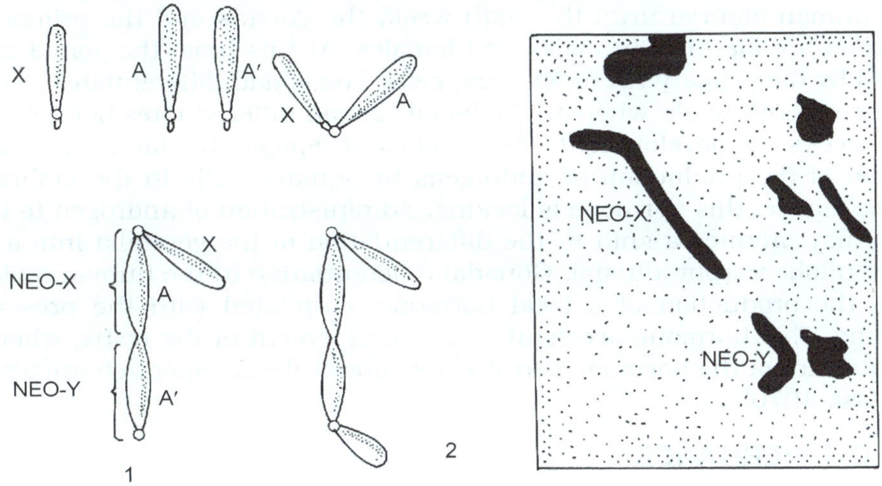

Fig. 10.4. 1, The mechanism of centric fusion between a member (A) of a pair of autosomes (AA') and the sex chromosome (X) the small fragment with the centromere at the right is lost. Below, left, the neo-x, neo y-chromosome at metaphase II right the same element, showing the neo-y with a submerging Centromere 2, the neo-x-neo y chromosome during the first meiotic metaphase.

There is evidence that the neo-Y chromosome gradually become heterochromatic. It was demonstrated by radioautography with H—thymidine that, as in the case of the C chromosome, the neo-Y chromosome replicates late in the synthetic period.

GYNANDROMORPHS

Gynandromorphism is a genetic mosaic of both male and female sexual characters, present in the same individual. A gynandromorphy possesses chromosomes of both sexes.

In *Drosophila* a gynandromorphy is produced by the elimination of one of the X chromosomes during division of the egg. Figure 10.5 shows an individual in which the right half is male and the left half, female. Gynandromorph are common among silk-worms and bees. The occurrence of gynandromorphism in vertebrates is difficult to assess because it depends on the hazardous distinction between gynandromorphism and intersexuality due to hormonal effect.

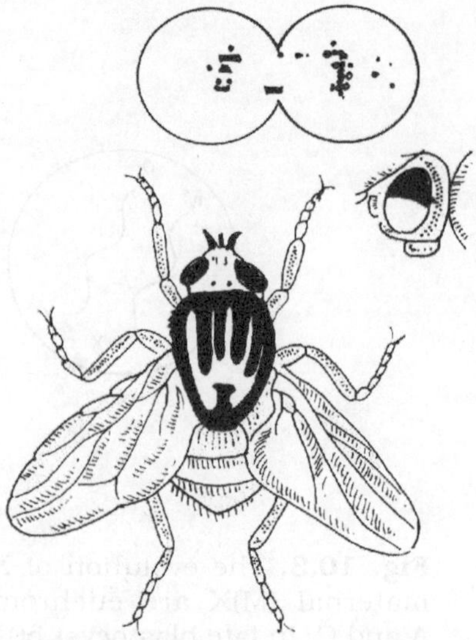

Fig. 10.5. Gynandromorph of *Drosophila*.

SEX DIFFERENTIATION

Although the primary determination of sex is made at fertilization, the embryo acquires its definite sex characteristics by a more complex mechanism.

An epigenetic factor (*i.e.*, hormonal) may assume control of the genetic determination during development, thereby changing the phenotypic direction of sex. Among vertebrates, a condition of bisexuality may exist (*e.g.*, the coexistence of structures of the functional sex, together with primordial of the heterologous sex). For example, male amphibians have a rudimentary ovary (Biddr's organ) and vestigial oviducts.

In the human embryo until the sixth week, the gonads and the primordial of the urogenital tract are identical in males and females. At this time, the gonad has already been invaded by the primary XX or XY germ cells. The gonad differentiates into a definite testicle at the seventh week, whereas the female gonad differentiates between the eighth and ninth weeks of development. An important epigenetic factor at the time of differentiation is the production of androgens by somatic cells in the embryonic male gonad; in the female, this hormone is lacking. Administration of androgen to the mother at this time may produce a shift in the differentiation of the genitalia into a male type (feminine pseudohermaphroditism). Gonadal differentiation in the human male probably depends on the production of a local hormone, correlated with the presence of the Y chromosome. This hormone accelerates the development of the testis, whereas in the female the absence of the hormone permits the shower development characteristic of the ovary (see *Jost.* 1970).

SEX-LINKED INHERITANCE

Sex-linked genes are those carried by the sex chromosomes and yet not involved in sex determination.

The following is an example of sex linked inheritance in *Drosophila*, when a homozygous red-eyed female (dominant) is crossed with a white-eyed male (recessive), all individuals in the F_1 are red-eyed, but when the cross is between a white-eyed female and a red-eyed male, male offspring in the F_1 have white eyes. These experiments demonstrate that the gene for red eye in this case is carried by the X chromosome, but not by the Y.

In organisms with an XY type of sex determination, genes may be present in the differential segments of the X and Y. Such genes are not alleles, since they are in a nonhomologous section of the chromosome. They are completely linked and crossing over cannot occur in them.

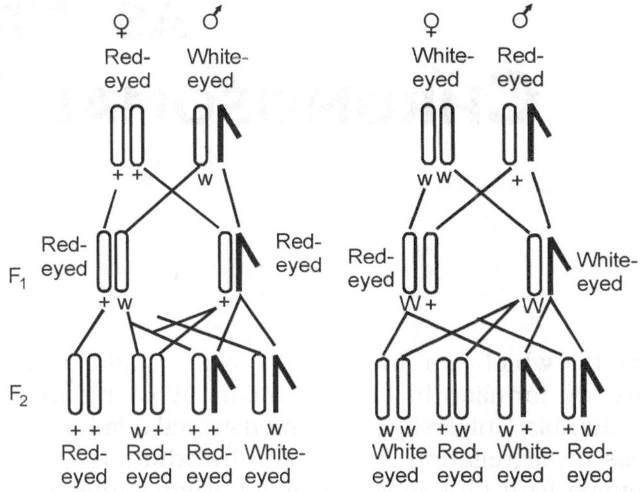

Fig. 10.6. Sex-linked Inheritance of eye colour in *Drosophila.*

There are three types of sex-linked inheritance: (1) X-linked, by genes localized in the nonhomologous section of X and which have no corresponding alleles in Y; (2) Y-linked, by genes localized in the nonhomologous section of Y and which have no alleles in X; and (3) XY-linked by genes localized in one chromosomal segment which is homologous in both X and Y (the so-called incomplete linkage).

Genes Linked to the X Chromosome

The classic example of this form of sex linkage is the red colour in *Drosophila.* In man the genes that determine *daltonism* (*i.e.*, red-green colour blindness) and hemophilia are linked to the X chromosome. Eight per cent of males have daltonism, whereas this is found in only 0.5 per cent of females. In the latter, both X chromosomes are altered at the same locus. Hemophilia (a defect of blood clotting) is inherited as a sex-linked recessive gene. Rarely, is female a hemophiliac. In such a case, the father is a hemophiliac and the mother is a carrier of hemophilia.

Other X-linked genes produce the following conditions in the human; *ichthyosis, myopia, Gower's muscular atrophy*, and one type of colour blindness. All these anomalies re-transmitted in the same way as the "white eyed" trait in *Drosophila*; the same reasoning can be followed to obtain F_1 and F_2.

Genes Linked to the Y Chromosome

Genes in the nonhomologous region of the Y chromosome pass directly from father to son. For example, *ichthyosis hystrix gravis* and other diseases follow the male line.

Genes Localized in the Homologous Segments of both X and Y Chromosomes

These genes are inherited as the autosomal genes. They are partially sex-linked. In the human, there are several defects of this type, among which are total colour blindness, two skin diseases (*xeroderma pigmentosum* and *epidermolysis bullosa*), *rectinitis pigmentosa, spastic paraplegia* and other diseases (See Fig. 11.3).

11 NON-DISJUNCTION AS PROOF OF CHROMOSOMAL THEORY

INTRODUCTION

The mechanism by which an individual's cells might come to have an extra chromosome was already familiar to cell biologists. Working with fruit flies in 1913, *Calvin B. Bridges*, of Columbia University had demonstrated a failure, called non-disjunction, in the normal process of segregation at meiosis. In addition to three pairs of somatic chromosomes, the female fruit fly has a pair of X chromosomes (the male has an X and a Y chromosome). Normal meiosis in the female produces an ovum with a single X chromosome; normal meiosis in the male, a sperm with an X or a Y chromosome. On rare occasions, Bridges found, the two X chromosomes of a female germ cell fail to "disjoin" in the first stage of meiosis. One of the ova produced by the next division accordingly receives both X chromosomes. Subsequent fertilization by a normal Y-bearing sperm gives rise to an exceptional fly, female in appearance but with a chromosome complement of XXY.

Apparently a similar accident during meiosis in the human germ cell can give rise to a mongoloid infant. In this case, however, the non-disjunction involves not the X chromosomes but a small somatic chromosome: Number 21. The ovum with two 21's is fertilized by a sperm also bearing a 21; the result is a zygote with three, instead of the normal pair, of these chromosome. This is called "trisomy" of chromosome 21.

NON-DISJUNCTION

The failure of homologous (at meiosis) or sister chromatids (at mitosis) to separate properly to opposite poles is called non-disjunction. The sex chromosomes pass through the phenomenon of non-disjunction and present various interesting situations. The non-disjunction may be of following types:

Primary Non-disjunction

The phenomenon of primary non-disjunction was discovered by *C.B. Bridges* (1916) during his classical matings of *Drosophila melanogaster.* Let us consider one of his classical crosses as follows:

The gene for wild type red eyes (+) is carried by the X chromosome; a recessive allele (v) produces vermilion eyes in homozygous females (X^vX^v) and in all males (X^vY). Normally vermilion-eyed females mated to red-eyed males produce only red-eyed daughters and vermilion eyed sons.

Fig. 11.1. The hen and cock feathering in domestic fowl. Cock-feathering is characterized by long, pointed, curving neck and tail feathers.

Parents	X^vX^v		×	X^+Y
	Vermilion female (or mother)			Red male (or father)
Gametes	**Eggs:**	(X^v)	(X^v)	
	Sperms:	(X^+)	(Y)	
F_1	1/2 X^+X^v	:	1/2 X^vY	
	Red female (or daughter)		Vermilion male (or son)	

Crosses of this type, however, in rare cases produce unexpected vermilion-eyed daughters and red-eyed sons with a frequency of one per 2000 to 3000 offsprings. Bridge speculated that these unusual progeny are due to a failure of the X chromosomes in an XX female to disjoin during meiosis 1 of oogenesis. This phenomenon was termed primary disjunction by Bridges and this event tends to produce three types of eggs; the majority of eggs with normal single X chromosome and a small number of eggs with their two C chromosomes or no X at all. These eggs on normal fertilization produce the following results:

The metafemales (AA XXX) are weak and seldom live beyond the pupal stage. The AADY individual die in the egg stage. Note that the AA XXY females indicates that the presence of a Y chromosome does not determine maleness itself, though males without it (AAXO) are sterile. From these results three facts became established : 1. In *Drosophila* sex determination occurs according to genetic balance theory of Bridges. 2. In *Drosophila*, Y chromosome contains genes for the spermatogenesis. 3. It provides strong support to chromosome theory of inheritance (*i.e.*, Mendelian factors of genes are carried on the chromosomes).

Parents:	AAX^vX^v	×		AAX^+Y
	Vermilion female			Red male
Gamets	Eggs : (AX^v) (numerous)	+ (AX^vX^v) (rare due to non-disjunction)	+	(AO) (recalled nullo)
	Sperms : (AX^+)	(AY)		
F_1	1. AAX^+X^v = Red female (numerous normal)			

2. $AAX^{+}X^{v}X^{v}$ = Red "metafemales" (rare, die)
3. $AAX^{+}O$ = Red male (rare, sterile)
4. $AAX^{v}Y$ = Vermilion male (numerous)
5. $AAX^{v}X^{v}Y$ = Normal
6. $AAOY$ = Vermilion female (rare, fertile)

   Die (rare)

Secondary Non-disjunction

In next cross, Bridges mated the exceptional vermilion-eyed females ($AAX^{v}X^{v}Y$) that arose as a result of primary non-disjunction to normal red-eyed males ($AAX^{+}Y$). The progeny of this cross included 96 per cent females with red eyes and only 4 per cent females had vermilion eyes indicating that there was 4 per cent secondary non-disjunction (Fig. 11.2).

Thus, primary non-disjunction may occur in either XX females or XY males. In the former it leads to the production of XX and O eggs. Occurrence in the first meiotic division of males produces XY and O sperms. When such non-disjunction takes place during the second meiotic division then XX, XY and O sperms result. Secondary non-disjunction, on the other hand, occurs in XXY females, when it gives rise to XX, XY, X and Y eggs (Fig. 11.2). As the term non-disjunction implies, these aberrant gametes are produced only as a result of failure of the sex chromosomes to disjoin after synapsis; they are not physically attached.

F_2		Sperms: AX^{+} (50%)	Sperms: AY (50%)	
Eggs: X-Y Pairing (16%)	$AX^{V}X^{V}$ (4%)	$AAX^{+}X^{V}X^{V}$ Metafemale (die) (2%)	$AAX^{V}X^{V}Y$ Vermilion female (2%)	4% secondary exceptional progeny phenotype
	AY (4%)	$AAX^{+}Y$ Red female (2%)	AAYY (die) (2%)	96% regular (*i.e.*, expected progeny phenotypes)
	AY^{V} (4%)	$AAX^{+}X^{V}$ Red female (2%)	$AAX^{V}Y$ Vermilion male (2%)	
	$AX^{V}Y$ (4%)	$AAX^{+}X^{V}Y$ Red female (2%)	$AAX^{V}YY$ Vermilion male (2%)	
Eggs: X-X pairing (84%)	AX^{V} (42%)	$AAX^{+}X^{V}$ Red female (21%)	$AAX^{V}Y$ Vermilion male (21%)	
	$AX^{V}Y$ (42%)	$AAX^{+}X^{V}Y$ Red female (21%)	$AAX^{V}YY$ Vermilion male (21%)	

Fig. 11.2. Results obtained from secondary non-disjunction.

Similar results were obtained by *Bridges* (1916) from the cross of red eyed male and white eyed female *Drosophila*.

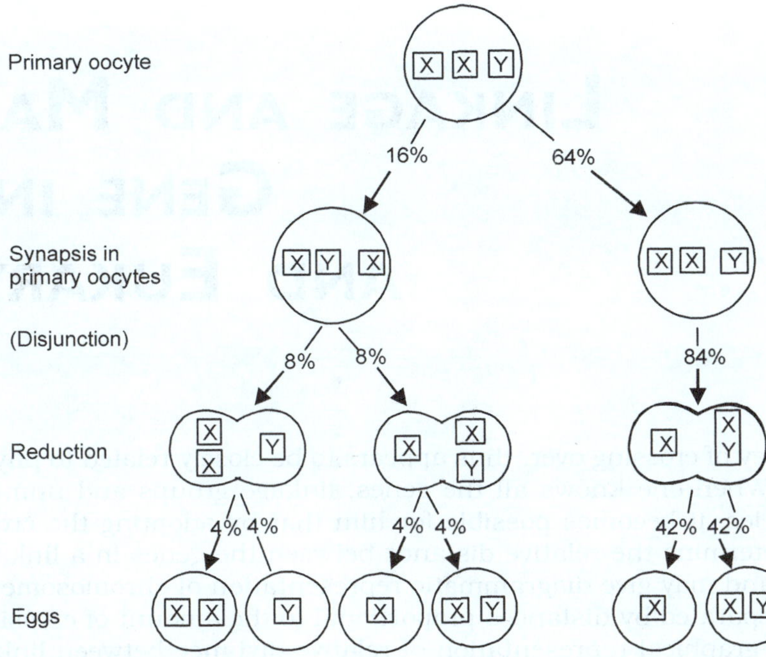

Fig. 11.3. Secondary non-disjunction in XXY female *Drosophila* resulting in 46 per cent X, 46 per cent XY, 4 per cent XX, and 4 per cent Y eggs.

12 LINKAGE AND MAPPING GENE IN PRO AND EUKARYOTES

The frequency of crossing over, thus appears to be closely related to physical distance between genes. When one knows all the genes, linkage groups and number of linkage groups of a species, it becomes possible for him that by adopting the crossing over as a tool he may determine the relative distance between the genes in a linkage group and also their order and may give diagrammatic representation of chromosomes showing the gene as points separated by distances proportional to the amount of crossing over. Such a diagrammatic, graphical representation of relative distance between linked genes of a chromosome is called linkage or genetic map.

MAPPING GENES IN BACTERIA

The most commonly used bacterium for genetic and molecular analysis is *Escherichia coli* (*E. coli*), which is found most commonly in animal (including human) intestines. It is a good subject for study since it can be grown on a simple defined medium and can be handled with simple microbiological techniques.

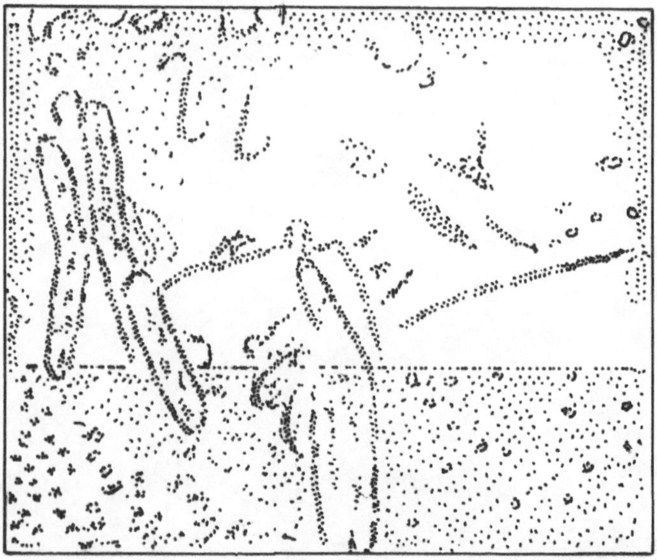

Fig. 12.1. Staining Electron micrograph of cells of the bacterium *Escherichia coli.*

Bacterial cells are relatively small compared with eukaryotic cells. Bacterium *E. coli* is a cylindrical organism about 1 μm long and 0.5 μm in diameter. Its cytoplasm is full of ribosomes (the organelles responsible for protein synthesis), while the central nucleoid region contains the genetic material, a single circular chromosome consisting of DNA. As in all prokaryotes, there is no membrane between the bacterium's nucleoid region and the rest of the cell.

Like other bacteria, *E. coli* can be grown both in liquid medium and on the surface of growth medium solidified with agar. Genetic analysis of bacteria typically is done by plating cells out on agar-solidified media. Wherever a single bacterium lands on the agar surface it will grow and divide repeatedly, resulting in the formation of a visible cluster of genetically identical cells called a colony. The concentration of bacterial cells in a suspension (the titer) can be determined by spreading known volumes of suspension or of a known dilution of the suspension on the agar surface, incubating the plates, and then counting the number of resulting colonies. By varying the nutrient composition of the media it is possible to detect and quantify the various genotypic classes in a genetic analysis.

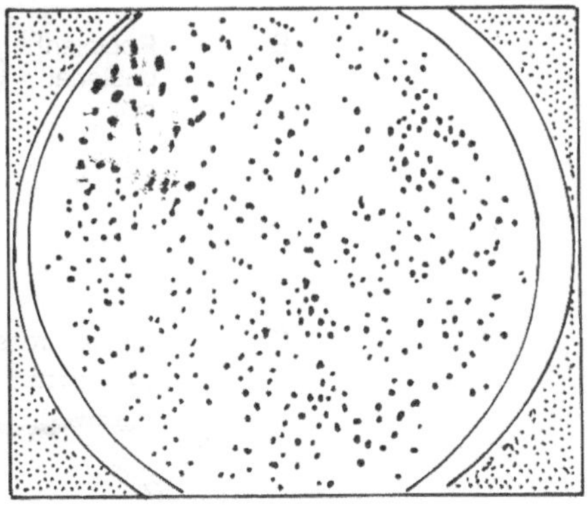

Fig. 12.2. Bacterial colonies growing on a nutrient medium in a pertri dish.

Genetic material can be transferred between bacteria by three processes. In each case: (1) transfer is unidirectional; (2) there is no true zygote; and (3) the cells are essentially always haploid. Thus, these genetic transfer processes are different from those in eukaryotes.

1. **Transformation** is a process in which there is a transfer of genetic information by means of extracellular pieces of DNA in bacteria. The DNA fragments are derived from donor bacteria and may be taken up by another living bacterium, the recipient. If the donor and the recipient bacteria differ genetically, then genetic recombinants are produced by crossover events involving the DNA fragments and the recipient's chromosome. The genetic recombinants generated by the transformation process are called transformants.
2. **Conjugation** is a process in which there is a unidirectional transfer of genetic information through direct cellular contact between a donor and a recipient bacterial cell. The contact is followed by the formation of a cellular bridge physically connecting both cells. Then a segment (rarely all) of the donor's chromosome may be transferred into the recipient and may undergo genetic recombination with a homologous chromosome segment of the recipient cell. The recipients inheriting donor DNA are called transconjugants.
3. **Transduction** is a process by which bacteriophages mediate the transfer of bacterial genetic information from one bacterium (the donor) to another (the recipient). Since the amount of DNA a phage can carry is limited, the amount of genetic material that can be transferred is usually less than 1 per cent of that in the bacterial chromosome. Once the donor genetic material has been introduced into the recipient, it may undergo genetic recombination with a homologous chromosome region. The recipients inheriting donor DNA in this way are called transductants.

It is possible to map bacterial genes by using anyone of these methods. However, not all methods can be used on each species of bacterium.

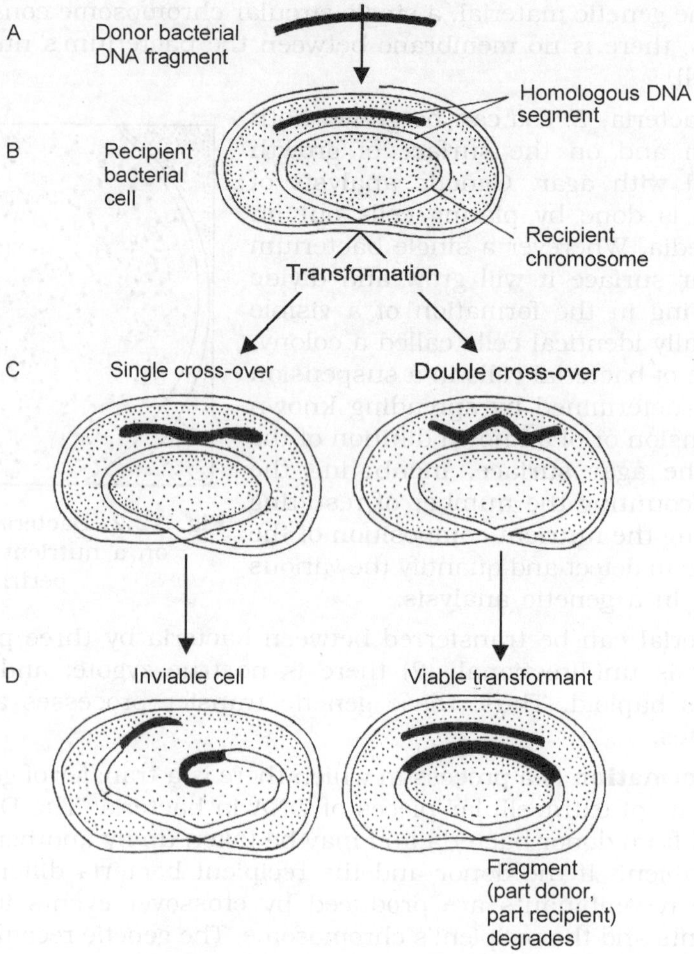

Fig. 12.3. Transformation in bacteria. A. Linear donor bacterial DNA fragment. B. Donor transforming DNA fragments find homologous region on the bacterial chromosome. C. Single Scross-over opens chromosomes produce linear molecule resulting in an inviable cell.

BACTERIAL TRANSFORMATION

Transformation of bacteria was first observed by *F. Grifith* in 1928 and in 1944 *O. Avery and his colleagues* showed that DNA was responsible for the genetic change that was observed. In transformation fragments of free DNA are taken up by a living bacterium. This process is generally used to map genes in bacterial species in which conjugation or transduction does not occur. In transformation mapping experiments, DNA from a donor strain is extracted and purified. The extraction process breaks the DNA into small fragments, and this genetic material is then added to a suspension of recipient bacteria with a different genotype. The donor DNA that is taken up by the recipient cell may undergo genetic recombination with the homologous parts of the recipient's chromosome to produce a recombinant chromosome. Those recipients whose phenotypes are changed by recombination event are called transformants.

While most bacterial species are probably capable of the recombination event of the transformation process, there is a great deal of variability in their ability to take up DNA. Wild-type *E. coli*, for example, is not readily transformable because the bacterium rapidly degrades incoming DNA. Other bacteria, such as *Bacillus subtilis* and *Streptococcus pneumoniae*, can be easily transformed in a test tube, and mapping has been done on these bacteria by using this technique. Even in a bacterial species that is capable of being transformed, only a small population will actually take up DNA. Over the years, though, scientists have devised experimental conditions for a number of bacterial species that produce a population of cells, called competent cell, most of which are able to take up DNA because they are leaky, meaning extraneous DNA is readily passed through the membrane of competent cells.

Once a donor DNA fragment has been taken up by a recipient bacterium with a different genetic constitution, we are in a position to detect the transformants produced by genetic exchange between the transforming DNA pieces and the recipient cell's chromosome. Since the bacterial chromosome is circular and the DNA fragment is linear, a single cross-over (or any add, number of crossover) will open up the chromosome and add the transforming DNA to make a long, linear chromosome. A linear chromosome is unable to replicate in the bacterium and will often degrade leading to the death of the cell. Thus the integration of part or all of the transforming DNA pieces requires an even number of cross-overs, for example, a double cross-over. First, the donor DNA migrates to the homologous region in the recipient chromosome, and then the cross-overs take place. These cross-overs leave an intact, circular, recombinant bacterial chromosome and a residual linear piece of DNA (part donor and part recipient) that degrades. Given highly competent recipient cells and an excess of DNA fragments, transformation of most genes will occur at the frequency of about 10^{-3} (*i.e.*, one cell per 10^3 cells).

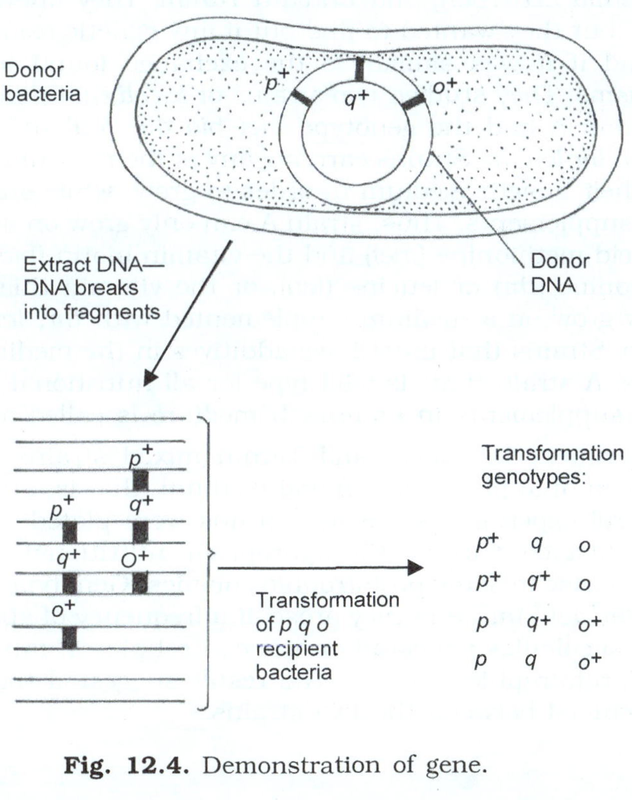

Fig. 12.4. Demonstration of gene.

Using transformation, it is possible to determine gene linkage, gene order, and map distance. The principles are as follows.

If two genes, x^+ and y^+ are far apart on the donor chromosome, they will always be found on different DNA fragments. Thus, given an x^+ y^+ donor and an x y recipient, the probability of simultaneous transformation (cotransformation) of the recipient to x^+ y^+ (from the product rule) is the square of the probability of transformation with one gene alone. If transformation occurred at a frequency of 1 to 10^6 recipient cells ($10^{-3} \times 10^{-3}$). Therefore, if two genes are close enough that they often are carried on the same DNA fragment, the cotransformation frequency would be close to the frequency of transformation by a single gene. To turn this argument around to the way it is used practically, if cotransformation occurs at a frequency that is substantially higher than the product of the two single-gene transformations, the two genes must be close together.

Gene order can also be determined from cotransformation data. For example, if genes p and q are often cotransformed, and genes q and o are often cotransformed, but genes o and p are never cotransformed, the gene order must be p-q-o.

Since it is possible to control the size of DNA fragment used in transformation expriments the probability of cotransformation of two genes can be related to the average molecular size of the transforming DNA. Using this approach of relating cotransformation frequency to average size of the transforming DNA, a physical map of the genes (*i.e.*, a map of the relative physical locations of genes along the DNA) can be derived.

CONJUGATION IN BACTERIA

Discovery of Conjugation in *E. coli*

In 1946, another method of genetic transfer between bacteria, conjugation, was discovered by *Joshua Lederberg* and *Edward Tatum*. They knew that meiosis does not occur in bacteria, but they wanted to find out if any genetic exchange between bacterial cells occurred and if it was similar to the exchange found in sexually reproducing eukaryotic organisms. They studied two strains of *E.coli* that differed in their nutritional requirements. Strain A had the genotype *met bio* thr^+ leu^+ thi^+, and strain B had the genotype met^+ bio^- *thr leu thi*. Strains carrying any of the mutant genes require nutritional supplements in their growth medium in order to grow, while strains carrying wild-type genes require no supplements. Thus, strain A can only grow on a medium supplemented with the amino acid methionine (*met*) and the vitamin biotin (*bio*) but does not need the amino acids threonine (*thr*) or leucine (*leu*), or the vitamin thiamine (*thi*). In contrast, strain B can only grow on a medium supplemented with *thr*, *leu*, and *thi* but does not require *met* or *bio*. Strains that must have additives in the medium in order to grow are called auxotrophs. A strain that is wild type for all nutritional requirement genes and thus requires no supplements in its growth medium is called a prototroph.

In their experiment Lederberg and Tatum mixed strains A and B together and spread (plated) them onto plates of minimal medium, that is, a medium with no growth additives. In control experiments the two strains were plated separately. As expected, neither strain could grow without the appropriate nutritional supplement. The mixed cells, however, gave rise to some prototrophic colonies. Genotypically, these prototrophic strains met^+ bio^+ thr^+ leu^+ thi^+, and they arose at a frequency of about 1 in 10 million cells (1×10^{-7}). Since no colonies appeared on the control plates, mutation was ruled out as the origin of the prototrophic colonies. The result suggested that some type of genetic exchange had occurred between the two strains.

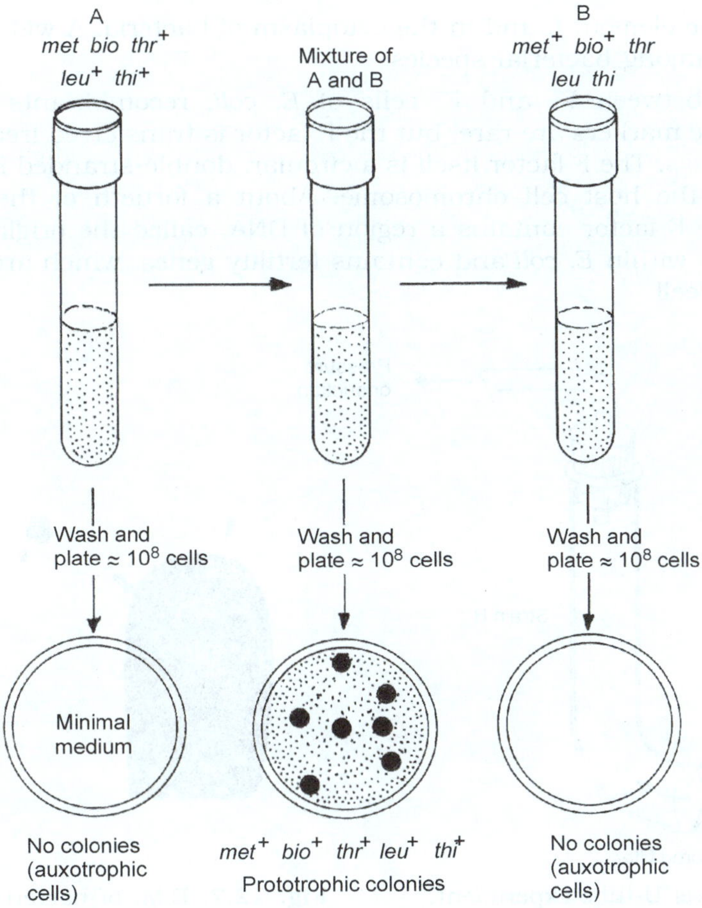

Fig. 12.5. Lederberg & Tatum experiments showing that sexual recombination occurs between cells of *E. coli.* After the cells have been mixed and the mixture plated, a few colonies grow on the minimal medium.

To determine whether cell-to-cell contact was needed for the genetic exchange to occur, the researchers performed an experiment using a U tube apparatus made by Bernard Davis, Strains A and B were placed in a liquid medium on either side of the U separated by a fine filter whose pores were too small to allow bacteria to move through it. After several hours of incubation during which the medium was moved between compartments by alternating suction and pressure, the cells were plated on a minimal medium to check for the appearance of prototrophic colonies. No prototrophic colonies appeared. The interpretation was that cell-cell contact was required in order for the previously observed genetic exchange to occur. In other words, *E. coli* has the mating system, called conjugation, in which genetic material can be transferred from one bacterium to another through a tube transiently linking the two cells.

THE SEX FACTOR F

In 1953, *William Hayes* obtained evidence that the genetic exchange in *E. coli* is not reciprocal. Instead, as he discovered, one cell acts as a donor and the other cell acts as a recipient. Hayes proposed that the transfer of genetic material between the strains is mediated by a fertility factor called F that the donor cell possesses (F^+) and the recipient cell lacks (F^-). The F factor is found only in *E. coli* and is an example of a plasmid, a self-replicating genetic element found in the cytoplasm of bacteria. A wide range of different plasmids occur among bacterial species.

In crosses between F^+ and F^- cells of *E. coli,* recombinants for chromosomal (nonplasmic) gene markers are rare, but the F factor is transferred frequently during the conjugation process. The F factor itself is a circular, double-stranded DNA molecule that is distinct from the host cell chromosome. About a fortieth of the size of the host chromosome, the F factor contains a region of DNA, called the origin, that is required for its replication within *E. coli* and contains fertility genes, which are required if a cell is to be a donor cell.

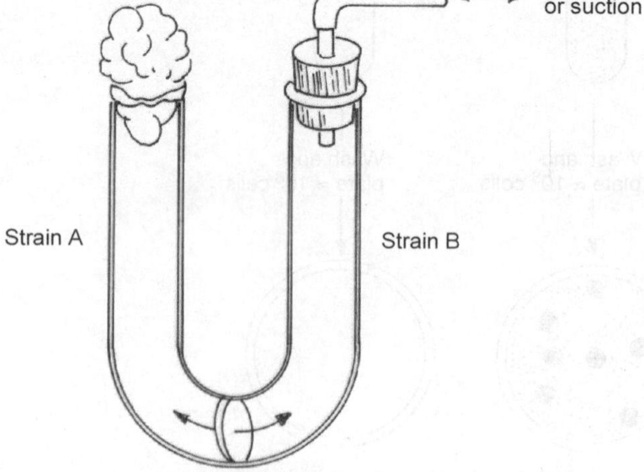

Fig. 12.6. Davis U-tube experiment.

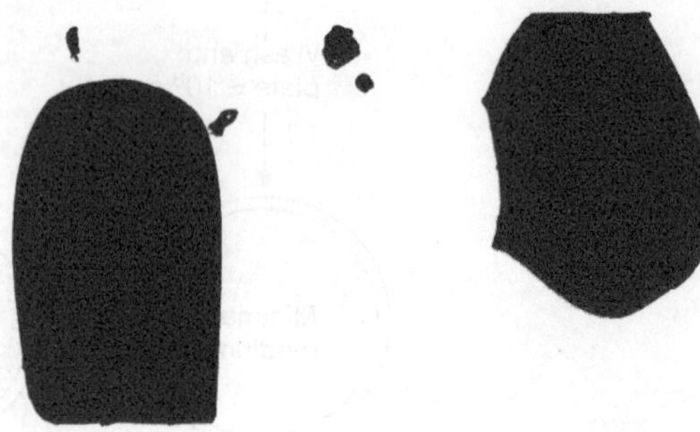

Fig. 12.7. E.M. of bacterial conjugation bet and an F^+ donor and F^- recipient of *E. coli.*

When F^+ and F^- cells are mixed, they conjugate. No conjugation occurs between two cells of the same mating type (*i.e.*, two F^+ bacteria or two F^- bacteria). One strand of the F factor is nicked at the origin, and DNA replication proceeds from that point. As replication proceeds, a copy of the F factor is transferred to the F^-cell as shown in (Figure 12.8). The complementary strand is synthesized in the F^- cell. There is a polarity of transfer, that is, the origin of replication is always transferred first, followed by the rest of the F factor in one particular orientation.

When the complete F factor has been transferred, the F^- cell becomes an F^+ cell as a result of the genes specific to the F factor. These genes include genes that specify hairlike cell surface components called F-pili or sex-pili, which allow the physical union of F^- and F^+ cells to take place. In a population of cells, only some of which are F^+ to start with the end result of the conjugation of F^+ and F^- cells and the subsequent transfer of the F factor is an epidemic conversion of the population to the F^+ state.

HIGH-FREQUENCY RECOMBINATION STRAINS

Although geneticists established that F^- cells can be converted to the F^- state by the transfer of the F factor, they did not know how conjugation resulted in the formation of recombinants for chromosomal genes. This puzzle was solved when *W. Hayes* and *L. Cavalli-Sforza* separately isolated a different type of donor strain from an F^+ stock. In a cross of this donor strain with an F^- strain that differed with respect to chromosomal gene markers, the frequency of recombinants for the chromosomal genes was about a thousand times greater than that normally seen for chromosomal genes in $F^+ \times F^-$ crosses. This new, highly recombinant type of F^+ donor strain is called an *Hfr*, or high frequency recombination, strain. The particular strains were later called *Hfr H* and *Hfr C* after their discoverers. The *Hfr* strains are the strains generally used in mapping *E. coli* genes through conjugation.

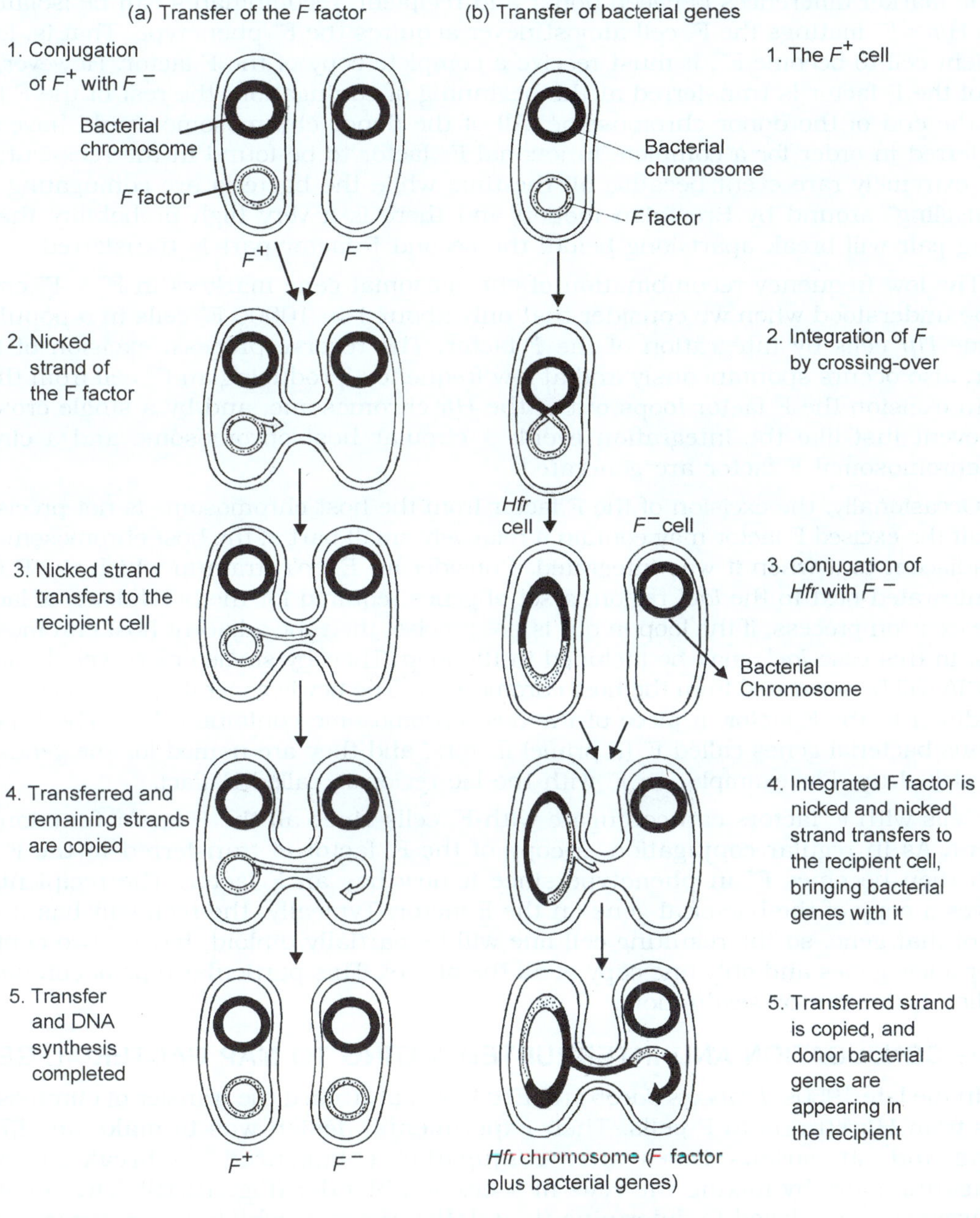

Fig. 12.8. Bacterial conjugation.

The H*fr* strains originate as a result of a rare event in which the F factor actually integrates into the bacterial chromosome by a single cross-over event (Fig. 12.8). When the F factor is integrated, it is no longer replicated independently; instead, it is replicated as part of the host chromosome. Since the F factor genes are still functional, H*fr* cells are able to mate with F^- cells. When mating happens, the same events occur as in the $F^+ \times F^-$ matings. The integrated F factor is nicked in one of the two DNA strands and replication begins. During replication part of the factor first moves into the recipient cell. With time some of the bacterial chromosome is also transferred into the recipient. So if there are genetic marker differences between donor and recipient, recombinants can be isolated. In these H*fr* $\times F^-$ matings the F^- cell almost never acquires the F^+ phenotype. That is, for the recipient cell to become F^+, it must receive a complete copy of the F factor. However, only part of the F factor is transferred at the beginning of conjugation; the rest of the F factor is at the end of the donor chromosome. All of the donor chromosome would have to be transferred in order for a complete functional F factor to be found in the recipient. This is an extremely rare event because all the time while the bacteria are conjugating, they are "jiggling" around by Brownian motion and there is a very high probability that the mating pair will break apart long before the second F factor part is transferred.

The low frequency recombination of chromosomal gene markers in $F^+ \times F^-$ crosses can be understood when we consider that only about 1 in 10000 F^+ cells in a population become *Hfr* cells by integration of the *F* factor. The reverse process, excision of the F factor, also occurs spontaneously and at low frequency, producing an F^+ cell from the *Hfr* cell. In excision the F factor loops out of the *Hfr* chromosome, and by a single crossing-over event just like the integration event, a circular host chromosome and a circular extrachromosomal F factor are generated.

Occasionally, the excision of the F factor from the host chromosome is not precise. As a result the excised F factor may contain a relatively small part of the host chromosome that was adjacent to it when it was integrated. Consider an *E. coli* strain in which the F factor has integrated next to the *lac*$^+$ region, a set of genes required for the breakdown of lactose. In the excision process, if the loop in out is not precise, then the adjacent host chromosomal genes, in this case *lac*$^+$, may be included in the loop. Then by single cross-over the looped out DNA will be separated from the host chromosome. The newly generated plasmid contains, in addition to the F factor, a piece of the host chromosome containing *lac*$^+$, The F factors contains bacterial genes called F' (F prime) factors, and they are named for the genes they have picked up. For example, an F' with the lac region is called F′ (lac).

Cells with F′ factors can conjugate with F^- cells since all the F factor functions are present. As in regular conjugation, a copy of the F′ factor is transferred to the F^- cell, which then becomes F^+ in phenotype since it now has an F factor. The recipient also receives a copy of the bacterial gene on the F factor. Typically, the recipient has its own copy of that gene, so the resulting cell line will be partially diploid, having two copies of one or a few genes and only one copy of all the others. This particular type of conjugation is called F duction or sexduction.

USING CONJUGATION AND INTERRUPTED MATING TO MAP BACTERIAL GENES

In the late 1950s, *Francois Jacob* and *Elie Wollman* studied the transfer of chromosomal genes from *Hfr* strains to F-cells. Their experimental design was to make an *Hfr* $\times F^-$ mating and, at various times after conjugation commenced, to break apart the conjugating pairs by mixing the cells in a kitchen blender (Fig. 12.10). The separated cells were then analyzed to determine the relative times at which donor genes entered the recipient's and produced genetic recombinants. That is, if gene a entered the recipient after 9 minutes and gene b after 15 minutes, genes a and b would be considered to be 6 minutes apart on the genetic map. In this way the distance between genes was measured as time units (minutes) between the two genes in terms of their appearance as recombinants in the recipient. The genetic map of *E. coli* was constructed in just this way, with map units in minutes, timed from an arbitrarily located origin.

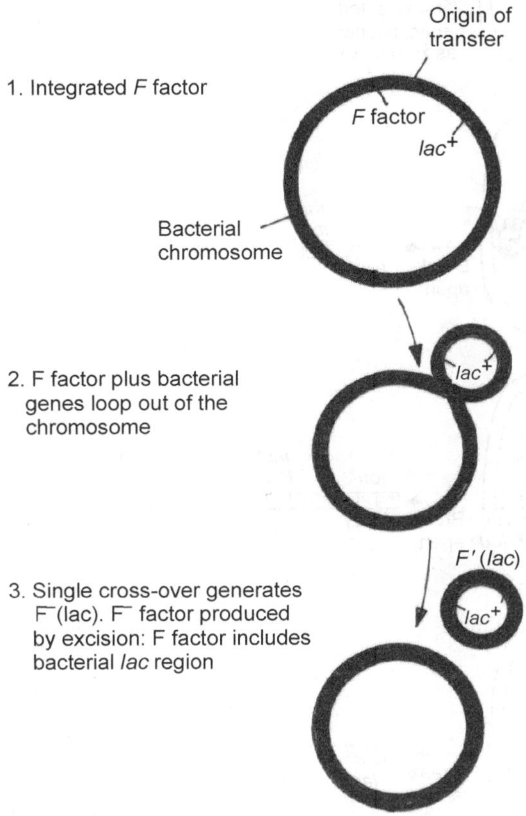

Fig. 12.9. Production of an F′ factor: 1. region of bacterial chromosome into which the F factor has integrated; 2. the F factor looping out incorrectly, so it includes a piece of bacterial chromosome, the lac^+ genes; 3. excision, in which a single crossover between the looped-out DNA segment and the rest of the bacterial chromosome (*i.e.*, the reverse of integration) results in an *F'* factor, called F^- (*lac*.).

The use of interrupted mating to map bacterial genes may be illustrated by considering one of Jacob and Wollman's experiments, which involved the following cross:

Donor

Hfr H str^S thr^+ *leu* azi^r ton^r lac^+ gal^+

Recipient

F^- str^r *thr* leu^- azi^s ton^s *lac gal*

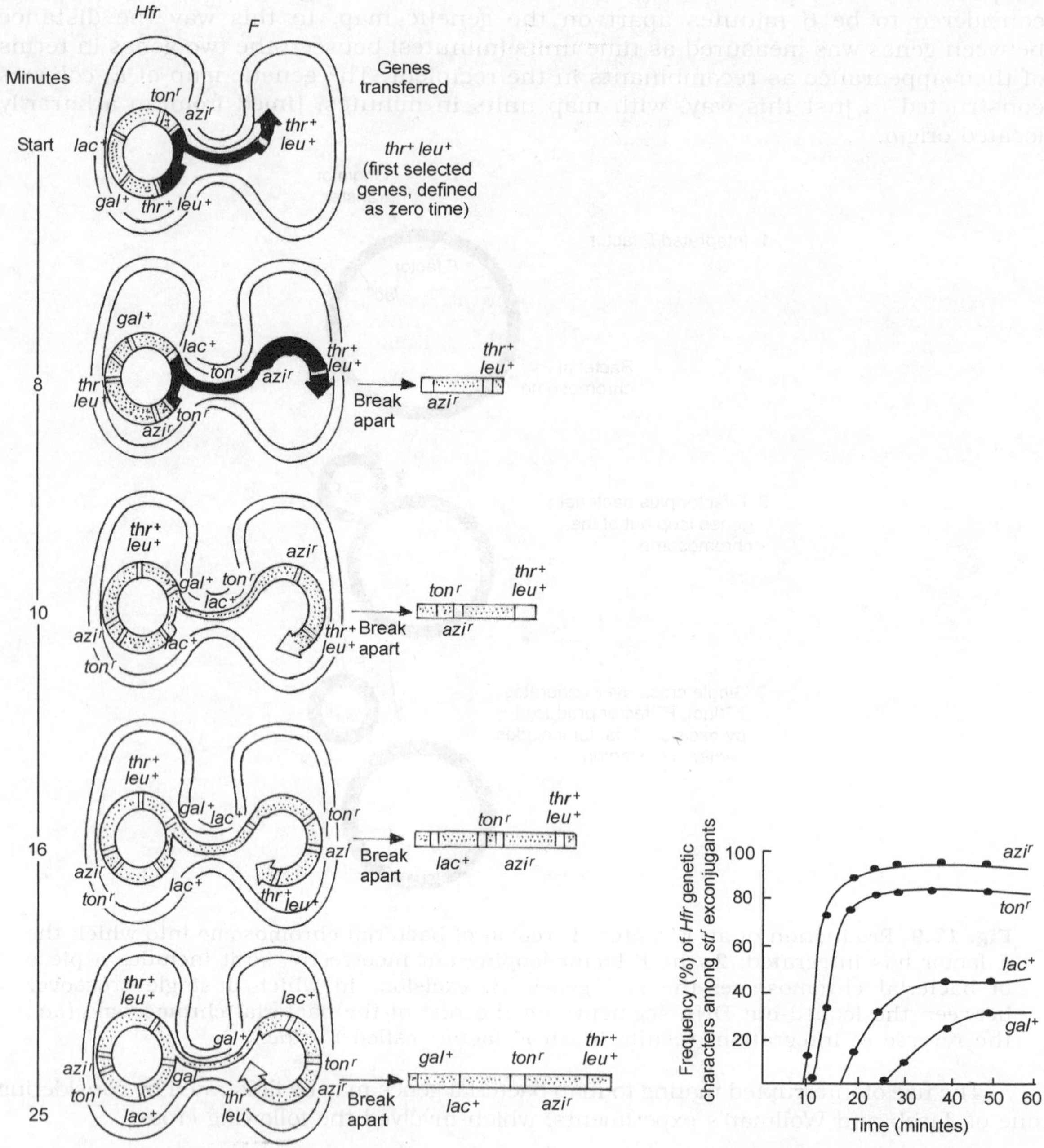

A. Progressive transfer of donor genes to recipient during *Hfr* × F$^-$ conjugation.

B. Appearance of donor genetic markers in recipient as a function of time.

Fig. 12.10. Interrupted-mating experiment involving the cross *HFr H str^s thr$^-$ leu$^-$ azi^r ton^r lac$^+$ lac$^+$* × *F$^-$ str^r thr leu azi^s ton^s Lac gal*: A. At various times after mating commences, the conjugating pairs are broken apart and the transconjugant cells are plated on selective media to determine which genes have been transferred from the *Hfr* to the F$^-$. The progressive transfer of donor genes with time is illustrated. B. The graph shows the appearance of donor genetic markers in the F$^-$ cell as a function of time that is, after the selected *thr$^-$* and *leu$^-$* genes have entered.

The *Hfr* H strain is an *Hfr* strain of *E. coli* distinct from that found by Cavalli-Sforza, and it was isolated by William Hayes, as already discussed. *Hfr* H is prototrophic and is sensitive to the antibiotic streptomycin. The F^- cell carries a streptomycin resistance gene and also a number of mutant genes. These genes cause the F^- to be auxotrophic for threonine (*thr*) and leucine (*leu*), to be sensitive to sodium azide (azi^s) and to be infected by bacteriophage T1 (ton^s), and to be unable to ferment lactose (*lac*) or galactose (*gal*) as a carbon source.

In a conjugation experiment, the two cell types are mixed together in a nutrient medium at 37°C, the normal growth temperature for *E. coli*. After a few minutes have passed so that the H*fr* and F^- cells may paired, the culture is diluted to prevent the formation of new mating pairs. This procedure ensure some synchrony in the chromosome transfer between cells. Samples are removed from the mating mixture at various times and are then blended to break the pairs apart. These transconjugants are plated on a selective medium designed to allow recombinant types to grow and divide while killing both the H*fr* H and F^- parental types.

For this particular cross, the medium contains streptomycin to kill the *Hfr* H and lacks threonine and leucine so that the parental F^-dies. In this cross, the threonine (thr^+) and leucine (leu^+) genes are the first donor chromosomal genes to be transferred to the F^-, so recombinants formed by the exchange of those genes with the *thr leu* genes of the F^- recipient are produced and will grow on the selective medium. This exchange process occurs in the same way described for transformation; that is, by double cross overs. Appropriate media can be used to test for the appearance of other (unselected) marker genes among the selected transconjugants.

Figure 12.10(B) shows an example of the results. The transconjugants here are thr^+, leu^+, and str^s. The first unselected marker gene to be transferred is azi^r and recombinants for this gene are seen at about 9 minutes. The second gene transferred is ton^r at 10 minutes, followed by lac^+ at about 17 minutes and gal^+ at about 25 minutes. This figure also shows the rates of appearance and the maximum frequencies obtained for each recombinant type. It is apparent that with increasing time after conjugation commences, both the rates of appearance and the maximum frequencies of recombinants decrease. The rate of appearance of recombinants decrease because the population of mating cells become asynchronous with respect to the transfer of genetic material that is, the rate of transfer from mating couple to mating couple is not constant. The maximum frequency decreases because with time there is an increasing chance that mating pairs will break apart.

In this experiment, then each gene from an *Hfr* appears in recombinants at a different but reproducible time after mating commences, and the time intervals between gene appearances are used as a measure of genetic distance. Also the maximal frequency of recombinants becomes smaller, the later the gene enters the recipient. From such data we can conclude that an *Hfr* chromosome transfer starts at a specific point and is integrated into the F^- cell in a linear way. The transfer starts at a specific point within the integrated F factor called the origin or O. Genes located far from the origin tend not to be transferred because of the high probability that the conjugating pairs will be broken apart by that time.

CIRCULATION OF THE *E. COLI* MAP

It would take about 100 minutes to transfer the entire chromosome plus F factor from an *Hfr* cell to an F^- cell. Normally however, the conjugating pairs break apart well before such a complete transfer. Then how can we construct a complete genetic map of the *E. coli* chromosome.

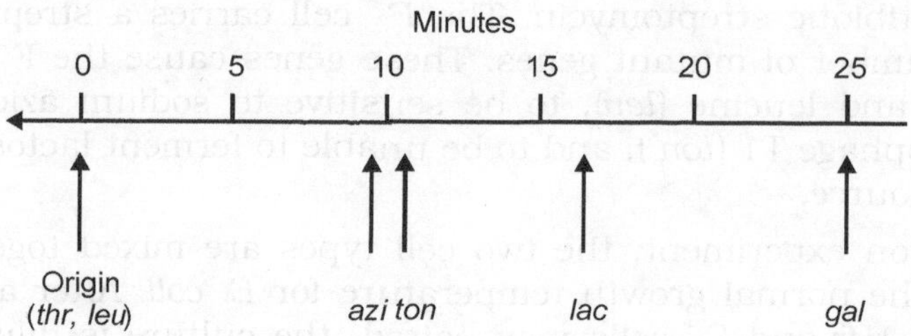

Fig. 12.11. Chromosome map of the genes in the interrupted-mating experiment of Figure 12.10. The marker positions represent the time of entry of the genes into the recipient during the experiment. Thus the map distance is given in minutes. The *azi* marker for example, appeared about 8 minutes after conjugation began, whereas the *gal* marker appeared after about 25 minutes.

Further experiments showed that different *Hfr* strains transfer donor genes into the recipient strain with different starting points and polarities of transfer. That is, since the F factor can integrate into the bacterial chromosome at a number of places, and since the F factor itself determines the direction of gene transfer, chromosomal transfer may be in different directions and from different places in the map. Fig. 12.11 shows the order of chromosomal gene transfer for *Hfr* strains H, 1, 2, and 3. In *Hfr* H, the genes are transferred in the order *O-thr-pro-lac-pur-gal*....in *Hfr* 1 the order is *O-thr-thi-gly-his-gal*...in *Hfr* 2 it is *O-pro-thr-thi-gly-his*... and in *Hfr* 3 it is *O-pur-lac-pro-thr-thi*.... Moreover, the genetic distance in time units between a particular pair of genes is constant, within experimental error, no matter which *Hfr* strain is used as donor, that is the genetic distance between *thr* and *pro*, for example, is the same in H, 1, 2, and 3.

The best way to construct a genetic map of the chromosome is to try to align the genes transferred by each H*fr*. In view of the overlap of the genes, the simplest map that can be drawn from these data is a circular one. This was a truly significant finding since all previous genetic maps of eukaryotic chromosomes were linear.

TRANSDUCTION MAPPING OF BACTERIAL CHROMOSOMES

Transduction, the process whereby pieces of bacterial DNA are carried between bacterial strains by a phage (phage vector) is a useful mechanism for bacterial genes. Of necessity the useful phage are temperate phages, *i.e.*, those phages able to lysogenize, the bacteria they infect. Examples are phages P_1 and λ for *E. coli* in phages P22 for *Salmonella typhimurium.* Two types of transduction occurs : In generalize transduction any gene may be transferred between bacteria, in specialized transduction only specific genes are transferred.

Generalized transduction. The discovery of generalized transduction is credited to Joshua and Esther Lederberg and Norton Zinder. In 1952 these researchers were testing to see whether a conjugation mechanism occurred in the bacterial species *Salmonella* typhimurium, which causes typhus in the mouse. Their experiment was essentially similar to the one that showed conjugation existed in *E. coli.* They mixed together two multiple auxotrophic strains and looked for the appearance of prototrophs. One strain was *phe*$^+$ *trp*$^+$ *tyr*$^+$ *met his*, which required methionine and histidine and the other strain was *phe trp tyr met*$^+$ *his*$^+$ which required phenylanine, tryptophan, and tyrosine. When they crossed these two auxotrophic strains, they found prototrophic recombinants phe^+ trp^+ met^+ his^+ at a low frequency.

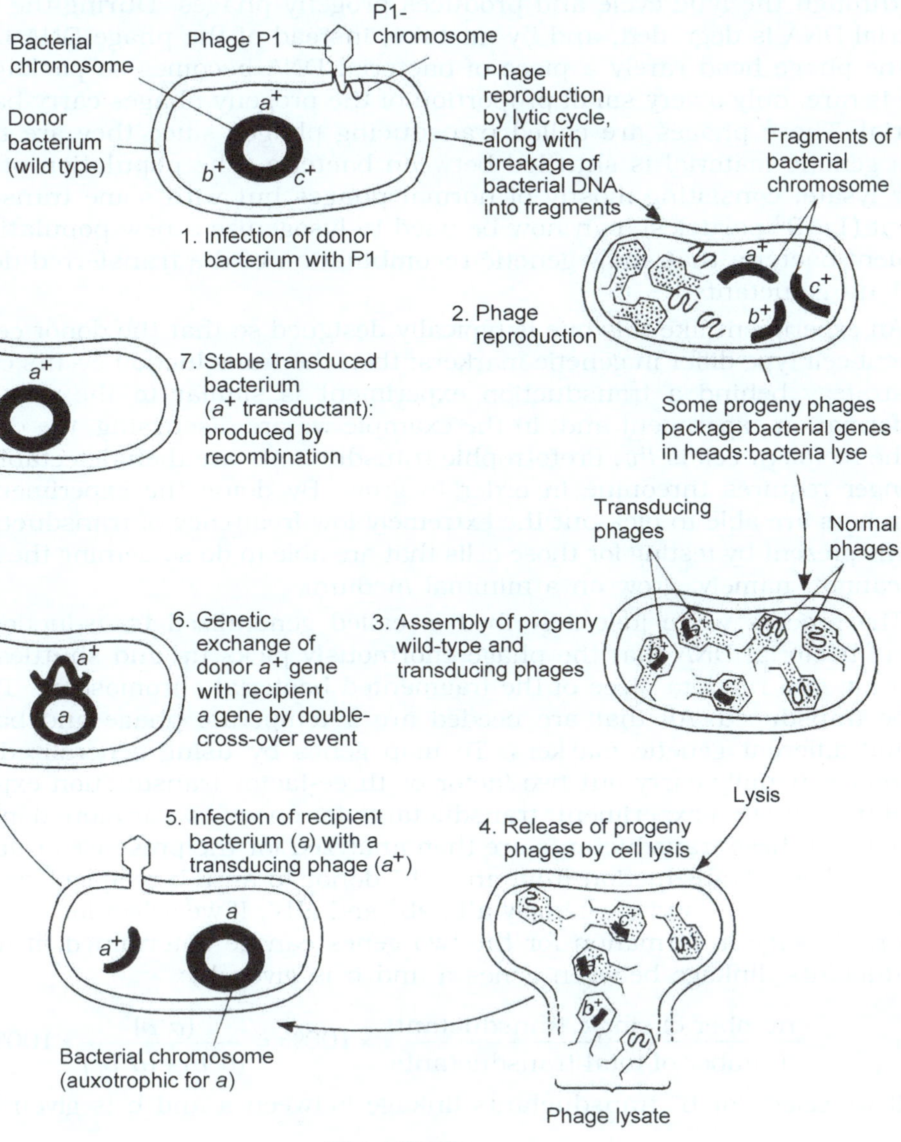

Fig. 12.12

When Zinder and the Lederbergs used the U-tube apparatus (which showed the existence of conjugation in *E. coli*). They still found prototrops. This result indicated that recombinants were being produced by a mechanism other than conjugation, since cell-to-cell contact, was not necessary. The interpretation was that some noncellular agent was responsible for the formation of recombinants. They called this agent a filterable agent since it could pass through a filter with pores small enough to block bacterial movement. In this particular case, the filterable agent was identified as the temperate phage P22.

The mechanism for generalized transduction by P22 is as follows:

Normally, the P22 phage enters the lysogenic state when it infects a bacterium, and it then exists in the prophage state in these cells (*i.e.*, the phages genes required for taking the phage through the lytic cycle are suppressed). As we have said, the mechanism that maintains this state breaks down occasionally, and the phage genome goes through the lytic cycle and produces progeny phages. During the lytic cycle, the bacterial DNA is degraded, and by mistake, instead of the phage DNA being packaged into the phage head rarely a piece of bacterial DNA becomes so packaged. Since this event is rare, only a very small proportion of the progeny phages carry bacterial genetic material. These phages are called transducing phages since they are the vehicles by which genetic material is shuttled between bacteria. The population of phages in the phage lysate, consisting mostly of normal phages but with some transducing phages present ($1/10^5$ particles), can now be used to lysogenize a new population of bacteria. Recipient bacteria that show genetic recombination of the transferred donor genes are called transductants.

An experiment like this one is typically designed so that the donor cell type and the recipient cell type differ in genetic markers; then the transduction events can be followed. The strategy behind a transduction experiment is similar to the strategy. Behind a transformation experiment and, in the example we are discussing, the donor cell is *thr*$^+$ and the recipient cell is *thr*. Prototrophic transductants are then detectable since the cell no longer requires threonine in order to grow. By doing the experiment in this way, researchers are able to pick out the extremely low frequency of transductants among all the cells present by testing for those cells that are able to do something the nontransduced cells cannot, namely, grow on a minimal medium.

The process we've just described is called generalized transduction because the piece of bacterial DNA that the phage enormously picks up and shuttles to a recipient bacterium is a random piece of the fragmented bacterial chromosome. Thus any genes can be transduced. All that are needed are a temperate phage and bacterial strains carrying different genetic markers. To map genes by using generalized transduction, researchers usually carry out two-factor or three-factor transduction experiments. In a typical transduction experiment, transductants for one of two or more donor markers are selected and these transductants are then analyzed for the presence or absence of other donor markers. Transduction from an a^+ b^+ donor to an a b recipient produces various transductants for a^+ and b^+, namely a^+b, ab^+ and a^+b^+. If we select for one or other donor markers, linkage in formation for the two genes can be determined. If we select for a^+ transductants, linkage between genes a and b is given by:

$$\frac{\text{number of single transductants}}{\text{number of total transductants}} \times 100\% = \frac{(a^-b)}{(a^+b)+(a^+b^-)} \times 100\%$$

If we select for b^+ transductants linkage between a and b is given by

$$\frac{(ab^-)}{(ab^-)+(a^+b^-)} \times 100\%$$

Since the production of a transducing phage is a rare event, and since the amount of bacterial DNA the phage can carry is limited by the capacity of the phage head, the chance of either a^+ or b^- being included individually in the phage head is proportional to the distance between them. If the genes are close enough to be included in the transducing phage, an a^+ b^+ transductant can be produced which is called a *cotransductant.* Thus, the cotransduction of two or more genes is a good indication that the genes are closely linked.

The map distances between cotransduced genes may be determined by using transduction and it is by this procedure that linkage maps of bacterial chromosomes have been constructed. For example, consider the mapping of some genes in *E. coli* by using transduction with the temperate phage P1. The donor *E. coli* strain is *leu*$^+$ *thr*$^+$ *azi*^{r} (able to grow on a minimal medium and resistant to the metabolic poison sodium azide), and the recipient cell is *leu thr azi*^{s} (need leucine and threonine as supplements in the medium and sensitive to azide). The P1 phages are grown on the bacterial donor cells, and the phage lysate is used to transduce the recipient bacterial cells. Transductants are selected for any one of the donor markers and then analyzed for the presence of the other two (in this case) unselected markers.

Transduction data or deducing gene are in order to this particular phage in the *E. coli* chromosome is between the gal (galactose fermentation) and the bio (biotin biosynthesis) genes.

The particular *E. coli* strain that λ lysogenized is *E. coli* K12, and when it contains the λ prophage, it is designated *E. coli* K12 (λ). Let us focus just on the gal gene and assume that the particular K12 strain that λ lysogenized is gal$^+$ that is it can ferment galactose as a carbon source. This phenotype is readily detectable by plating the cells on a solid medium containing galactose as a carbon source and containing a dye that changes colour in response to the products of galactose fermentation. The gal$^+$ colonies are pink, while gal colonies are white on this medium. If we induce the prophage functions—then the lytic cycle is initiated. Induction may be accomplished by any procedure that destroys the protein repressor molecules responsible for the inhibition (ultraviolet light irradiation is one commonly used procedure).

After the lytic cycle is initiated, the phage chromosome loops out, and a separate circular λ chromosome is generated by a single cross-over event at the att λ/b2 sites (Fig. (12.13(B). This sequence of events is exactly the reverse of the integration process and parallels the events of F factor integration and excision in conjugation. In most cases the excision of the phage chromosome is precise so that the complete λ chromosome and only the λ chromosome is produced. In rare cases the excision events are not precise so that by crossing over between the phage chromosome and the bacterial chromosome at sites other than the homologous recognition sites, an abnormal circular DNA product results. In the case diagramed a piece of λ chromosome has been left in the bacterial chromosome, while a piece of bacterial chromosome, including the *gal*$^+$, gene, has been added to the rest of the λ chromosome. When a bacterial gene (or genes) is included in a progeny phage, we have a transducing phage, called, in this case, λd gal. The d stands for "defective" since not all phage genes are present, and the gal indicates that the bacterial host cell gal gene has been picked up. This event is similar to F$^+$ production by defective excision of the F factor.

Since the abnormal looping-out phenomenon is a rare event, the phage lysate produced from the initial infection diagramed contains mostly normal phages and a few transducing phages ($1/10^3$). The phage lysate can be used to infect nonlysogenic bacteria (bacteria that do not contain phages) that are gal. Nonlysogenic bacteria must be used because one a bacterium contains a prophage, it is immune to infection by a second phage of the same type. The phage lysate contains mostly wild-type phages and relatively few (about one in a million) gal$^+$ transducing phages. Because of the small proportion of transducing phages, the lysate is called a low frequency transducing (LFT) lysate.

Infection of the gal bacterial cells with the LFT lysate produces two types of transductants. One type is unstable. The wild type λ integrates at its normal att λ site, and then the λd gal$^+$ phage integrates by a crossing over event within the common λ sequence. In the bacterial chromosome the bacterium is heterozygous gal^+/gal and hence can ferment galactose. This transductant is unstable because by induction phage growth can be initiated. (The event will be described a little later). The second type is stable : These tranductants are produced when only a λd gal^+ phage is present. The gal^- gene carried by the phage may be exchanged for the bacterial gal gene by a double cross-over event. Such a transductant is stable because the bacterial chromosome contains only one type of gal gene and no phage genes are integrated.

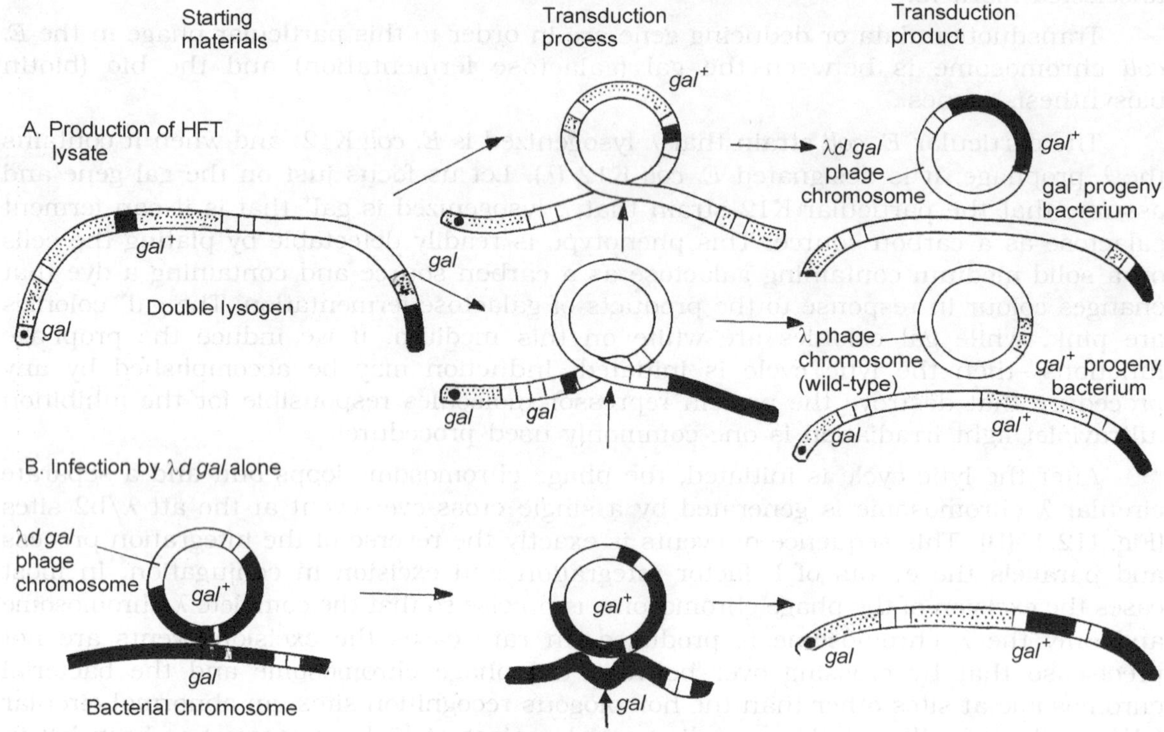

Fig. 12.13. Specialized transduction by bacteriophage λ : A. The unstable double lysogen containing λ and λ*d* gal producing about equal number of λ*d* gal and wild-type λ phages, thus giving rise to a high-frequency transducing (HFT) lysate. B. Low concentrations of phage from the HFT lysate used to infect bacteria so that a cell can be infected by λ*d* gal alone.

A high frequency transducing (HFT) lysate may be obtained with this system. The starting point is the doubly lysogenic, unstable transductant in which both a wild-type λ and λd gal^+ phage are produced at high frequency. When the HFT lysate is used to infect new nonlysogenic gal bacteria so that only one phage infects each bacterium, the defective phage genome is integrated by a single cross-over in the gal gene region. This integration produces a gal^+/gal transductant that is stable since it lacks sufficient λ genes to lyse the cells. Thus by manipulating the number of phages that infect a bacterium, and using appropriate selection event, we can detect specialized transduction and cause marker genes to be transduced at high frequency. Then we can make a detailed analysis of the genes transduced in this way.

LINKAGE AND MAPPING GENE IN EUKARYOTES

Genes on the same chromosome are called linked genes and are said to belong to a linkage group. The number of linkage groups in an organism equals the haploid number of chromosomes. Through test crosses we can determine which genes are linked to each other and can then construct a linkage map or genetic map of each chromosome.

Genetic mapping provides information that is useful in many aspects of genetic analysis. Genetic mapping can tell us, for example, whether genes that work together to produce a given phenotype are located together on the chromosome. Such information can be used to construct models that explain the regulation of gene expression in an organism. Also, the knowledge of the locations of genes on chromosome has been useful in recombinant DNA research and in understanding the DNA sequences in and around genes and at parts of the chromosomes (such as the centromeres) that are of particular functional interest.

In this chapter, we will learn about the discovery of gene linkage, about the evidence that gene recombination is the result of a physical exchange of chromosome parts (crossing over) and about how a genetic map of a chromosome is constructed.

DISCOVERY OF GENETIC LINKAGE PARTIAL LINKAGE IN THE SWEET PEA

In 1905, *W. Bateson, E. R. Saunders*, and *R. C. Punnett*, in their studies of heredity in the sweet pea, *Lathyrus odoratus L*, discovered the first exception to the law of independent assortment of gene pairs. They crossed two true breeding strains, one that produced purple flowers and long pollen and another that produced red flowers and round pollen. All F_1 plants had purple flowers and long pollen, indicating that purple flower colour was dominant to red flower colour and that long pollen was dominant to round pollen. The 381 plants of the F_2 consisted of 305 plants with purple petals and 76 plants with red petals. (This is not a good fit to the 3 : 1 ratio predicted by the law of segregation for a cross of two heterozygotes, so presumably there was a difference in viability for the two phenotypic classes). For the pollen shape trait the F_2 plants segregated into 305 with long pollen and 76 with round pollen. These results indicated that both the flower colour trait and the pollen shape trait are controlled by a single pair of alleles at their respective genes.

A natural extension of the data analysis was to see whether the two character pairs assorted independently in the cross. From the $F_1 \times F_1$ cross the F_2 progeny displayed the following phenotypes.

284 (74.6 per cent) purple flowers, long pollen

21 (5.5 per cent) purple flowers, round pollen

21 (5.5 per cent) red flowers, long pollen

55 (14.4 per cent) red flowers, round pollen.

More precise frequencies determined by Punnett in 1917 and based on 6932 F_2 plants derived from the same type of purple and long X red and round cross gave these results:

4831 (69.5 per cent) purple, long
390 (5.6 per cent) purple, round
393 (5.6 per cent) red, long
1338 (19.3 per cent) red, round.

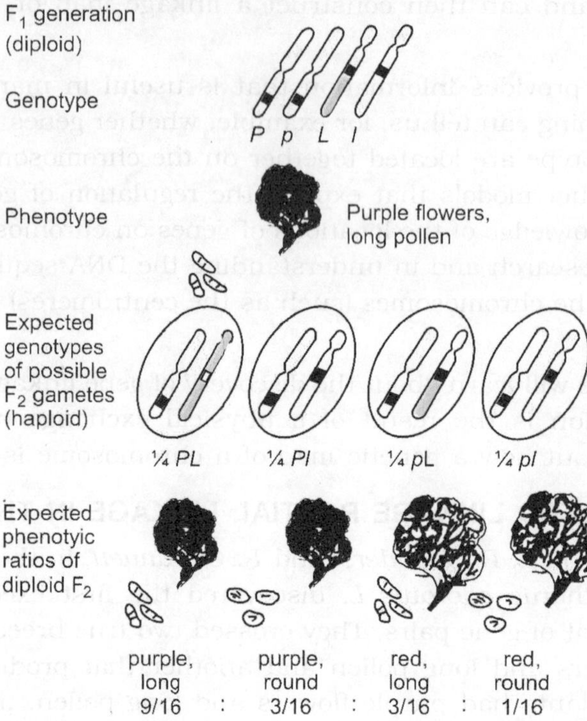

Fig. 12.14. Expected results of F_1 and F_2 cross it genes.

If the genes assorted independently, the expected proportions would be 9/16 : 3/16 : 3/16 : 1/16 for the four phenotypes in the order given, or in percentages, 56.25 per cent : 18.75 per cent : 18.75 per cent : 6.25 per cent, respectively. The observed values differ significantly from those expected according to the independent assortment hypothesis. What constitutes a "significant" difference between observed results and results expected on the basis of a particular hypothesis may be determined by a chi-square test, a statistical procedure described under the heading CRI-DU-CHAT.

Bateson, Saunders, and Punnett recognized that the dominant purple and long characters occurred together more often than predicted by Mendel's second law but could not explain why this result occurred. They tried to explain their exceptional results in terms of modified Mendelian ratios but without success. We know now that the genes they were studying are located on the same chromosomes. Genes located on the same chromosome are said to exhibit linkage and are considered to be linked genes. Thus Bateson and colleagues results can be explained as follows:

If we symbolize the purple allele as P, the red allele as p, the long allele as L, and the round pollen allele as 1, the cross is PP LL X ppll. The F_1 double heterozygote is PpLl. If independent assortment were occurring (Fig. 12.14), equal proportions of PL, Pl, pL, and pl gametes would be produced by the F_1 plants, and the resulting F_2 generation would exhibit the expected 9 : 3 : 3 :1 ratio.

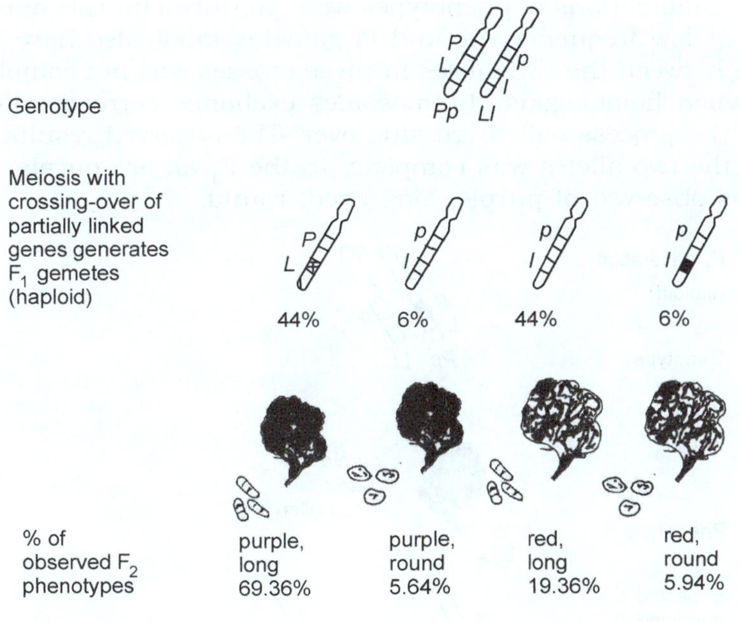

A

Gametes from F_1 heterozygote (%)	Pollen: *PL* 44	*Pl* 6	*pl* 44	*pL* 6
Eggs	Genotype frequencies in F_2 (%)			
PL 44	19.36	2.64	19.36	2.64
Pl 6	2.64	0.64	2.64	0.36
pl 44	19.36	2.64	19.36	2.64
pL 6	2.64	0.36	2.64	0.36

B

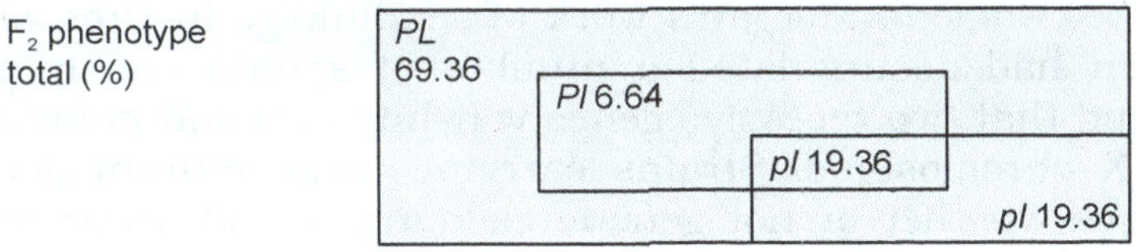

C

Fig. 12.15 A. F_1 generation (diploid). B. Random union of F_1 gametes to produce F_2. C. Phenotype frequencies in F_2 (%).

Instead, the observed ratio of F_2 phenotypes can best be explained if 44 per cent of the gametes are PL, 44 per cent are pl and 6 per cent each are Pl and pL. At the time of pollen and egg formation in the F_1 sweet peas, there was a tendency for the parental

P and L alleles to stay together and for the parental p and l alleles to stay together. Since two nonparental combinations of phenotypes were produced (purple and round, and red and long) albeit at low frequency, pL and Pl gametes must also have been generated. Thus the linkage between the two alleles in these crosses was not complete. This partial linkage occurs when homologous chromosomes exchange corresponding parts during meiosis through the process called crossing over. The observed results would be if the linkage between the two alleles was complete. In the F_1 all are purple, long and in the F_2 a 3 : 1 ratio is observed of purple, long : red, round.

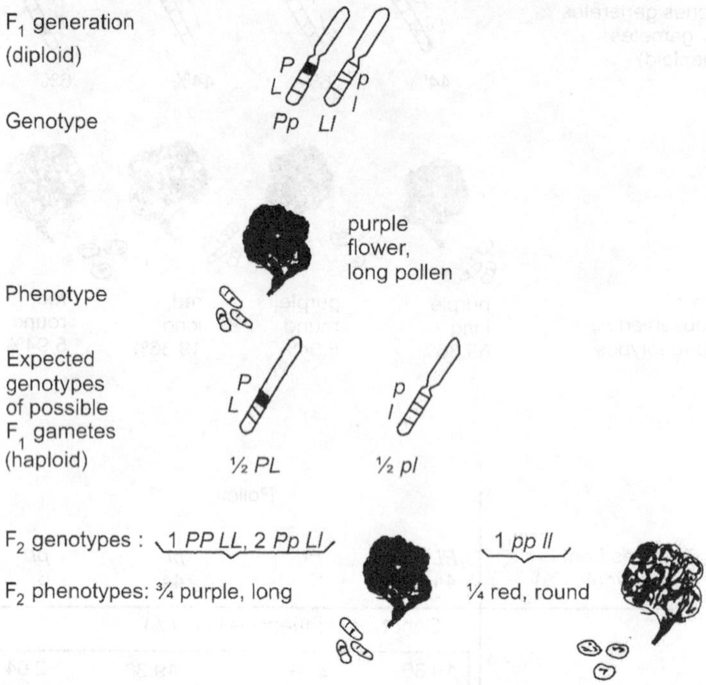

Fig. 12.16. Expected genes result of F × F_1 cross if genes are completely linked.

MORGAN'S LINKAGE EXPERIMENTS WITH *DROSOPHILA*

Our modern understanding of genetic linkage stems from the work of *T. H. Morgan* and his colleagues with linkage in *Drosophila melanogaster*, which was done around 1911. We described some of Morgan's work of sex linkage in *Drosophila* in Chapter 3. By 1911, Morgan had accumulated a number of spontaneously arising sex-linked mutants. He found that two recessive genes w (white eye) and m (miniature wing) were located on the X chromosome. Strains carrying these mutant genes were used in experiments to see whether or not genetic exchange could occur between the linked genes on the chromosomes.

Morgan crossed a white miniature (wm/wm) female with a wild-type male (w^-m^+/Y). As expected, all the F_1 males were white-eyed and had miniature wings, while all females were wild type for both the eye colour and wing size traits. The F_1 flies were interbred and the 2441 F_2 flies were analyzed. In results, similar to these of Bateson, Saunders, and Punnett, the most frequent phenotypic classes in both sexes were the grandparental phenotypes of mutant white eyes plus, miniature wings or wild-type red eyes plus large wings. Conventionally, we refer to the original genotypes of the two chromosomes as parental genotypes, parental classes, or more simply, parentals. The term is also used to describe phenotypes, as the original white miniature females and wild-type males in the crosses under discussion are defined as the parentals.

Morgan was surprised to find that significant number of flies had the nonparental combinations of the eye and wing phenotypes of white eyes plus normal wings and red eyes plus miniature wings. The two nonparental combinations of linked genes are called recombinants. In all, 900 of 2441 F_2 flies, or 36.9 per cent exhibited recombinant phenotypes. To explain the recombinants, Morgan proposed that in 36.9 per cent of the meioses an exchange of genes had occurred between the two X chromosomes of the F_1 females. (No such exchange needs to be postulated for sperm production since the males are hemizygous and no genetic exchange occurs between the nonhomologous X and Y chromosomes).

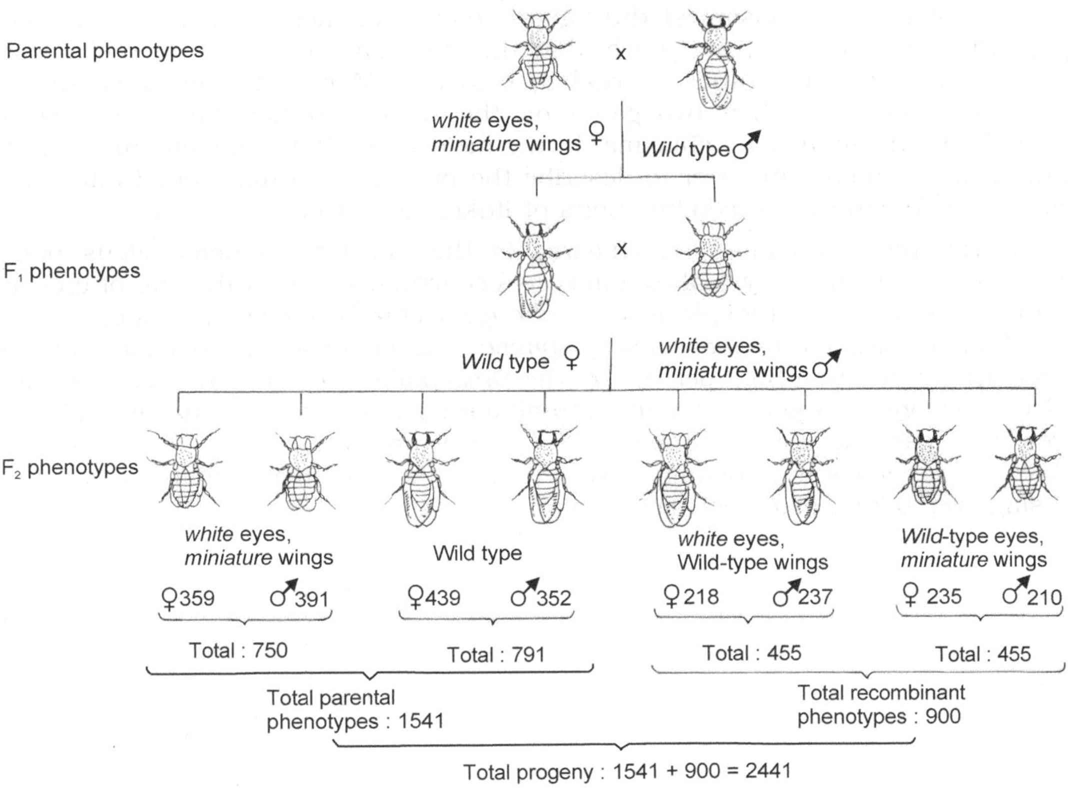

Fig. 12.17. Morgan experimental crosses of white eye and miniature wing variants of *D. melanogaster.*

Next Morgan extended the work to include another sex-linked character, which was yellow body colour. He crossed a true-breeding white-eyed, yellow-bodied mutant female with a wild type red-eyed grey-bodied male (where red eyes and a grey body are dominant traits). As expected, the F_1 progeny consisted of wild-type females and white-eyed, yellow-bodied males; that is the two parental phenotypic classes. Among the 2205 F_2 flies obtained the original grabed-parental combinations of white eyes plus yellow body and red eyes plus grey body were the most frequent phenotypic classes. While only 1.3 per cent of the flies (29 of 2205) showed recombinant phenotypes. Morgan concluded that the genes that controlled the eye colour and the body colour traits were more closely linked than were the genes for eye colour and wing size.

A large number of other crossed of this type were done and the conclusions were always the same. In each case, the parental phenotypic classes were the most frequent while the recombinant classes occurred much less frequently. Approximately equal numbers of such of the two parental classes were obtained, and similar results were obtained for the recombinant classes. Morgan's general conclusion was that linear to each other on the same chromosome. To take this conclusion one step further, characters remain together to a greater extent the closer the corresponding genes are on the chromosome.

Morgan was also able to relate the production of the phenotypic classes in the F_2 progeny of the cross to chromosomal segregation. Earlier, in 1909, *F. Janssens* had described chiasmata (singular = chiasma; cytologically observable physical exchange between homologus chromosomes) during prophase I of meiosis in the salamander. Janssens thought (but could not prove) that the chiasmata might be sites of physical exchange between maternal and paternal homologues. Morgan hypothesized that the partial linkage occurred when two genes on the same chromosome were separated physically from one another by chiasma during meiosis. In 1912, *Morgan* and *E. Cattell* introduced the term crossing over to describe the process of chromosomal interchange by which recombinants (new combinations of linked genes) arise.

To clarify the terminology: A chiasma is the place on a homologous pair of chromosomes at which a physical exchange is occurring, *i.e.*, it is the site of crossing-over. *Crossing-over is the actual process of exchange of chromosome segments of nonsister chromatids of homologous chromosomes by symmetrical breakage and crosswise rejoining.* Therefore, the reason that 36.9 per cent of the *Drosophila* progeny were recombinant for W and m, and only 1.3 per cent were recombinant for w and y, is that w and m are relatively far apart on the X chromosome while w and y are quite close. Thus the probability of crossing over occurring between w and m is much higher than the probability of crossing over taking place between w and y.

13 Interference and Coincidence in Pro and Eukaryotes

In most higher organisms it has been found that one chiasma formation reduces the probability of mother chiasma formation in an immediately adjacent region of the chromosome, probably because of physical inability of the chromatids to bend back upon themselves within certain minimum distances. *The tendency of one cross-over to interfere with the other cross-over is called interference.* Thus, the proximity of one cross-over to another decreases the probability of another very close by. The centromere has a similar interference effect; frequency of crossing over is also reduced near the ends of the chromosome arms. The net result of this interference is the observation of fewer double cross-over types than would be expected according to map distances. The strength of interference varies in different segments of the chromosome and is usually expressed in terms of a coefficient of coincidence, or the ratio between the observed and the expected double cross-overs.

Coefficient of coincidence = % observed double crossovers

The coincidence is the complement of interference, so:

Coincidence + Interference = 1.0

When interference is complete (1.0), no double cross-over will be observed and coincidence becomes zero. When interference decreases, coincidence increases. Coincidence values ordinarily vary between 0 and 1. Coincidence is generally quite small for short map distance. There is no interference across centromere.

Example: For explaining interference and coincidence, we can consider the results of one of the experiment of *Hutchison* (1922). He reported the map distance for three genes, c (colourless aleurone), Sh (shrunken grains) and wx (waxy endosperm) of corn and observed following crossing over frequencies between these genes :

Table 13.1. Crossing over Frequencies between Genes c, sh and wx of Corn

Regions	*Genes*	*Percentage Crossovers*	*Map Distance (in map units)*
I	c-sh	3.4	3.4 + 0.1 = 3.5
II	Sh-wx	18.3	18.3 + 0.1 = 18.4
Double cross-over	c-Sh-wx	0.1	

Table 13.2. Some Linkages Groups and Chromosome Assignment in Human Beings

Number of the Chromosomes	*Gene*	*Number of the Chromosomes*	*Gene*
1	Anylase (pancreatic)	13	Ribosomal RNA
1	Actin, skeletal muscle alpha chain	13	Collagen-IV, alpha-I chain
1	Xeroderma pigmentosum A	13	Collagen-IV, alpha-II chain
1	Rhesus blood antigen (Rh)	14	Ribosomal RNA
1	Histone cluster B : H3, H 4	14	T-cell leukemia-I
1	Cystic fibrosis antigen	14	Immunoglobulin heavy chain gene cluster
1	Acid phosphatase		
2	Elastin	15	Ribosomal RNA
2	Interferon-I	16	Haemoglobin alpha
2	Collagen-III	16	Haemoglobin zeta
2	Collagen-IV	16	Nonhistone chromosomal protein-I
2	Glucagon	17	Growth hormone
2	Tubulin, alpha, testis specific	18	Gastric releasing peptide
3	Rhodopsin	19	Bombay phenotype
3	Somatostatin	19	Lewis blood group
4	Huntington's disease	19	Green/blue eye colour
4	MN blood group	20	Growth hormone releasing factor, somatocrinin
4	Diabetes insipidus		
5	Fibroblast growth factor, acidic	21	Ribsomal RNA
6	Tubulin, beta M40	22	Myoglobin
6	Insulin-dependent diabetes mellitus	22	Ribosomal RNA
7	Actin, cytoskeletal beta	X	Polymerase DNA, alpha
7	Collagen I alpha 2	X	Gonadal dysgenesis, XY female type
7	Histone cluster A : H1, H2A, H2B	X	Haemophilia A, factor V
7	Non histone chromosomal protein-2	X	Haemophilia B, factor IX
8	Polymerase, DNA beta	X	Deutan (green) colour blindness; green cone pigment
8	Carbonic Anhydrase cluster	X	Protan (red) colour blindness; red cone pigment

9	ABO blood group	X	Ocular albinism
10	Hexokinase-1	Y	Azoospermia third factor
11	Catalase	Y	Stature
11	Insulin	Y	H-Y antigen
12	Salivary protein complex	Y	Testicular determining factor (TDF)
12	Collagen II, alpha-I	Y	Pseudoautosomal segment (PAS)

If crossing over in region I and II were independent, we should predict 0.035 × 0. 184 = 0.6 per cent double cross-overs, whereas only 0.1 per cent was observed.

So, $$\text{coincidence} = \frac{0.1}{0.6} = 0.167$$

LINKAGE MAPS OF DIFFERENT ORGANISMS

By adopting the above mentioned techniques, geneticists have constructed the linkage or genetic maps of various organisms, such as viruses, bacteria, fungi, tomato, barley, wheat, rice, sorghum, morning glory, garden pea, maize, *Drosophila*, chickens, mice, man (Table 13.2), etc. The first linkage map has been constructed for two chromosomes of *Drosophila* by Strutevent in 1911. The linkage maps of other chromosomes of Drosophila have been constructed by *C.B. Bridges*. The linkage or genetic mapping in maize has been done by *McClintock* under the leadership of *R. A. Emerson*.

SYNTENIC GENES

If two or more specific human gene products and a given human chromosome are both present in the same hybrid cells, then those genes are located in the same chromosome; that is, they are *syntenic*. The term *synteny* refers to genes that are located on the same chromosome, whether or not they show recombination; linkage refers only to genetic loci that have been shown by recombination studies to be in the same chromosome. Syntenic genes may be so far apart in their chromosome that they seen to segr. The independently; that is, they may show as much as 50 percent recombination as would be exhibited by non-syntenic genes.

14 HEREDITARY DEFECT—KLINEFELTER, TURNER, CRI-DU-CHAT AND DOWN'S SYNDROME

INTRODUCTION

In humans, additions and deletions of chromosomes particularly the large chromosomes almost always result in lethals. Some new borns with extra chromosomes of the smaller groups such as G (Example, chromosome 21) survive but show multiple physical and mental abnormalities.

KLINEFELTER'S SYNDROME (47, XXY)

The very same anomaly that Bridges found in the chromosomes of the fruit fly actually showed up in another constitutional disorder of human beings that was linked to an abnormality of chromosome complement. This was Klinefelter's syndrome, named for *Harry Fitch Klinefelter*, the U.S. physician who described it. Individuals with this syndrom are phenotypically males but some tendency toward femaleness can also be recognized, particularly in secondary sex characteristics. It occurs once in about every 400 live male births and is characterized in the adult by tallness, enlargement of the breasts, small prostrate glands, testicular dysgenesis (defective development of testicles) and virtual nonproduction of sperm. The syndrome is sometimes also associated with severe mental retardation. Most of the patients show a chromosome complement of two X chromosomes and one Y chromosome: XXY. Bridge's XXY flies were female; an XXY human is a sterile male.

An XXY zygote could result, of course, from non-disjunction during meiosis in either the father or the mother; an abnormal ovum would supply two X chromosomes and the normal sperm a Y, or an abnormal sperm would contribute an X and a Y to the X of the ovum.

In one case it has been possible to establish in which parent the failure occurred. This victim of Klinefelter's syndrome was colour-blind, it was also determined that he had a colour-blind mother and a father with normal vision. The gene for colour-blindness, is sex-linked (*i.e.*, it lies on the X chromosome) and recessive (*i.e.*, its effect is negated by another X chromosome with normal gene). The colour-blindness gene must therefore have been carried by both of the mother's X chromosome, it could not have been carried by the father's X chromosomes or the father would have been colour-blind. It followed that the patient inherited both X chromosomes from his mother. In other instances, however, it is likely that the patient inherits an X chromosome as well as a Y from his father as the result of non-disjunction in meiosis of the sperm, and only the normal one-X complement from his mother. Additional cases have been identified, with XXXY and

XXXXY cells the genes on the little Y chromosome should not be underestimated; even when it is outnumbered four to one, the patient still has masculine characteristics.

If non-disjunction occurs in the mother, the XX ovum need not, of course, be fertilized by a Y-bearing sperm. What would be the consequence of fertilization by an X-bearing sperm, yielding an XXX zygote? When this occurs in fruit flies, the result "super females" appear remarkably normal. When it happens in the human species, some of the XXX females are normal and fertile but many are mentally defective and infertile.

The few individuals that have been studied with tetrasomic X chromosomes (48, XXXX) are all mentally defective females. In the general population, a trisomic X chromosome occurs in the proportion of about one per 1000. Individuals with the "triple X syndrome" are comparable in some ways to *Drosophila* metafemales (XXX). In *Drosophila*, however, such individuals are strikingly abnormal and sterile, whereas human XXX individual, are visibly indistinguishable from individuals with the normal XX arrangement.

The best known symptoms associated with this syndrome are abnormalities associated with functional processes such as menstruation.

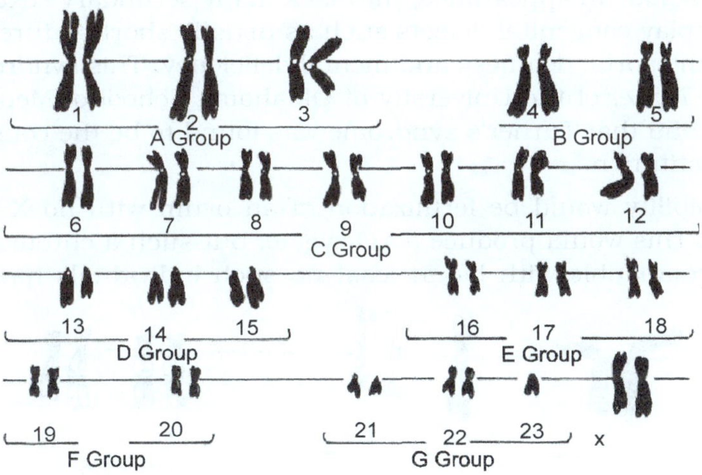

Fig. 14.1. Karyotype of a young man with Klinefelter syndrome, showing an extra X chromosome (47, XXY) (Courtesy of *R.M. Fineman*).

47, XYY and Behaviour

P.A. Jacobs and her associated reported in 1965 that seven XYY males were detected in population of 197 males, mentally subnormal inmates of a penal institution in Scotland. The XYY men were unusually tall, with an average height of 73.1 inches, compared with 64 inches for XY men in the same prison. Numerous, have since confirmed that a high proportion of XXY individuals are tall, subnormal in intelligence (with Iqs ranging from 80 to 95), and antisocial. The aggressive behaviour that brought them into conflict with the law was usually against property rather than people.

XXY trisomy occurs about once in 1000 live male births in the general European population. Only a few of these can be accounted for in the criminal population. Furthermore, most XYY men have been described as perfectly normal in behaviour. In studies carried out in the 1970's, *E.B. Hook* showed that only 3.6 per cent of all XYY men are institutionalized for any reason.

Environmental factors are presumed to be involved in the development of aggressiveness. Since some XXY men are subnormal in intelligence and excessively tall in stature, particular environmental situations in childhood or adulthood may lead to withdrawal from society or adulthood may lead to withdrawal from society or aggressive behaviour. Unfavourable social conditions such as frustration in personal accomplishment and tauning from associates may encourage physical aggression as a means of adaptation. Males with this sex trisomy have not been found to transmit the extra Y chromosome to their sons. This extra chromosome seems to be weeded out in gametogenesis. A wide range of physical and mental abnormalities has been detected in institutionalized XXY men, but most of these are irregular in occurrence and do not form a syndrome. Tallness of stature and mental dullness, however, are fairly characteristics, among institutionalized XYY men.

Turner's Syndrome (45, X)

If non-disjunction can result in ova with two X chromosomes, it can also produce ova with no X chromosome at all. When such a gamete is fertilized by an X-bearing sperm, the zygote will have only one sex chromosome—it is called XO. Such individuals have been found female in appearance, they lack many secondary sexual characteristics; they may also display congenital defects such as usually short stature, a "webbed" neck, malformation of the aorta, deafness and mental deficiency. This syndrome was described by *Henry Hubert Turner* of the University of Oklahoma School of Medicine in 1938, but it was not until 1959 that Turner's syndrome was found to be the consequence of an XO chromosome constitution.

A final possibility would be fertilization of an ovum with no X chromosome by a Y-bearing sperm. This would produce a YO zygote, but such a chromosome complement is apparently incompatible with life at least no such individuals have been found.

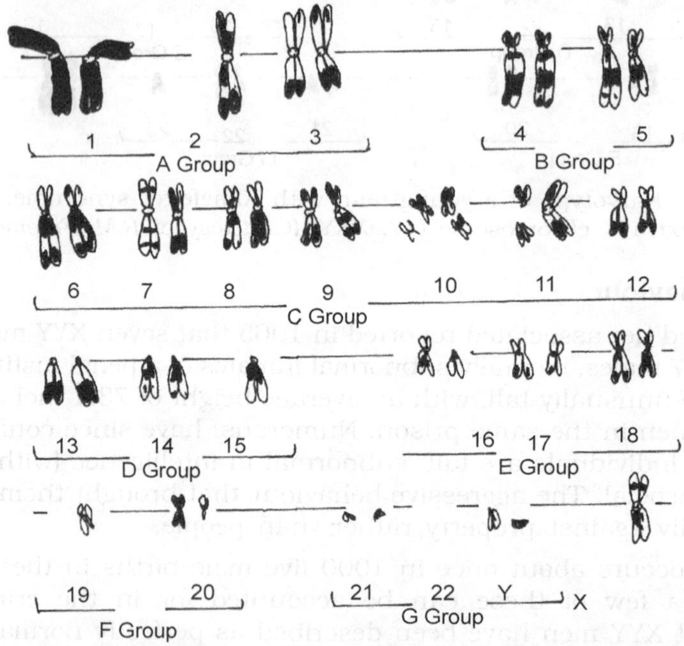

Fig. 14.2. Karyotype of a boy showing an extra Y chromosome (47, XYY); (Courtsey of *S.W. Rogers*).

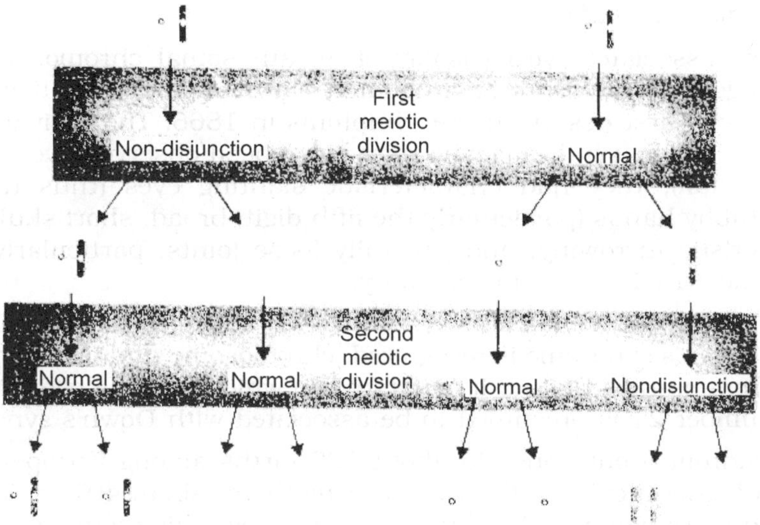

Fig. 14.3. Consequences of non-disjunction at the first and second meiotic divisions. Non-disjunction at meiosis I produces no normal gametes. Non-disjunction at meiosis II produces two gemetes containing (or lacking) two identical chromosomes, both derived from the same member of the homologous pair, and two normal gametes. It is assumed that other chromosome pairs not shown in this diagram behave normally in the first and second meiotic divisions.

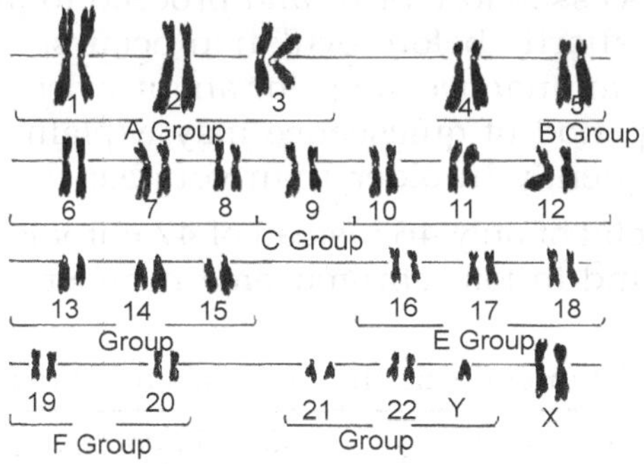

Fig. 14.4. Karyotype of a newborn with trisomy-13 syndrome, showing three number 13 chromosomes 47, XX, + 13. (Courtesy of *R.M. Fineman*).

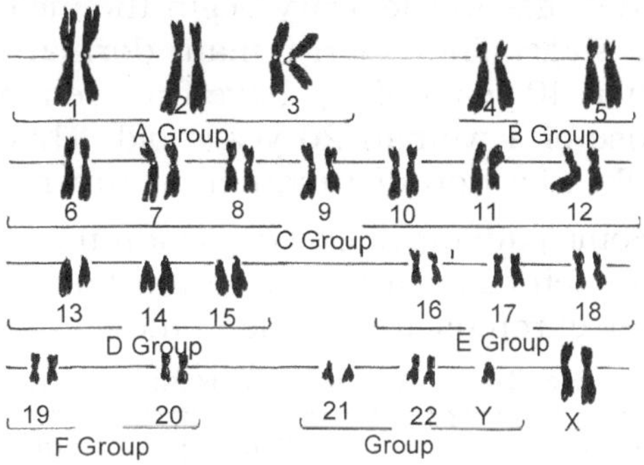

Fig. 14.5. Karyotype of a child with trisomy-18 syndrome, showing three number 18 chromosomes 47, XX, + 18. (Courtesy of *R.M. Fineman*).

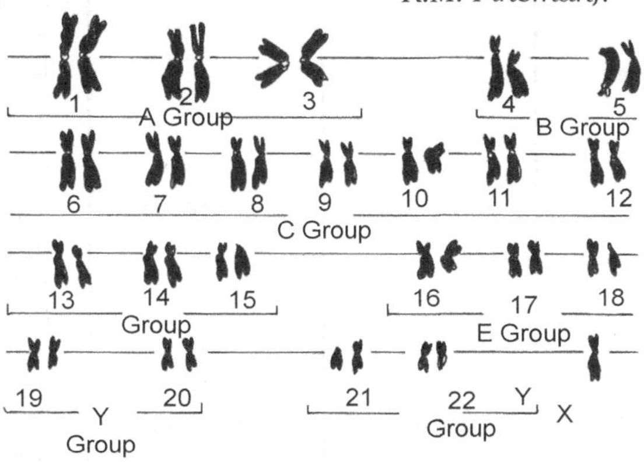

Fig. 14.6. Karyotype of a child with Turner syndrome showing only one sex (X) chromosome (45 X).(Courtesy *R.M. Fineman*).

Down's Syndrome (47, +21)

A syndrome associated with trisomy of an autosomal chromosome was formerly known as "mongolism" but is now designated Down's syndrome. It was named after *Langdon Down*, who first described the symptoms in 1866. The individuals he studied, and those with the same syndrome who have since been observed, were short in stature (about four feet tall); they had characteristic slanting eyes (thus the earlier name "mongolian"), stubby hands (particularly the fifth digit, broad, short skull, large tongues, with a characteristic furrowing, and generally loose joints, particularly in the ankles. They were characterized as low in mentality.

A small chromosome presumably number 21, is added to the normal complement in all cases where Down's syndrome is recognized. chromosome number 21 has been difficult to distinguish from number 22 and it is not known which of these two is involved. On the basis of size, number 21 is presumed to be associated with Down's syndrome.

Down's syndrom occurs once in about 700 births among European people. Older mothers have a higher incidence than younger mothers. About 40% of Down's syndrome victims are born to women over 40 years of age, whereas this group of mothers produces only about 4% of all babies born in the general population. The mean maternal age for all babies born in the population is estimated to be 28.5 years compared with a mean age of 36.6 years for mothers of babies with Down's syndrome. The greater incidence among older mothers may be associated with the long delay in completion of meiosis. Potential eggs in the ovary begin the meiotic process before birth and proceed to primary oocyte stage. They then remain dormant until shortly before ovulation occurs. Cells of a woman 40 years of age have thus remained in an inactive stage for about twice as long as those of a woman 20 years old. This long period of quiescence may explain atleast partially why non-disjunction occurs more frequently in older than younger mothers.

Some patients of Down's syndrome have a total of only 46 instead of 47 chromosomes, but in such cases a translocation has been found to join chromosome number 21 with another chromosome in the same complement.

Table 14.1. Aneuploidy Resulting from Nondisjunction in the Human Population

Chromosome Nomenclature	*Chromosome Formula*	*Clinical Syndrome*	*Estimated Frequency at Birth*	*Main Phenotypic Characteristic*
47, + 21	2n + 1	Down	1/700	Short broad hands with simian-type palmar crease, short stature, hyper-flexibility of joints, mental retardation, broad head with round face, open mouth with large tongue.
47, + 13	2n + 1	Trisomy-13	1/20000	Mental deficiency and deafness, minor muscle seizures, cleft lip and/or palate, polydactyly, cardiac anomalies, posterior heel prominence.
47, + 18	2n + 1	Trisomy-18	1/8000	Multiple congenital mal-formation of many organs, low-set, malformed ears, receding mandible, small

				mouth and nose with general elfin appearance, mental deficiency, Seahorse shape of double kidney, short sternum. 90% die in the first 6 months.
45, X	2n − 1	Turner	1/250 female births	Female with retarded sexual development, usually sterile, short stature, webbing of skin in neck region, cardiovascular abnormalities, hearing impairment.
47, XXY 48, XXXY 48, XXYY 49, XXXXY 50, XXXXXY	2n + 1 2n + 2 2n + 2 2n + 3 2n + 4	Klinefelter	1/500 male births	Male subfertile with small testes, developed breasts, feminine pitched voice, long limbs, knock knees, rambling talkativeness.
47, XXX	2n + 1	Triple X	1/700	Female with usually normal genitalia and limited fertility. Slightly mental retardation.

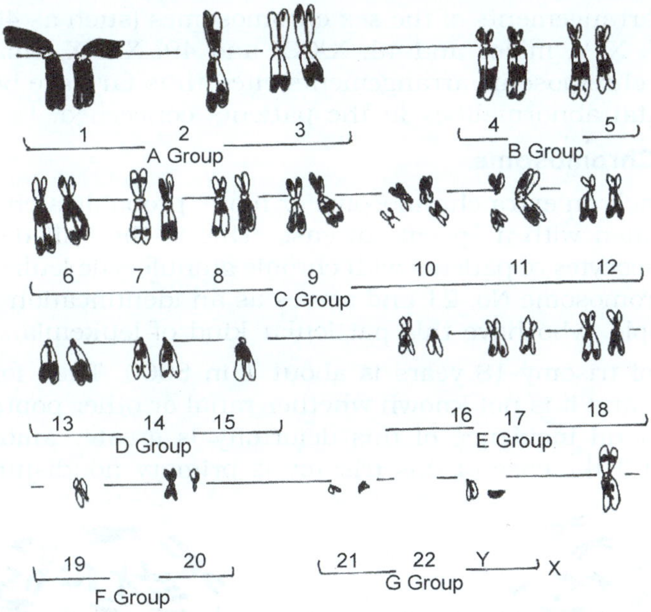

Fig. 14.7. Karyotype of a child with Down syndrome, showing trisomy for chromsome 21 (47, XX, +21). (Courtesy of *R.M. Fineman*).

Other Genetic Defects in Human Development

Much of the information about the regulation of gene expression during development has come from studies of nonhuman organisms. The reason for the relative dearth of information about human development is that studies of early human embryos is both technically and legally difficult. Nevertheless, several human developmental mutants have been identified. These mutants presumably result from defects in the regulation of expression of genes that are crucial for the developmental processes affected or from defects in structural genes whose products are needed for the developmental defects.

1. **Achondroplasia.** Phenotypically, people with achondroplasia have a normal-sized trunk, enlarged head, and unusually short of long-bones in the arms and legs. As a result, adults with achondroplasia average about 4 ft in height. Often other skeletal abnormalities can be seen upon X-ray examinations.

   The disorder is inherited as an autosomal dominant mutation. Most probably the mutation involved in a gene that functions in bone formation.

2. **Nail-Patella Syndrome.** The phenotypes of an individual with nail-patella syndrome include deformed or absent fingernails and toenails and patella (knee cap) which is abnormally small or absent. This syndrome is also the result of an autosomal dominant mutation.

3. **Cleft Lip and Cleft Palate.** Cleft lip and cleft palate are also the result of gene mutations. In embryonic development, the palate (roof of the mouth) halves and do not fuse since they are not in the correct position at the appropriate time. Unlike the achondroplasia and nail-patella syndrome, most forms of cleft lip and palate are not the result of a single gene mutation. Rather, there appearance to be a number of rare mutations that as part of the phenotypic abnormalities they produce, included left lip, cleft palate or both.

   One additional chromosome in the D group is associated with a cleft lip.

More complex arrangements of the sex chromosomes (such as 48, XXXY, 48, XXYY; 49, XXXXY, and 47, XYY, males and 48, XXXX and 49, XXXX females) are on record. Irregular numerical chromosome arrangements cited thus far have been associated with structural and mental abnormalities in the patients concerned.

The Philadelphia Chromosome

A special case in which an extra chromosome or more, probably a chromosome fragment of No. 21, is associated with a specific disease. This is the Philadelphia chromosome, which is found in leucocytes of patients with chronic granulocytic leukemia. It is apparently a deleted part of chromosome No. 21 and serves as an identification marker in the bone marrow cells of people, who have this particular kind of leukemia.

The incidents of trisomy-18 years is about 1 in 8000. Thus for, only a few cases have been observed, and it is not known whether ratial or other population groups differ in incidence. In general incidence of this deformity is greater among infants of older women, as expected if the case of this trisomy is primary nondisjunction in meiosis.

A B

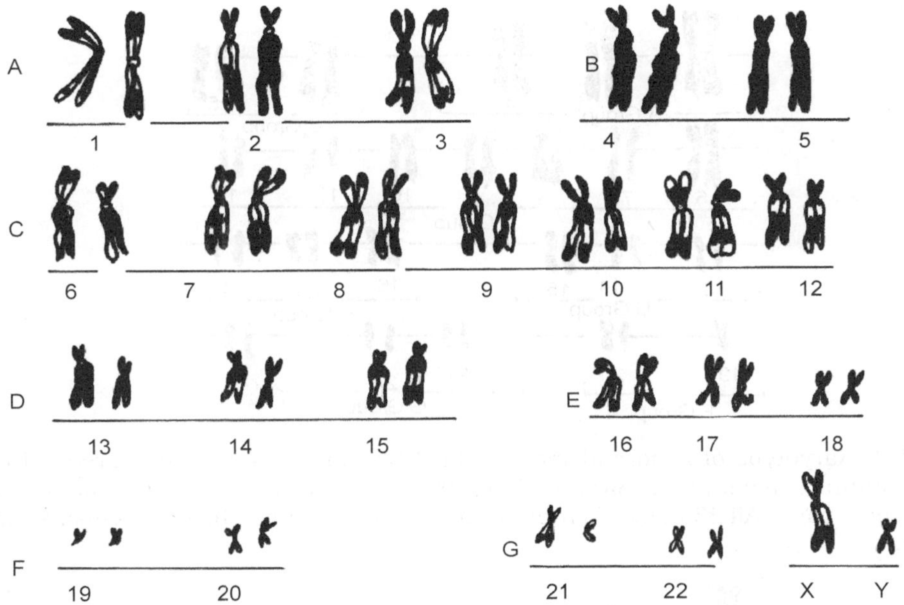

Fig. 14.8. Human metaphase chromosomes of A. a normal female and B. a normal male as they appear in microscope preparations, and C. as they are classified (kary, Karyotyped) in seven major groups according to criteria established at the Denver (1960), London (1963), and Chicago (1966) conferences (Photos A and B by John P. Limbach. Ripon Microslides Laboratory; photo C by Peter Arnold, Inc.).

CRI-DU-CHAT

Chromosome deletions are usually lethal even as heterozygotes, resulting in zygotic loss, still births or infant deaths. Sometimes infants with small chromosome deficiencies, however, survive long enough to permit observation of some of the abnormal phenotypes they express. *J. Lejeune and his colleagues*, for example, discovered a chromosome deficiency in humans that has been associated with the cri-du-chat (cat cry) syndrome. The name of this syndrome came from a plaintive cat like mewing cry from small weak infants with the disorder. Other characteristics are microcephaly (small head), broad face and saddle nose, widely spaced eyes with epicanthic folds, unique facial features, and physical and mental retardation. IQs of cri-du-chat children studied are in the range of 20–40. The chromosome deficiency is in the short arm of chromosome deficiency 5 and is designated 5p-A karyotype (46, XX, 5p-) of a newborn.

Cri-du-chat patients die in infancy or early childhood and do not transfer the chromosome deletion to offsprings. This chromosome deficiency, however, has been shown by *Lejeune and others*, to become involved sometimes in a reciprocal translocation and thus to be transmitted. When the short arm of chromosome 5 become translocated to chromosome 15, the heterozygous translocation was carried in a normal healthy parent. Some gametes, however, carried only the deficient member of the translocation pair. Children inheriting the $5p^-$ chromosome expressed the cri-du-chat syndrome.

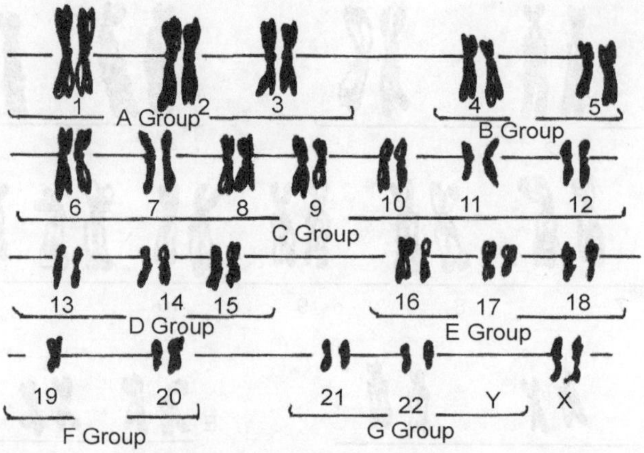

Fig. 14.9. Karyotype of a normal male child. Chromosomes were prepared with the aiesma-staining technique. Bands along the chromosomes identity members of chromosome pairs. All 23 pairs of chromosome can be distinguished by aiemsa-staining method.

Fig. 14.10. Karyotype of a normal young woman. Chromosomes were prepared with the C-technique. Hetrochromatids regions of chromosomes are darkly stained.

Like deficiencies, chromosome duplications are usually lethal even as heterozygotes, but sometimes, they are sufficiently viable to permit observations of the abnormalities they produce. Like deficiencies, they may be associated with translocations and thus transmitted by normal healthy parents. Remember that a translocation is an exchange of parts between nonhomologous chromosomes or a transfer from one chromosome to a nonhomologue. Broken parts may be further divided and some may be lost or gained in a transfer. Thus, deficiencies and duplications may accompany a translocation carried in a parent. Such a parent may be a translocation heterozygote, with the long arm of one chromosome 21 attached at the centromere to the long arm of chromosome 14 along with normal chromosomes 14 and 21. Some gametes will contain a normal 21 and translocation 14q 21q. When such a gamete is fertilized with a gamete from a normal individual, a translocated 21, along with two normal chromosomes 21, results in a viable infant with trisomy-21 and Down syndrome.

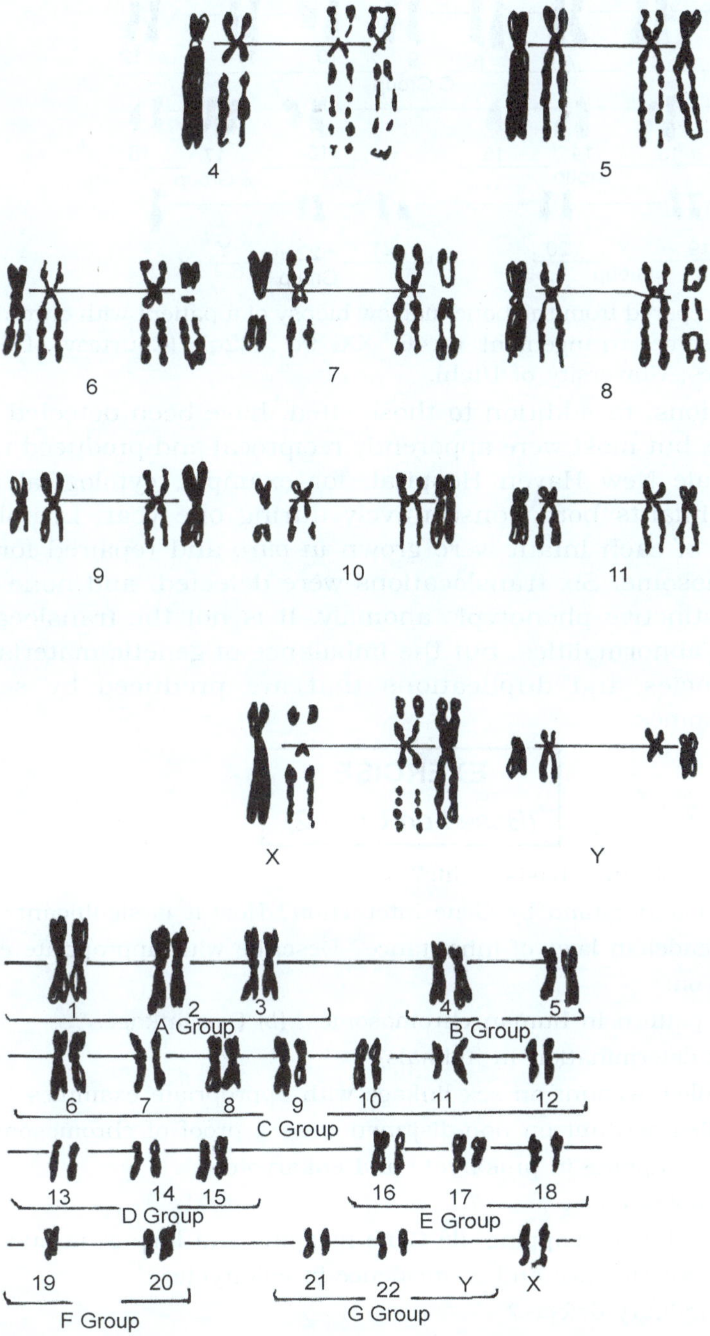

Fig. 14.11. Karyotype of newborn with the cri-du-chat syndrome. A heterozygoteous deficiency is shown in chromosome 5. The chromosome arrangement is 46, XX.

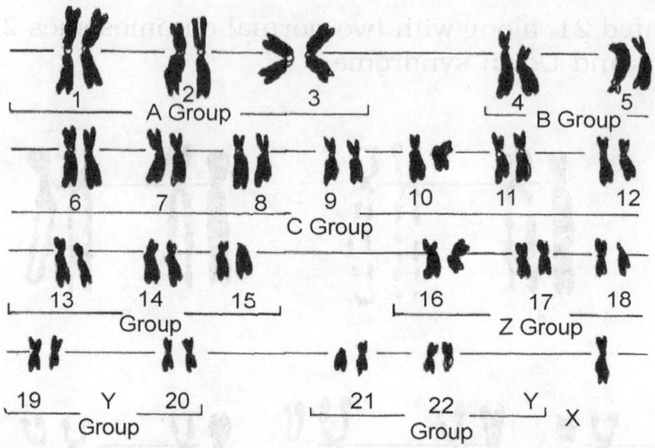

Fig. 14.12. Karyotype prepared from the bone marrow biopsy of a patient with chronic myelogenous leukemia. The chromosome arrangement is 46, XX, $9q^+$, $22q^-$. (Courtesy of R.M. Fineman, Department of Pediatrics, University of Utah).

Many translocations, in addition to those cited, have been detected in studies of human chromosomes, but most were apparently reciprocal and produced no phenotypic anomalies. At the Yale New Haven Hospital, for example, cytological studies were conducted on 4500 infants born consecutively during one year. Lymphocytes from umbilical cord blood of each infant were grown *in vitro* and repaired for microscopic observation of chromosome. Six translocations were detected, and none of these was associated with a distinctive phenotypic anomaly. It is not the translocation process per se that produces abnormalities, but the imbalance of genetic material reflected in chromosome deficiencies and duplications that are produced by segregation of translocated chromosomes.

EXERCISE

(Based on Unit –2)

Q.1 What is the molecular basis of life?

Q.2 What do you understand by Gene interaction? How it is significant?

Q.3 What are Mendelian laws of inheritance? Describe with appropriate examples.

Q.4 Write notes on:

(*a*) Banding pattern in human chromosomes. (*b*) Sex linkage.

Q.5 What is sex determination in animals?

Q.6 Write a detailed account on sex linkage with appropriate examples.

Q.7 Give a detailed account on non-disjunction as a proof of chromosomal theory.

Q.8 What is gene mapping in prokaryote and eukaryotes?

Q.9 Write short notes on:

(*a*) Linkage and gene mapping. (*b*) Gene mapping and linkage in eukaryotes.

Q.10 Describe the interference and co-incidence in eukaryotes.

Q.11 What are hereditary defects?

Q.12 Write an essay on hereditary defects.

Q.13 Write detailed account on

(*i*) Klinefelter syndrome (*ii*) Cri-du-Chat syndrome

(*iii*) Turner syndrome (*iv*) Down syndrome.

UNIT-3

15 Structure of Prokaryotic Gene, Prokaryotic Transcription, Translation and Gene Expression (Lac, His, Trp, Catabolic Repression)

INTRODUCTION

Gene is defined as a *unit of hereditary material located on a chromosome* that, by itself or with other genes determines a characteristic in an organism. Genes may exist in a number of forms, termed as alleles. *Genes are the hereditary units which are transmitted from one generation to next generation.* The term gene was coined by *Johannsen. Watson* gave various definitions for gene as follows:

—as a fragment of gene
—as a complete protein
—as a heredity and variation.

The exact definition for gene is a discrete region of self-replicating molecule *i.e.*, chromosome which is responsible for specific cellular production. Watson gave units of gene as follows:

Recon—the smallest unit of gene for transcription and translation.

Muton—the unit of gene for mutation.

Cistron—the functional unit of gene for transcription and translation.

Cistron is a unit, element of which exhibits cistrons phenomenon. *Cistron* is not a synonym for gene and it is difficult to use this term in practice.

At the intragenic level, with the help of recombination studies, micromaps have been prepared for different genes in the same fashion as chromosome maps for different chromosomes were initially prepared at the intergenic level.

A *recon* is further subdivided into units of mutations called *mutons* and several mutons in a recon will not be separable due to recombination. Since mutons in a recon cannot be identified due to recombination test, they will have to be recognized through the changes in amino acid sequence of a polypeptide chain, synthesized by or through the nucleotide sequence of cistron. Since a mutation can takes place by single base replacement, a single nucleotide pair is the ultimate limit of muton. Therefore, cistron, recon and muton are units in descending order of size and since gene is not synonymous to cistron etc., it will be at top of this heirarchy. Thus, a gene can consist of several cistrons and a cistron of several recons, and a recon of several mutons. The recon and muton may be of same size, so that in such a case a recon may not consist of several mutons.

TRANSCRIPTION IN PROKARYOTES

Transcription is a process in which DNA directs the synthesis of RNA in a polymerization reaction, catalyzed by RNA polymerase. This is the I step of protein synthesis in which genetic messages are enclosed in it. DNA is transcribed into mRNA.

Components

It requires a template—activated precursors, an equivalent metal ion, and RNA polymerase.

1. **Template.** Preferred template double stranded DNA, of 2 strands only 1 strand called as sense strand which transcribed mRNA. Single stranded DNA can also serve as template.
2. **Activated Precursors.** Synthesis of mRNA strand requires all four ribonucleoside triphosphate—ATP, GTP, UTP and CTP.
3. **DMI.** Mg^{++}/Mn^{++} are effective in transcription. *In vitro* synthesis requires Mg^{++}.
4. **RNA polymerase.** Transcription of RNA requires the enzyme RNA polymerase. RT are joined together by this with elimination of pyrophosphate. The entire enzyme (holoenzyme) consists of a core enzyme and a sigma (σ) and a omega (ω) factors.

Functions

1. It recognizes a promoter on double stranded DNA.
2. It causes DNA to denature and unwind into single strand at promoter.
3. As a result of reading, promoter sequence RNA polymerase orients itself properly and transcribes the entire sense strand of gene.
4. It stops transcribing when it reaches and recognizes the terminator.

RNA Polymerase in Prokaryotes

It doesn't have any proofreading function characters, able to initiate new RNA chains. In prokaryotes a single RNA polymerase enzyme controls synthesis of all different types of cellular RNA. It catalyses synthesis of m, r and t RNA. Entire enzyme consists of a core enzyme and σ factor.

In *E. coli,* consisting of four polypeptide chains β, β', α, α' and designated as $\beta\beta'$ α, it has elongated structure.

The α chain is present twice while, σ factor not very firmly attached to core enzyme and can be easily isolated.

The β' subunit is required for binding of RNA polymerase to DNA template.

The β subunit is required for binding with σ factor.

The 2α subunit is not clear but it has a possible role in promoter recognition.

The σ factor recognizes start signal on DNA template and direct binding of RNA polymerase to promoter region. This increases specificity of core polymerase binding to DNA. Once the RNA synthesis has started σ dissociates and core enzyme brings out increase in length of mRNA. The dissociated σ factor again combines with core enzyme to form RNA polymerase.

Initiation

1. **Role of RNA Polymerase.** σ factor interacts with core enzyme to form complete enzyme. β subunit is required for interaction with σ factor. Core enzyme by itself is ineffective in transcribing RNA because it transcribes both strands. The σ factor recognizes start signal on DNA and directs binding of enzyme to promoter region and enhances specificity of transcription.
2. **Recognition, Binding and Initiation Sites on DNA.** RNA polymerase initially binds to an RNA polymerase recognition site on DNA promoter region. It then diffuses to a binding site consisting of sequences of 7 bases. The start of mRNA transertion takes place at initiation site. This consists of '1 or 2' bases, near binding site.
3. Prokaryotes mRNA are polycistronic and can initiate protein synthesis at internal position.
4. **Unwinding of DNA Double Helix.** Binding of RNA polymerase to DNA results in local unwinding of DNA double helix. Short segment of DNA molecule opens up. Synthesis of mRNA now begins on one of the two strands of DNA.
5. **RNA Chains Start with pppG/PPPA.** One base of RNA is always purine, either pppG/pppA. In *E. coli* most chains start with pppG. After synthesis, however terminal triphosphate group nucleotides may be cleaved by endonucleases. RNA chains don't require primers.

Pribnow Box

The Pribnow box is the sequence TATAAT of six nucleotides that is an essential part of a promotor site on DNA for transcription to occur in prokaryotes. It is located roughly 10 base pairs upstream from the site of initiation of transcription. It is considered a cosensus sequence that it is the most common base sequence to appear at such points on the DNA helix there may be variations in various organisms.

T	A	T	A	A	T
77%	76%	60%	61%	56%	82%

Fig. 15.1. Probability of occurrence of each nucleotide.

Elongation

Chain elongation takes place at 5'-3' direction by formation of phosphodiester bond. It takes place by addition of activated ribonucleoside triphosphates (ATP, UTP, CTP and GTP) to 1 strand of DNA template for each nucleotide added to growing RNA chain a pyrophosphate (pp_i) is given off, which is rapidly hydrolysed to inorganic phosphate (P_i) for each nucleotide moreover added to chain 2 high energy phosphates are expanded.

$$\text{DNA template} + \begin{array}{c}\text{Activated}\\ \text{ribonucleoside}\\ \text{triphosphates}\end{array} \xrightarrow[\text{Polymerase}]{\text{RNA}} \text{DNA template} + \begin{array}{c}\text{RNA polymer}\\ \text{(AMP, GMP,}\\ \text{MP, MPT)}\end{array} + 4n\ \text{PnP}$$

This step takes place by means of core enzyme which moves along DNA template

1. **Dissociation of σ Factor.** After commencement of chain elongation, the σ factor dissociates from core enzyme. It can now be used by another polymerase molecule. σ factor is released because either it is no longer required or its prcscncc may kinder effective movement of enzyme of mpDNA template.

2. **RNA Chains Grow in 5'-3' Direction.** This is shown that chain growth takes place in 5'-3' direction as a triphosphate group is attached to I nucleotide. During chain elongation, the region of DNA that has been transcribed regions takes its double helical conformation as next region of DNA unwinds. The newly formed RNA chain has a triphosphate group. at 5' end and a hydroxyl group at 3' end.
3. **Only 1 DNA Strand of a Gene Transcribes mRNA.** In dissociation DNA a given gene is transcribed from only 1 of 2 strands. The transcribed RNA is complementary to only 1 of 2 strands. 1 gene may transcribe mRNA from 1 strand while another transcribes from other strand, when both strands of DNA double helix transcribed mRNA the transe is said to be symmetrical.
4. **Termination.** Three key events occur at termination, such as:
   (*i*) RNA synthesis stops.
   (*ii*) RNA chain is released from DNA.
   (*iii*) RNA polymerase is released from DNA.

There are stop signals to mark the end of a gene or sequence of genes and stop transcription by RNA polymerase. Prokaryotic terminators often contain a sequence coding for an RNA stretch that can make hydrogen bond to form a hairpin shaped loop and stem structure. This structure appears to cause RNA polymerase to pause in transcribing DNA.

Termination of RNA chain synthesis appears to be brought by two types of mechanisms. In the first type termination, signal appears to be recognized by DNA itself. In second type termination signal involves an additional protein known Rho (ρ) factor. In the first type, it contains a stretch of about uridine residues following them RNA hairpin and causes polymerse to stop transcription and release mRNA without aid of any accessory factor.

In second type, it lacks a poly uridine region and often the hairpin. The ρ factor, in this, binds to mRNA and moves along the molecule until it reaches the RNA polymerase that has halted at a termination. The ρ factor then causes polymerase to dissociate from mRNA probably by unwidning mRNA-DNA complex.

MECHANISM OF PROKARYOTIC TRANSCRIPTION AND TRANSLATION BY PROTEIN BIOSYNTHESIS

Transcription

The process in which single stranded RNA with a base sequence complementary to the template strand of DNA or RNA synthesized.

Protein biosynthesis is the process in which cells build proteins. The term is sometimes used to refer only to protein translation but more often it refers to a multistep process, beginning with amino acid synthesis and transcription which are then used for translation.

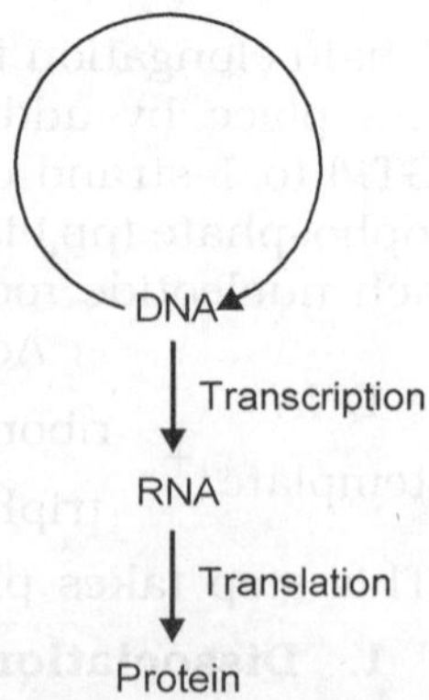

Fig. 15.2. Relation between DNA, RNA and protein

Some biologist have long recognized relationship between DNA, RNA and protein. DNA is carefully copied during its synthesis or replication. The expression of the information encoded in the base sequence of DNA begins with the synthesis of an RNA copy of the DNA sequence making up a gene. Although DNA has two comlementary strands only the template strand is copied at any particular points on DNA. If both strands of DNA were transcribed, two differentiate mRNAs would result and causes genetic confusion. Different gene may be encoded on opposite strands. This process of DNA-directed RNA synthesis is called transcription because the DNA base sequence is being written into an RNA base sequence. The RNA that carries information from DNA and direct protein synthesis is messenger RNA (mRNA). The last phase of gene expression is translation or protein synthesis.

The RNA Polymerase Enzyme

In *E. coli* and other prokaryotes, a single RNA polymerase (RNA-P) enzyme is responsible for the synthesis of all kinds of RNAs (such as mRNA, tRNA and rRNA). RNA-P is in fact one of the largest enzymes known (MW 490,000). It consists of six subunits (*i.e.*, polypeptide chains), two identical alpha (α) subunits, and one chain of each of beta (β), beta dash (β'), omega (ω) and sigma (σ) subunits.

MECHANISM OF TRANSCRIPTION

Transcription is the process by which an mRNA template, carrying the sequence of the protein, is produced for the translation step from the *genome*. Transcription makes the template from one strand of the *DNA* double helix, called the *template strand.*

Transcription takes place in three stages.

1. Transcription starts with the process of *initiation. RNA polymerase*, the enzyme which produces RNA from a DNA template, binds to a specific region on DNA that designates the starting point of transcription. This binding region is called the *promoter*. As the RNA polymerase binds on to the promoter, the DNA strands are beginning to unwind.
2. The second process is *elongation.* RNA polymerase travels along the template (noncoding) strand, synthesizing a ribonucleotide polymer. RNA polymerase does not use the coding strand as a template because a copy of any strand produces a base sequence that is *complementary* to the strand which is being copied.

   Therefore DNA from the noncoding strand is used as a template to copy the coding strand.
3. The third stage is *termination.* As the polymerase reaches the termination stage, modifications are required for the newly transcribed mRNA to be able to travel to the other parts of the cell, including *cytoplasm* and *endoplasmic reticulum.* for translation. A 5' *cap* is added to the mRNA to protect it from degradation. In eukaryotes, poly-A-polymerase adds a *poly-A tail* onto the 3' end for stabilization, protection from cytoplasmic *hydrolytic enzymes* and as a template for further processes. Also in *eukaryotes* (higher organisms), the vital process of *splicing* occurs at this stage by the spliceosome enzyme. It removes the introns (non-coding bits of genetic material) and glues together the exons (the segments that code for a specific protein).
4. The mRNA now exit the nuclear pore to be translated.

Translation

Protein synthesis is the process by which the genetic message carried by mRNA directs the synthesis of polypeptides with the aid of ribosomes and other cell constituents.

Protein translation involves the transfer of information from the mRNA into a peptide composed of amino acids. This process is mediated by the ribosome with thc adaptation of the RNA sequence into amino acids mediated by transfer RNA. Numerous initiation and elongation factors also play an important role. Translation requires lots of energy, with the hydrolysis of approximately 4NTP $\longrightarrow$ NDP per amino acid added. Thus, gene expression is highly regulated to ensure that only proteins which required are translated.

Translation involves three processes such as;

(*a*) Initiation (*b*) elongation (*c*) termination

(a) Initiation in Prokaryotes

The initiation of protein translation involves the assembly of the ribosome and addition of the first amino acid, methionine.

1. The 30S ribosomal subunit attaches to the mRNA, mediated by IF-1 and IF-3 (*initiation factors*). The 30S ribsome brings with it the P and A site, but the A site is blocked by IF-1 to prevent binding of tRNA. It aligns to the *Shine Dalgarno sequence*, which positions the first codon (AUG) in the P site.
2. Next, the specific aminoacyl-tRNA for *N-formylmethionine* (F-Met) is brought into the P site by IF-2. The anticodon of this tRNA will bind to the AUG codon on the mRNA.

   Note: this is the only tRNA brought into the P site; all successive aminoacyl-tRNAs will be brought to the A site for peptide elongation.
3. The 50S ribosomal subunit is then brought into complete the ribosome, and with it, IF-1, IF-2, and IF-3 come off the complex. The A and P site are completed, and the 50S subunit also brings the E (exit) site.

(b) Elongation in Prokaryotes

1. Elongation proceeds after initiation with the binding of an aminoacyl-tRNA to the A site, which is the next codon in the mRNA. The aminoacyl-tRNA is brought to the ribosome through a series of interactions with EF-Tu (an *elongation factor*). This step involves the hydrolysis of GTP: EF-Tu-GTP $\rightarrow$ EF-Tu-GDP (The hydrolyzed GDP is switched for GTP through another series of reactions with EF-Ts.)
2. The next aminoacyl-tRNA binds to the codon, and the C-terminus of the F-Met undergoes nucleophilic attack by the N-terminus of the second amino acid. The F-Met is now connected to the second amino acid through a *peptide bond.*
3. The first tRNA (for F-Met) is now uncharged. The entire ribosome complex moves along the mRNA through the action of another elongation factor (EF-G) and the hydrolysis of GTP $\rightarrow$ GDP.
4. The first tRNA is now in the E site and comes off from the ribosome, while the second tRNA, with the nascent peptide chain, is in the P site. Step 1–4 will repeat as successive amino acids are added.

(c) Termination

Termination of protein biosynthesis occurs when the ribosome comes across a *stop codon*, for which there is no tRNA. At this point, protein biosynthesis halts and one of three *release factors* will bind to the stop codon. This induces a nucleophilic attack of the C-terminus of the nascent peptide by water-this hydrolysis releases the peptide from the ribosome. The ribosome, release factor, and uncharged tRNA then dissociates and translation is complete.

Polysomes and Coupled Transcription—Translation

The unit of translation is never simply a ribosome traversing an mRNA molecule, but is a more complex structure. After about 25 amino acids have been joined in a polypeptide chain, the AUG initiation site of the encoding mRNA molecule is completely free of the ribosome. A second initiation complex then forms. The overall configuration is of two 70S ribosomes moving along the mRNA at the same speed. When the second ribosome has moved along a distance similar to that transversed by the first, a third ribosome is able to attach. The process, *i.e.*, movement and reinitiation, continues until the mRNA is covered with ribosomes at a density of about one 70S ribosome per 80 nucleotides. This large translation unit is called a polyribosome or simply a polysome. This is the usual form of translation unit in all cells. The use of polysomes is advantageous to a cell, since the overall rate of protein synthesis is increased compared to the rate that would occur if there were no polysomes.

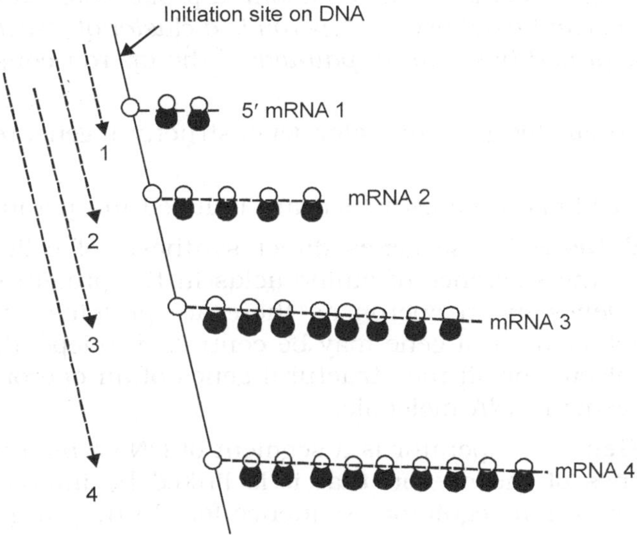

Fig. 15.3. Coupled transcription-translation. Transcription of a section of the DNA of *E. coli* and translation of the nascent mRNA. The dashed arrows show the distance of RNA polymerase from the transcription initiation mRNA 4 probably becomes shorter due to partition of its 5' end by a RNase (after *Freifelder*).

Further, an mRNA molecule being synthesized has a free 5' terminus and translation also occurs in the 5' $\rightarrow$ 3' direction, so each cistron contained in the mRNA immediately starts its translation. As a result, the ribosome binding site is transcribed first, followed in order by the AUG codon, the region encoding the amino acid sequence, and finally the stop codon. Thus, in bacteria in which no nuclear membrane separates the DNA and the ribosomes, there is no obvious reason why the 70S initiation complex should not form before the mRNA is released from the DNA. With prokaryotes (*E. coli*) this does indeed occur; the process is called coupled transcription-translation. This coupled activity does not occur in eukaryotes; because the mRNA is synthesized and processed in the nucleus and later on transported through nuclear membrane to the cytoplasm where ribosomes are located.

GENE EXPRESSION IN PROKARYOTES (THE OPERON CONCEPT)

The total amount of DNA present in a cell may contain from a few to thousands of genes. Although the different types of cells in the body of a multicellular organism differ in structure and function; their genes are identical, since all the cells are ultimately derived from the zygote. As development proceeds, certain genes become active at one time while others at that time become inactive *i.e.*, genes are "switched on" and "switched off" at different times. This process is called as differential gene action. When genes are active, they direct the formation of enzymes which affect certain traits. The metabolic products formed may repress synthesis of enzymes. Thus enzyme synthesis is induced and repressed at different times.

Although a cell has the genes to produce hundreds of enzymes, only the enzymes required at a particular time are produced. This control mechanism ensures that the cell is not flooded with unnecessary enzymes. A hypothesis to explain induction and repression of enzyme synthesis was first put forward by *Francois Jacob* and *Jacques Monod* in 1961 of the *Institute Pasteur* in Paris. Both Jacob and Monod proposed operon hypothesis to explain gene regulation mechanism in prokaryotes at transcriptional level. The operon hypothesis can be defined as *"operon is a cluster of structural genes that work in a concert in an integrated biochemical pathway,"* the operon consists of the following components;

Regulator gene, promoter gene, operator gene, structural gene, repressor, corepressor and inducer.

The are different kinds of genes which are induced in operon system.

The Structural Gene. These genes direct synthesis of cellular protein through mRNA and determine the sequence of amino acids in the protein synthesis. There are as many structural genes in an operon as there are proteins or polypeptides under common control. Each structural gene may be controlled independently and transcribe a separate mRNA molecule or all the structural genes of an operon may form one long polycistronic or polygenic mRNA molecule.

The Operator Gene. An operator is a segment of DNA which regulates the activity of the structural genes of the operon that it is linked by interacting with a specific repressor or activator. It is a regulatory sequence for shutting a gene down or turning it "on".

The Promoter Gene. A promoter is a sequence that enables a gene to be transcribed. The promoter is recognized by RNA polymerase which then initiates transcription. In RNA synthesis, promoter are a means to demarcate which genes should be used for messenger RNA creation, control which protein the cell manufacture.

Regulator Gene. The operon may also contain regulatory genes such as a repressor gene which codes for a regulatory protein that binds to the operator and inhibit transcription. Regulatory genes not be part of the operon itself, but may be located elsewhere in the genome. The repressor molecule will reach the operator to block the transcription of the structural genes.

Types of Operon

The two types of operon found in prokaryotic gene regulation are:

1. Inducible operon 2. Repressible operon

1. Inducible Operon. A category of enzymes called inducible enzymes are normally absent from the cell or are present in very small quantities. They increase in quantity only in presence of an inducer. Such a system is called as *Inducible system* or *Inducible operon system.* In this system, the substrate acts as inducer of transcription. The most common inducible system is *lac* operon which was traced in *E. coli.* In *lac* operon system the lactose is converted into allolactose which is the actual inducer in operon while lactose is the apparent inducer. An artificial inducer commonly used in experimental work is Isopropyl thiogalactoside (IPTG). When the inducer is absent, the active repressor protein produced by the regulator gene associates with the operator gene and blocks it. No transcription of mRNA by the structural gene can therefore take place and hence there is no enzyme synthesize. Absence of inducer may be due to the fact that it is not present in the growth medium or is not synthesized by the microorganisms.

When the inducer is present, the active repressor binds to the inducer molecule to form a repressor-inducer comples (R-I). A conformational change takes place in the repressor molecule, which is inactivated. This prevents binding to the operator, which is therefore not blocked. The structural genes transcribe mRNA, and enzyme synthesis takes place.

(*i*) ***Inducer Absent:*** Regulator gene → Active repressor → Binds to and block operator gene.

→ No mRNA transcription by structural genes → No enzyme synthesis.

(*ii*) ***Inducer Present:*** Regulator gene → Active repressor → + Inducer → Inactive repressor → No blocking of operator gene → mRNA transcription by structural genes → Enzyme synthesis.

2. Repressible Operon. The synthesis of an enzyme may be inhibited by a product of metabolism. The process is called enzyme repression. In this type of operon, the proteins produced by the regulator gene is an inactive repressor also called as the aporepressor. The aporepressor on combining with aporepressor molecules become activated. It undergoes a conformational change, which enables it to bind to the operator gene.

When the corepressor is absent, the inactive aporepressor does not bind to operator gene. Because the operator gene is not blocked, the structural genes transcribe mRNA and protein synthesis takes place.

When the corepressor is present, the aporepressor binds to form a complex, and is activated. The complex blocks the operator gene. The structural genes are therefore unable to synthesize mRNA and consequently no protein synthesis takes place. The corepressor may be a product of one of the enzymes produced by the structural genes.

(*i*) ***Corepressor Absent:*** Regulator gene → Inactive repressor (Aporepressor) → No blocking of operator gene → mRNA transcription by structural genes → Enzyme synthesis.

(*ii*) ***Corepressor Present:*** Regulator gene → Inactive repressor (Aporepressor) → +corepressor → Active repressor → Binds to and blocks operator gene → No mRNA transcription by structural genes → No enzyme synthesis.

LAC OPERON

Outline of Lac Operon

The lac operon of the bacterium *Escherichia coli* was the first operon discovered by, and provide a typical example of operon function. It consists of three adjacent structural genes, a promotor, terminator, and an operator. The lac operon is regulated by several factors including the availability of glucose and of lactose.

1. Type—It is a type of Inducible lac operon generally off.
2. Substrate—Lactose
3. When lactose is added to the medium the lactose is converted into its isomeric form allolactose and binds to the repressor protein produced by regulator gene and inhibits the binding of repressor to the operator site. Thus, allowing RNA polymerase to bind to the promoter site which results in transcription of structural genes coding for:

   X — β-Galactosidase

   Y — Permease

   Z — Transacetylase
4. When the medium lacks lactose repressor protein binds to the operator site inhibiting the binding of RNAP to the promoter site "switching off" transcription.
5. The enzymes in lac operon are responsible for catabolism of lactose.

$$\text{Lactose} \xrightarrow{\text{enzymes}} \text{Glucose} + \text{Galactose}$$

Process of Lac Operon

The Lac operon is generally "switched off" so that there is no transcription and so no synthesis of enzymes. It can be turned on if a metabolite is supplied to the bacterium from outside. The metabolite combines with active repressor bound to operator. This results in switching on the operon.

This mechanism was discovered in 1961 when Francois Jacob and Jacques Monod in *E. coli* supplied it with different types of sugars. They found that *E. coli* was grown on medium containing lactose, it showed synthesis of three new enzymes *viz* β-galactosidase, which breaks lactose into glucose and galactose; permease, which facilitate the entry of lactose into the cell for the medium; and transacetylase, which helps in the action of other two enzymes.

Lac operon is normally off due to binding of repressor protein produced by action of regulator gene "R" to operator site. The process begins when lactose is supplied in medium where *E. coli* is growing. Lactose binds to active repressor causing a change in structure so that it cannot bind to operator. Once the operator is free of repressor, RNA polymerase starts transcription of operon by binding to promoter site "P". mRNA corresponding to three genes enzymes synthesized which gets translated to produce three enzymes *viz* β-galactosidase, permease and transacetylase. With the production of these enzymes metabolism of lactose molecules are consumed by the cell. When the last molecules of lactose bound to repressor are also consumed, the inactive repressor becomes active and binds to operator site to switch off operon as usual.

TRYPTOPHAN OPERON

Discovered in 1953 by *Jacques Mono and Colleges*, the trp operon in *E. coli* was the first repressible operon to be discovered. While the lac operon can be activated by a chemical (lactose), the tryptophan (trp) operon is inhibited by a chemical (Tryptophan).

This operon contains five structural genes trp E, trp D, trp C, trp B, and trp A which encodes *tryptophan synthetase.* It also contains a promoter which binds to RNA polymerase and an operator which blocks transcription when bound to the operator.

Outline of Tryptophan Operon

1. Type—It is type of "Repressible operon" generally on.
2. Substrate—Tryptophan.
3. In this operon system the presence of tryptophan in the medium switches off transcription.
4. If tryptophan is present in the medium it bands to the repressor produced by repressor gene or regulator gene and acts as a co-repressor forming an active aporepressor that now possess the ability to bind to the operator gene making RNAP unable to bind to promoter thus, stopping transcription of structural genes.
5. If tryptophan is lacking in the medium then the repressor protein is unable to bind to operator site. Thus, RNAP binds to the promoter site causing transcription to occur.
6. The structural genes consists of five genes A, B, C, D, and E coding for three enzymes

   A and B = Tryptophan synthetase

   C gene = Indole glycerp; synthetase

   D and E gene = Anthranellic synthetase.
7. The enzymes of tryptophan operon are responsible for anabolism of tryptophan from chorismic acid.

Process of Tryptophan Operon

The operon is generally on so that transcription and translation are on to synthesize enzymes needed for synthesis of a certain metabolite by the *coli.* The operon can be switched off when cell does not require metabolite, or metabolite has been produced in excess or over need of all.

Tryptophan is a heterocyclic amino acid whose synthesis in bacterial cell requires action of five enzymes. The synthesis of tryptophan is on because regulator gene "R" produce an inactive repressor c/a aporepressor which doesn't bind to operator site. Since operator site of operon is free of repressor, the operon remains on synthesizing all five enzymes needed for tryptophan synthesis. If tryptophan synthesis is in excess or over its utilization by cell, tryptophan accumulates. Some molecules of tryptophan act as corepressor and bind to inactive repressor which becomes active and binds to operator site immediately switching off the operator and so there is no transcription and so no synthesis of enzymes required for tryptophan production. Thus, synthesis of tryptophan stops until all available molecules are utilized by cell for its metabolic activities. It is a set in train by the ability of glucose to reduce the level of cyclic AMP (CMP) in the cell. There are some promoters at which RNA polymerase cannot initiate transcription without assistance from an ancientary protein, such proteins constitute positive regulators because their presence is necessary to switch on the transcription unit.

This term Catabolite Repression describes the regulatory mechanism which allows the *lac* operon to sense the presence of glucose as an alternating energy source for lactose. If glucose and lactose both are present, cell will use up glucose first and will not spend energy in splitting of lactose into its complement sugars. The presence of glucose in the cell switches off the *lac* operon by a mechanism called as catabolite repression which involves a regulatory protein called as Catabolite Activator Protein (CAP). CAP binds to a DNA sequence upstream of the LAC promoter and enhances the binding of the RNAP leading to enhanced transcription of the operon. However CAP binds only in the presence of a derivative of ATP called as cyclic adenosine monophosphates.

EFFECTS OF GLUCOSE ON LAC OPERON (CATABOLIC REPRESSION) OR GLUCOSE EFFECT

The function of β-galactosidase enzyme in lactose metabolism is to form glucose by cleaving lactose. (The other cleavage product, galactose is also ultimately converted to glucose by the enzymes of galactose operon). Thus, if both glucose and lactose are present in the growth medium activity of *lac* operon is not needed, and indeed, no β-galactosidase is formed until virtually all of the glucose in the culture medium is consumed. The lack of synthesis of β-galactosidase is a result of lack of synthesis of *lac* mRNA. No *lac* mRNA is made in the presence of glucose, because in an addition of an inducer to inactivate in *lac i* repressor, another element (*i.e.*, cAMP-CAP) is needed for initiating *lac* mRNA synthesis; the activity of this element is regulated by the concentration of glucose. However, the inhibitory effect of glucose on expression of the *lac* operon is quite indirect.

Small molecules, the cyclic AMP (cAMP), are present in *E. coli* and may be other bacteria. Cyclic AMP is synthesized enzymatically by adenyl cyclase and its concentration is regulated indirectly by glucose metabolism. When bacteria are growing in culture medium containing glucose, the cAMP concentration in the cell is quite low. In a medium containing glycerol or any carbon source that cannot enter the biochemical pathway used to metabolize glucose (the glycolytic pathway), the cAMP concentration becomes high. The mechanism by which glucose controls the cAMP concentration is poorly understood; the significant point that cAMP regulates the activity of the *lac* operon (and other several operons as well) *E. coli* contains a protein called the Catabolic Activator Protein (CAP), which is encoded in a gene called *crp*. Mutants of either *crp* or the adenyl cyclase gene are unable to synthesize lac mRNA, indicating that both CAP and cAMP are required for *lac* mRNA synthesis. CAP and cAMP bind to one another to form a unit, called cAMP-CAP which is an active regulatory element of the *lac* system. The cAMP CAP complex must be bound to a base sequence in the DNA in the promoter region in order for transcription to occur. Thus, cAMP-CAP is a positive regulator, in contrast with the repressor, and *lac* operon is independently regulated both positively and negatively.

There are two types of mutation abolish catabolite repression. The first type of mutation is *cya*$^-$, that introduces a defect in the enzyme adenylate cyclase that synthesizes cyclic AMP. (The reaction uses ATP as substrate and consists of making a 3'–5' link *via* phosphodiester bonds, the result is a monophosphate joined to both of the positions at which phosphates are usually to be found).

The second type of mutation, known alternatively as *cap* or *crp*, identifies the regulator proteins that acts directly on the target operons. The protein is known as CAP (for catabolic between protein) or CRP (for cyclic AMP receptor protein).

16 STRUCTURE OF EUKARYOTIC GENE, EUKARYOTIC TRANSCRIPTION, TRANSLATION AND GENE EXPRESSION, TRANSCRIPTION FACTORS

INTRODUCTION TO EUKARYOTIC GENES

Eukaryotes

The living cells are differentiated into two kinds *viz* prokaryotic cells and eukaryotic cells. Eukaryotic cells are advanced and are basically those cells which have genetic material *i.e.,* DNA enclosed by membrane to form a nucleus.

Structure of Eukaryotic Genes

Gene is a unit of hereditary material, located on a chromosome that by itself or with other genes, determines a characteristic in an organism. These are simply defined as *smallest hereditary unit, capable either for recombination or of mutation or of controlling a specific function.* In the classical sense, each gene corresponds to that part of DNA molecule, which, if active in a cell, codes for one functional polypeptide chain. *Benzer* coined the term cistron for this unit of hereditary function. In the light of Benzer's assertion, a gene may be defined as "*distinct chromosomal region responsible for a single cellular function and consisting of a linear array of potentially mutable units between which recombination can occur.*"

One of the most starting findings of recent years is that the most eukaryotic genes are split genes (discontinuous); the gene comprises two kinds of alternating segment—exons and introns. The exons contain coding nucleotides sequences which are ultimately translated into polypeptides. Conversely, the introns contain non-coding nucleotide sequence which are not translated into polypeptides. Due to this pecularity of gene structure, messenger RNA initially transcribed at a gene template is a large heterogenous RNA molecule. Before being transported to the cytosol for translation, this primary RNA transcript is processed in nucleus into a functional messenger. During this processing the intervening sequences transcribed by introns are removed (splicing) leaving a much shorted functional mRNA transcribed by exons.

INTRODUCTION TO EUKARYOTIC TRANSCRIPTION

Transcription is the means by which the genetic information in chromosomal DNA is converted into a single strand of messenger RNA. The process is carried out in the nucleus and produces a complementary copy of genetic message in mRNA molecule. It is the process in which DNA directs synthesis of RNA; a polymerization reaction catalyzed by RNA polymerase. This is first step of protein synthesis in which genetic messages are encoded in DNA to transcribe into mRNA.

COMPONENTS OF EUKARYOTIC TRANSCRIPTION

It requires a template, activated precursors, a divalent metal ion and RNA polymerase.

1. Template. For eukaryotic transcription the preferred template is double stranded DNA. Out of two strands of DNA only one strand called as sense strand transcribe mRNA. Single stranded DNA can also serve as template.

2. Activated Precursor. Synthesis of mRNA strand requires all four ribonucleoside triphosphates—ATP, GTP, UTP, CTP.

3. Divalent Metal Ion. Mg^{++} (magnesium ion) or Mn^{++} (manganese ion) are effective in transcript. The individual synthesis requires Mg^{++}.

4. RNA Polymerase. Transcription of RNA requires enzyme RNA polymerase. In Eukaryoitc three major classes of RNA polymerases are required, designated as I, II and III found in nucleus. Each of these is a large protein with large and several smaller subunits.

Enzymes	*Location*	*Product*
RNA poly I	Nucleolus	rRNA
RNA poly II	Nucleoplasm	mRNA
RNA poly III	Nucleoplasm	tRNA

The RNA polymerase II contains certain genes called as promoters at distinct regions which are centered at sites lying between – 25 bp and –100 bp.

The three regions are:

(*i*) TATA/goldberg Hogness box — Least effective.
located upstream to starting
–20 bp to –25 bp point.

(*ii*) CAAT/CAT Box — necessary for initiation
located between –70 to –80 bp
GGT/ACAATCT base composition

(*iii*) GC Box — Found in 1/more copies
located at –60 to –100 bp upstream.

CAT/GC Boxes — determine efficiency of transcription.

TATA Box — aligns RNA polymerase at proper site with the help of proteins called as transcription factors.

Eukaryotic promoters have certain sites called as enhancer genes. These genes are DNA sequence which affect transcription mechanism and increase rate of transcription.

There are other regulatory sites called silencers or silencer genes which repress gene expression, both enhancer and silencer can function at great distance from genes they enhance and repress.

STEPS OF EUKARYOTIC TRANSCRIPTION

Initiation

The regulatory DNA sequence for genes are transcribed by each of three RNA polymerases. Various transcription factors are involved in formation of transcription complex which are needed for initiation of transcription. The transcription factors are eukaryotic proteins that are RNA polymerase to recognize promoters. These are analogous to pro sigma factor that IIF needs *viz* TFIIA, TFIIB, TFIID, TFIIE, TFIIF, TFIIH, TFIIJ. The TF may be defined as proteins which are needed for initiation of transcription but are not part of RNA polymerase. They help in DNA binding of RNA polymerase to constitute the pre initiation complex Transcription complex.

After formation of this complex initiation, transcription occurs.

Formation of Transcriptosome with RNA poly II

A promoter sequence is responsible for constitutive expression of common genes (also called as house keeping genes) in all cells called generic promoter. Initiation of transcription on generic promoter by RNA poly II requires action of Tfs.

TFIID binds at TATA box.

Above step permits association of TFIIA and TFIIB.

TFIIB forms DB complex and RNA poly II associated to promoter site.

Poly II accompanied to promoter by TFIIF to form transcription complex.

Addition of TFIIE, TFIIH and TFIIJ helps in initiation.

Elongation

There are certain accessory proteins of transcription called as elongation factors which enhance overall activity of RNA poly II and lead to increase in elongation rate. Two such proteins are:

1. **TFIIF**—accelerates RNA chain growth.
2. **TFIIS**—helps in elongation of RNA chain, also acts by I causing hydrolytic cleavage at 3' end of RNA chain therefore helping in forward movement of RNA polymerase through any block to elongation.

Termination

In eukaryotes, actual termination occurs through termination sites similar to those found in prokaryotes.

MECHANISM OF EUKARYOTIC TRANSCRIPTION

Eukaryotic transcription is more complex than prokaryotic transcription, because eukaryotes have evolved much more complex transcriptional regulatory mechanisms than prokaryotes, for instance, in eukaryotes the genetic material (DNA), and therefore transcription is primarily localized to the nucleus, where it is seperated from the cytoplasm (where translation occurs) by the nuclear membrane. DNA is also present in mitochondria in the cytoplasm and mitochondria utilize a specialized RNA polymerase for transcription. This allows for the temporal regulation of gene expression through the sequestration of the RNA in the nucleus, and allows for selective transport of RNAs to the cytoplasm, where the ribosomes reside.

The basal eukaryotic transcription complex includes the RNA polymerase and additional proteins that are necessary for correct initiation and elongation.

Initiation

Among eukaryotes that regulate the transcription of individual genes, the *core promoter* of protein-encoding gene contain binding sites for the basal transcription complex and RNA polymerase II, and is normally within about 50 bases upstream of the transcription initiation site. Further transcriptional regulation is provided by *upstream control elements* (UCEs), usually present within about 200 bases upstream of the initiation site. The core promoter for Pol II sometimes contains a *TATA box*, the highly conserved DNA recognition sequence for the TATA box binding protein, *TBP*, whose binding initiates transcription complex assembly at the promoter.

Some genes also have *enhancer* element that can be thousands of bases upstream or downstream of the transcription initiation site. Combinations of these upstream control elements and enhancers regulate and amplify the formation of the basal transcription complex.

Transcription Process

Eukaryotes have three nuclear RNA polymerases, each with distinct roles and properties.

Name	*Location*	*RNA transcribed*
RNA polymerase I (pol I, Pol A)	Nucleolus	ribosomal RNA
RNA polymerase II (Pol II, Pol B)	nucleus	messenger RNA and most small nuclear RNAs.
RNA polymerase III (Pol III, Pol C)	nucleus	transfer RNA and other small RNAs

There are many eukaryotes that differ from the cononical presentation of the roles of RNA polymerases. Certain organisms possess four distinct RNA polymerases. Other organisms utilize RNA polymerase 1 to transcribe certain protein coding genes in addition to rRNAs.

Transcription Regulation

The gregulation of gene expression is achieved through the interaction of several levels of control including the regulation of transcription initiation. Most eukaryotes possess robust methods of regulating transcription initiation on gene-by-gene basis. The transcription of a gene can be regulated by cis-acting elements within the regulatory regions of DNA, and transacting factors that include transcription factors and the basal transcription complex.

Eukaryotic translation is the process by which messenger RNA is translated into proteins in eukaryotes.

Initiation—the cap-dependent Initiation

Initiation of translation involves the interaction of certain key proteins with a special tag bound to 5'-end of an mRNA molecule. The protein factors bind the small *ribosomal* subunit (also referred to as the 40S subunit), and these *initiation factors* hold the mRNA in place. The eukaryotic Initiation Factor 3 (eIF3) is associated with the small ribosomal subunit, and plays a role in keeping the large ribosomal subunit from prematurely binding. eIF3 also interacts with the eIF4F complex which consists of three other initiation factors: eIF4A, eIF4E and eIF4G. eIF4G is a scaffolding protein which directly associates with both eIF3 and the other two components. eIF4E is the cap-binding protein. It is the rate-limiting step of cap-dependent initiation, and is often cleaved from the complex by some viral proteases to limit the cell's ability to translate its won transcripts. This is a method of hijacking the host machinery in favor of the viral (cap-independent) messages. eIF4A is an ATP-dependent RNA helicase, which aids the ribosome in resolving certain secondary structures formed by the mRNA transcript. There is another protein associated with the eIF4F complex called the Poly-A Binding Protein (PABP), which binds the *poly-A tail* of most eukaryotic mRNA molecules. This protein has been implicated in playing a role in circularization of the mRNA during translation.

This pre-initiation complex (43S subunit, or the 40S and mRNA) accompanied by the protein factors move along the mRNA chain towards its 3'-end, scanning for the 'start' codon (typically AUG) on the mRNA, which indicates where the mRNA will begin coding for the protein. In *eukaryotes* and *archaea*, the *amino acid* encoded by the start codon is *methionine*. The initiator tRNA charged with Met forms part of the ribosomal complex and thus all proteins start with this amino acid (unless it is cleaved away by a protease in subsequent modifications). The Met-charged initiator tRNA is brought to the P-site of the small ribosomal subunit by eukaryotic Initiation Factor 2 (eIF2). It hydrolyzes GTP, and signals for the dissociation of several factors from the small ribosomal subunit which results in the association of the large subunit (or the 60S subunit). The complete ribosome (80S) then commences translation elongation, during which the sequence between the 'start' and 'stop' codons is translated from mRNA into an amino acid sequence-- thus a protein is synthesized.

Eukaryotic Initiation Factor

There exist many more eukaryotic initiation (eIF) than prokaryotic initiation factors due to greater biological complexity. Processes eIF is involved in include: formation of initiation complex with 5'mRNA and complexing with Met-tRNA;, binding mRNA factor to Met-tRNA;, scanning mRNA for the intiator codon AUG, locating the binding site of initiator tRNA to the AUG start site, and joining of the 60S subunit to create the 80S subunit.

The eIF4 initiation factors include eIF4A, eIF4B, eIF4E, and eIf4G. eIF4F is often used to refer to complex eIF4A, eIF4B, and eIF4G.

EIF4G is a scaffolding protein that interacts with eIF3 as well as the other members of the eIF4F complex. eIF4B contains two RNA binding domains—one specific interacts with mRNA, while second specifically binds the 16S portion of the small ribosomal subunit' eIF1, eIF1A, and eIF3, all bind to the ribosome subunit-mRNA complex. They have been implicated in preventing the large ribosomal subunit from binding the small subunit before it is ready to commence elongation.

eIF2 is a GTP-binding protein responsible for bringing the initiator tRNA to the P-site of the pre-initiation complex. It has specificity for the methionine-charged initiator tRNA, which is distinct from other methionine-charged tRNAs specific for elongation of the polypeptide chain. Once it has placed the initiator tRNA on the AUG start codon in the P-site, it hydrolyzes GTP into GDP, and dissociates. This hydrolysis, also signals for the dissociation of eIF3, eIF1, and eIF1A, and allows the large subunit to bind. This signals the beginning of elongation.

Modification of Released Protein

The released polypeptide chain contains the formylated methionine at its one end. An enzyme deformylase removes the formyl group of methionine. The exopeptidase enzyme may remove some amino acids from N-terminal end or the C-terminal end of polypeptide chain. At this stage the polypeptide (protein) possesses its primary and probably its secondary structures. The linear sequences of amino acids forms the primary structure; at least some portion of many proteins has a secondary structure in the form of an alpha helix. The protein chain may then fold back upon itself, forming internal bond (including hydrogen bonds, ionic bonds, and strong disulphide bonds) which stabilize its tertiary structure into a precisely and often intricately folded pattern. Two or more tertiary structures may unite into a functional quarternary structure. For example, haemoglobin consists of four polypeptide chains two identical α-chains and two identical β-chains. A protein does not become an active enzyme until it has assumed its tertiary or quarternary pattern. Some quarternary proteins consist of identical subunits (*e.g.*, the enzyme β-galactosidase consists of four identical polypeptide chains). Such proteins are called homopolymers. Other quarternary proteins, such as haemoglobin, consists of non-identical subunits and are called heteropolymers. Some polypeptide chains are subject to posttranslational modification before they assume their biologically active form. For example, chymotrpsinogen is cut at one position by an enzyme and the active split product is chymotrypsin.

Role of Chaperones in Protein Folding

Molecular chaperones represent a diverse class of protein molecules which play a significant role in folding of individual polypeptides and in assembly of multimeric proteins. They act by inhibiting incorrect molecular interactions which are not possible in their absence (In fact, chaperone is an English word which means a 'person' usually a married or elderly women who for the sake of propriety, accompanies a young unmarried lady in public as guide and protector). One example of chaperone is ATPase.

Expression in Eukaryotes

It is believed that in eukaryotes, only 2–15% of the genes are expressed at a particular time obviously, a regulation mechanism must be operating. Contrary to prokaryotes, genes in eukaryotes can be broadly of following two types:

1. First category of genes contains those, which have multiple copies like genes for rRNA, 5sRNA and tRNA, while rRNA genes are transcribed by RNA polymerases I (or A), the genes for 5sRNA & tRNA are transcribed by RNA polymerase III (or C).
2. The second category of genes are found in single copies or unique sequence and are transcribed by RNA polymerase II (or B). It has been possible to demonstrate that the three RNA polymerases are specific in transcribing specific DNA sequences. Thus, a gross level of regulation in eukaryotes is exercised through the availability of multiple RNA polymerase. The more specific regulation of specific genes meant for specific protein is exercised through the chromosomal proteins, most of which are now known to be the transcribing factors working as activators, repressor, w-activators and corepressors.

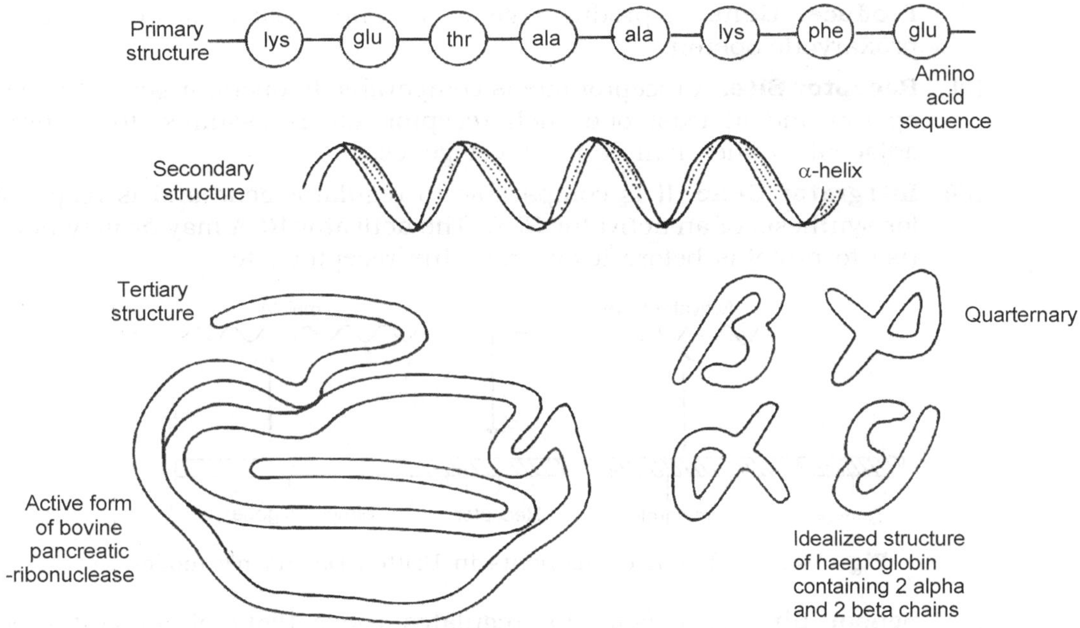

Fig. 16.1. Stages in the development of a functional protein.

CHROMOSOMAL PROTEINS AND GENE EXPRESSION

As it is known that chromosomal proteins are mainly of two kinds (*i*) histones and (*ii*) non-histones hence the chromatin has three components DNA, histones and non-histones. While histones are non-specific and may be involved in repressing gene activity in general, non-histones are specific, so that specific non-histones activate specific genes. Such a role of non-histones was demonstrated by chromatin reconstitution experiments, in which the chromatin was reconstituted using DNA, histones and non-histones from different tissues, and the synthesized proteins using this reconstituted chromatin was examined. The observation made in these chromatin reconstituted experiments suggested a positive regulatory role of non-histones in eukaryotes. This role of non-histones is also suggested by following additional information:

(*i*) The non-histones are present in increased quantity in tissue active in RNA synthesis, while quantity of histones is constant.

(*ii*) The non-histones exhibit 2–3 times more diversity than histones.

(*iii*) The non-histones have tissue specificity and DNA binding specificity.

(*iv*) Certain non-histones are known to induce RNA synthesis *in vitro*.

(*v*) Synthesis of specific non-histones is associated with induction of gene activity.

But in absence of histones, non-histones cannot induce RNA synthesis.

MODELS FOR UNIT OF GENE EXPRESSION

1. **Britten-Davidson Model.** Although various models for regulation of protein synthesis in eukaryotes have been proposed, the most popular model, known as "Britten-Davidson model" or the "gene battery model" was proposed by Britten and Davidson in 1969. It was subsequently elaborated by them in 1973. The model assumes the presence of four classes of sequences:

(*i*) **Producer Gene.** A producer gene is comparable to structural gene of prokaryotic operon.

(*ii*) **Receptor Site.** A receptor site is comparable to operator gene of bacterial operon and at least one such receptor site is assumed to be present adjacent to one or more producer genes.

(*iii*) **Integrator Gene.** It is comparable to regulator gene and is responsible for synthesis of an activator RNA. The activator RNA may or may not give rise to proteins before it activates the receptor site.

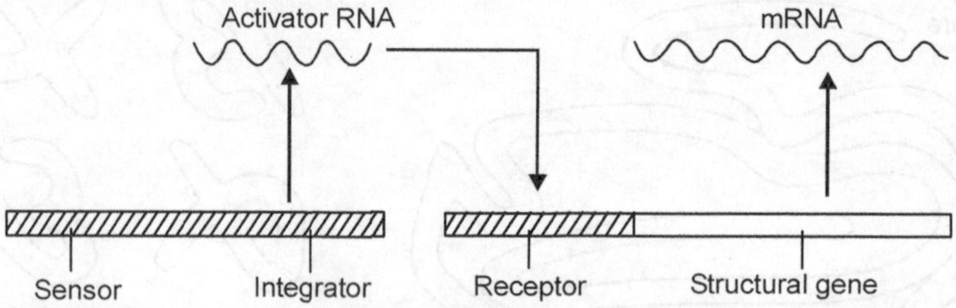

Fig. 16.2. Different components in Britten-Davidson's model.

(*iv*) **Sensor Site.** A sensor site regulates the activity of integrator gene. Integrator gene can be transcribed only when the sensor site is activated. The sensor sites are recognized by agents which change the patterns of gene expression like hormones and proteins etc. For instance hormone protein complex or transcription factor may bind to sensor sites to cause the transcription of integrator.

The interrelationship of the four classes of sequences given above are illustrated in Fig. 16.2. In this model the genes (producer gene and integrator gene) are those sequences, which are involved in RNA synthesis and sites (receptor and sensor sites) are those sequences which help only in recognition without taking part in RNA synthesis.

It is also proposed that receptor sites and integrator genes may be repeated a number of times, so as to control the activity of a large number of genes in a cell. Repetition of receptor ensures that same activator recognizes all of them and several enzymes of one pathway are simultaneously synthesized.

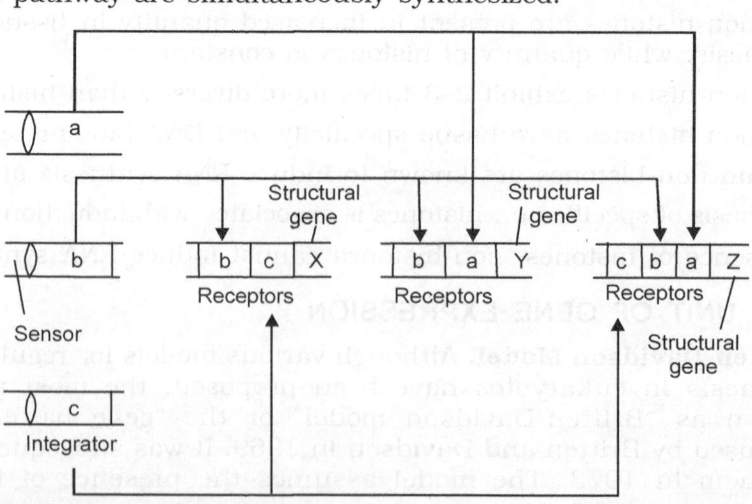

Fig. 16.3. Redundancy of receptor sites (as proposed in Britten-Davidson's model).

It is also possible that the transcription of same gene is needed in different states. This can be achieved by multiplicity of receptor sites and integrator genes. Each structural gene may have several receptor sites. Each responding to one activator so that different activators may activate the same gene at different times. Several integrator genes may also be associated with same sensor site.

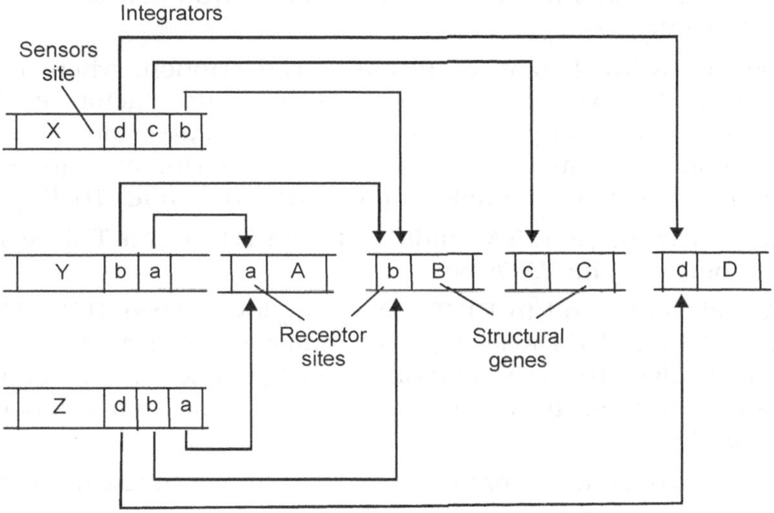

Fig. 16.4. Redundancy of integrator genes (as proposed in Britten-Davidson's model).

2. **A Gene Battery.** In Britten-Davidson model a set of structural genes controlled by one sensor site is termed a "battery" sometimes, when major changes are needed, it is necessary to activate several sets of genes. If one sensor site is associated with several integrators, it may cause transcription of all integrator at the same time thus causing transcription of several producer genes through receptor sites.

The repetition of integrator and receptor sequences is consistent with the knowledge that sufficient repeated DNA occurs in eukaryotes.

TRANSCRIPTION FACTORS

The initiation of transcription in eukaryotes involves the formation of a pre-initiation "transcription complex" or "transcriptosome". At least seven transcription factors (TFIIA, TFIIB, IFIID, TFIIE, TFIIF, TFIIH, TFIIJ), take part in the formation of transcription complex for RNA polymerse II, at the promoter sites of 'house keeping genes' (genes functional in all cells at all times for essential activities). These promoters are called 'generic promoters'. There are genes called "smart genes", whose expression is regulated by additional transcription factors. Hundreds of these transcriptional factors have now been isolated from higher plants and animals. Most of these transcriptional factors have each a 'DNA binding domain' and an 'activation domain'. On the basis of similarity, DNA binding domains have been grouped into the following main families:

(*i*) heliz-turn-helix (HTH);

(*ii*) helix-loop-helix (HLH);

(*iii*) zinc fingers;

(*iv*) leucine zippers. Similarly activation domains have been grouped into:

(*i*) acidic domains,

(*ii*) glutamine rich domains and

(*iii*) proline rich domains.

Detailed structure and function of a number of transcription factors have been studied in recent years.

Transcription by RNAP II may be (*i*) basal transcription, when it involves house keeping genes and makes use of only general transcription factors (GTFs), or may be (*ii*) activator dependent transcription, when regulated expression requires activator(s) and/or coactivator(s), in addition to GTFs, basal transcription requires the action of several transcription factors in the following defined order (Fig. 16.4).

(*i*) TFIID, consisting of TATA binding protein (TBP) and TBP associated factors (TAFs), binds at the TATA box;

(*ii*) TFIIA and TFIIB bind to FTIID-DNA complex to form TFIID-TFIIA-TFIIB-DNA complex. (also called DAB complex); in intermediate DA complex or DB complex may be involved in the formation of DAB complete (TFIIA is involved only in activator dependent transcription and TFIIB provides a link between promoter and RNAPII)

(*iii*) RNAPII is escorted to the promote site by TFIIF, forming a transcription complex;

(*iv*) orderly addition of TFIIE, TFIIH and TFIIJ help in the initiation.

The carboxy-terminal domain (CTD) of RNAP II gets phosphorylated with the help of TFIIH (Fig. 16.5), which is an essential requirement for transcription by RNAII. After initiation of transcription, RNAPII leaves promoter but TFIID remains bound to start another cycle of transcription initiation. The RNAPII after initiation keeps on synthesizing RNA, thus elongating the RNA chain. Atleast two elongation factors are used, (*i*) TFIIF accelerates RNA chain growth uniformly and (*ii*) TFIIS removes obstruction in the path of elongation. Termination of RNA chain synthesis is brought about by termination sites similar to those in prokaryotes.

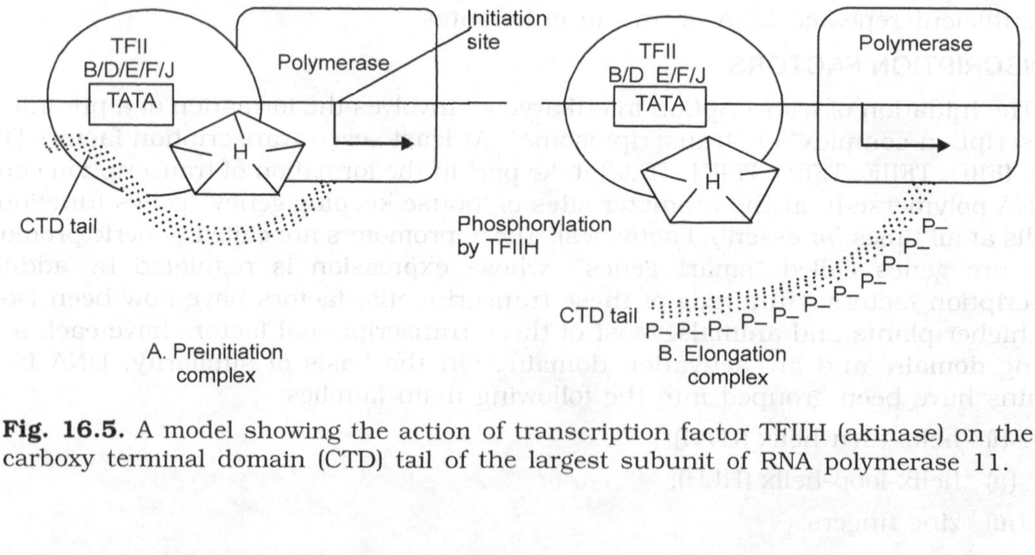

Fig. 16.5. A model showing the action of transcription factor TFIIH (akinase) on the carboxy terminal domain (CTD) tail of the largest subunit of RNA polymerase 11.

17 One Gene-One Enzyme Hypothesis

INTRODUCTION

Archibald E. Garrod (1909), the English Physician-Biochemist was the first to establish a relation between genes and enzymes. When mutations occur in genes, they result in the production of defective or inactive enzyme. This causes "inborn errors in metabolism" *e.g., alkaptonuria* and *phenylketonuria*, in humans.

The one gene-one enzyme hypothesis was then proposed by *Beadle* and *Tatum*. They proposed that *single gene controls the synthesis of one enzyme*, derived while working on *Neurospora crassa* (an omycetous fungus). The wild varieties of *Neurospora* can grow on minimal medium *i.e.*, the medium containing inorganic salts, a simple sugar and vitamin biotin. This means that wild varieties of *Neurospora* can synthesize all other essential *metabolities viz* purines, pyrimidines, amino acids and other vitamins.

MECHANISM OF HYPOTHESIS

According to this theory, each gene in an organism controls the production of specific enzyme. Beadle and Tatum irradiate a sexual condition stage of wild type *Neurospora* to X-rays and isolated mutants (auxotrophs) for specific *metabolities*. Only mutant strains that yielded 1:1 mutant to wild type ratio when crossed to wild type were studied further so as to select strains with mutation in only one gene. Mutant clones which could grow on complete medium but could not grow on minimal medium were further analyzed for their ability to grow on medium supplemented with just amino acids or just vitamins or just purines and pyrimidines. The strains that grow when just vitamins were added, were in turn analyzed for their ability to grow on media supplemented with each of the vitamins separately. This way Beadle and Tatum isolated a mutant that required only one vitamin. By correlating a genetic analysis with a biochemical study, they could demonstrate that one mutation had resulted in loss of one enzyme activity which in turn, resulted in loss of a function. Soon it became confirmed that DNA controlls the metabolic activity in the cells by allowing the synthesis of enzymes. The enzymes are generally proteinaceous or they have a protein part and they may be composed of more than one polypeptide formed by joining of amino acids for instance, haemoglobin of our blood is a tetramer having two α-polypeptides and two β-polypeptides. Thus, the concept of one gene, one enzyme hypothesis gradually changed to one gene, one polypeptide hypothesis.

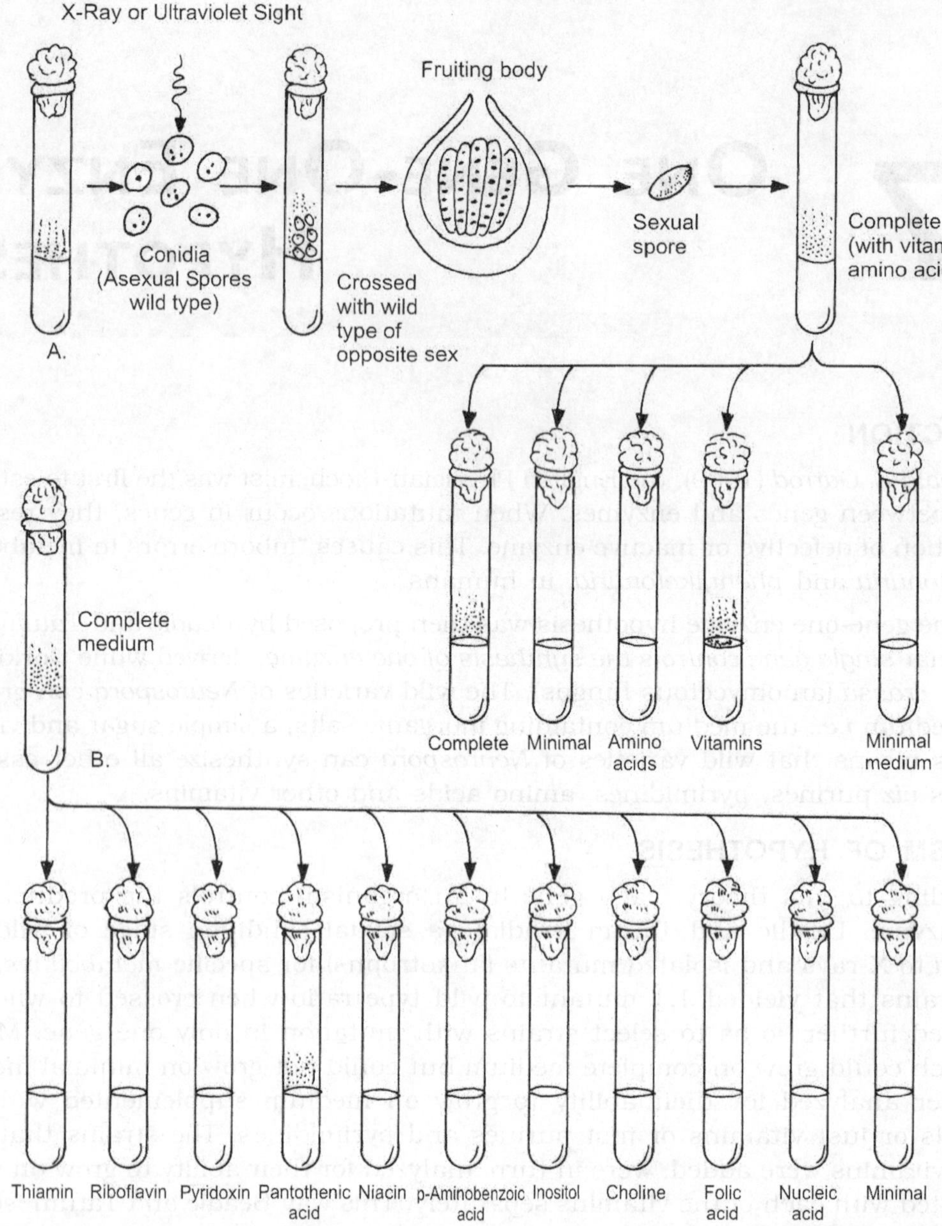

Fig. 17.1. Diagram of the classic experiment of *Beadle* and *Tatum*, showing that one mutation results in a new requirement for one growth in *cospora conidia* were mutagenized with X-rays or ultraviolet light and then screened for their ability to grow on "minimal medium" (containing inorganic salts, a simple sugar, and biotin) and "complete medium" (supplemented with all the vitamins, purines, pyrimidines, and amino acid) t-type *Neurospora* will grow on either minimal or complete medium. Mutant strains that could grow on complete medium, but not analyzed further to determine their exact growth-factor requirements (A and B). A back cross to wild type (shown at top) was done with into clone to select mutant strains carrying only one mutation. Single mutants should yield a 1:1 mutant to wild type segregation pattern related biochemical analyses demonstrated that each mutation resulted in the loss or inactivation of one enzyme. (From G.W. Beadle, "Genes Chemistry of the Organism," *Amer. Scientist* 34: 31–53, 1946. Reprinted by permission of *American Scientist*, Journal of Sigma XL "The Scientific Reserch Society").

IMPORTANCE OF HYPOTHESIS

This hypothesis was able to explain inborn errors of metabolism described by *Archibold Garrald:*

1. *Alkaptonuria.* In this, urine of a person suffering from this disease turns black on exposure to air because of the presence of a substance (alkapton, homogentistic acids).
2. Because in these persons, the enzyme homogentistic acids oxidase required in the conversion of homogentistic acid to malcylocytic acids present.
3. *Phenykotonuria.* The disease is caused due to the metabolism of phenylalanins. As a result, it gets accumulated in the blood causing mental retardation and epilepsy.

MODIFICATION OF ONE GENE-ONE ENZYME HYPOTHESIS

One gene-one polypeptide concept: *Example*-enzymes are made up of proteins. These proteins are made up of polypeptide chains. Thus, a theory was modified to one gene one-polypeptide concept. In sickle cell anaemia, Hb is a protein of a molecular weight 64000 containing about 600 amino acids. These amino acids are arranged in +polypeptide chain, two identical α-chains and two identical β-chains. The sickle shaped Hb differs from normal Hb in the presence of valine in place of glutamic acid at the 6th position in the β-chain. The sickle cell anaemia is produced by a single change in the two β-polypeptide chains. Thus, one gene controls the synthesis of one polypeptide.

18 DNA Recombination: Molecular Mechanism in Prokaryotic and Eukaryotic Organisms

INTRODUCTION

During the last three decades, techniques for manipulating eukaryotic as well as prokaryotic DNA have witnessed a remarkable development. There are at least three main phases of the development of these techniques, which include the following:

(*i*) Recombinant DNA and gene cloning (development in 1970s).

(*ii*) Polymerase chain reaction or PCR (developed in 1980s).

(*iii*) DNA chips and microarrays (developed in 1990s).

RECOMBINANT DNA

Recombination is defined as *exchange of DNA sequence between different molecules, occurring either naturally or as a result of DNA manipulation.* Recombinant DNA technique includes all the techniques involved in construction, study and use of recombinant DNA molecules.

The recombinant DNA molecule is that DNA molecule which is created in test tube by ligating together, pieces of DNA that are not normally contiguous. Recombinant DNA involves breakage of a DNA molecule at desired places to isolate a specific DNA segment and then insert it in another DNA molecule at a desired position. The product thus obtained is called *recombinant DNA* and the technique is often called *genetic engineering.* With this technique, genes can be isolated, cloned and characterized, so that the technique has led to significant progress in all areas of molecular biology. Using the recombinant DNA technology, we can isolate and clone single copy of a gene or a DNA segment into an indefinite number of copies, all identical. This became possible, because bacteria, phages and plasmids reproduce in their usual style, even insertion of foreign DNA, so that the inserted DNA also replicates faithfully with the parent DNA. This technique is called *gene cloning.*

RESTRICTION ENZYMES FOR CLONING

For cloning of DNA, often we need to cut the DNA at specific sites, which are recognized and cleaved by specific enzymes (endonucleases, which cleave DNA at internal sites), described as restriction enzymes. These restriction enzymes recognize short sequences of double stranded DNA as targets for cleavage. Different enzymes recognize different, but specific sequences, each ranging in length from 4 to 8 base pairs. Each enzyme is named by three letters (or four letters) abbreviation (letters are italicized) that identifies its origin. Roman numerals (I, II, III, etc.) are added to distinguish several enzymes with same origin. For instance *Eco* RI is derived from *E. coli* and *HpaI* is derived from *Haemophilus parainfluenzae*.

TECHNIQUES USED IN RECOMBINANT DNA TECHNOLOGY

There are several techniques, which are regularly used, while employing recombinant DNA technology. Some of these techniques will be briefly described in this section.

1. Separation of DNA Fragments using Agarose or Polyacrylamide Gel Electrophoresis. When genomic DNA, extracted from any tissue of a plant or animal species, is digested with a restriction enzyme, it is cleaved into fragments. The fragments of different sizes can be separated through gel electrophoresis, which involves movement of fragments or molecules from a groove created on one edge of the gel. The gel may be a cylinder or a slab (usually a slab for cloning experiment) about 10 cm long and 5 cm thick. The rate of movement of fragments is inversely correlated with the size of fragments or molecules, so that heavier fragments will move away.

Fragment of different sizes will appear as bands on the gel and can be examined or isolated for further study. More often agarose gels are used, but for separation of fragments differing by few base pairs, polyacrylamide gels are used.

2. Southern, Northern and Western Blotting. A mixture of DNA, fragments can be separated by gel electrophoresis and the separated bands can be stained and visualized directly in the gel. Similar methods are also available for RNA and proteins. However, to confirm the identity of bands obtained due to electrophoresis or to find similarity of one or more of these bands with a known and available molecular probe, it is possible to hybridize these bands with the labelled probe. Although hybridization of separated DNA fragments with a prove of small size (10 – 20 bases) can be conducted within the gel (in-gel hybridization), to facilitate this hybridization with relatively bigger probes, the bands are often transferred to a nitrocellulose membrane through a technique described as blotting. When DNA bands are thus blotted, it is called *Southern blotting* (after the name of *E.M. Southern*); when RNA bands are thus transferred it is described by the *JARGON* term *Northern blotting* and similarly when protein bands are transferred, the technique is described as *Western blotting*.

CLONING VECTORS FOR RECOMBINANT DNA

One of the most important uses of recombinant DNA technology is the clonning of (*i*) random DNA or cDNA fragments, often used as probes or (*ii*) specific genes, which may be either isolated from the genome or synthesized either organochemically or in the form of cDNA from mRNA. This cloning of DNA is possible only with the help of another DNA molecule which is capable of replication in an appropriate host. This other DNA molecule is often used in the form of a vector, which could be a plasmid, a bacteriophage, a derived cosmid (see later in this section), a phagemid (phage + plasmid) or even a virus. Techniques should also be available, which will allow selection of chimeric vector obtained after insertion of foreign DNA from a mixture of chimeric and the original vector (see later). Another critical desired feature of any cloning vector is that it should possess a site at which foreign DNA can be inserted without disrupting any essential function. Sometimes vectors are modified by inserting a DNA segment to create unique site (s) for one or more restriction enzymes to facilitate its use in gene cloning. This inserted DNA is sometimes called 'polylinker'.

Recombination occurs when there is an exchange of genetic information among different DNA molecules that result in a reshuffling of genes.

1. This process can produce numerous new combination of genetic information.
2. In eukaryotic cells, process takes place during sexual reproduction and in prokaryotes cells, it takes place during asexual and parasexual mating. As a result, allelic form of genes are exchanged producing new combination of alleles.

TYPES OF RECOMBINATIONAL PROCESS

Homologous Recombination

1. In this, a recombination process occurs between the regions of DNA containing the same or nearly the same nucleotide sequence.
2. Scene in the crossing over of chromosome, containing the same gene loci pair and exchange allelic proteins of the same chromosomae.
3. It takes place between the alleles of the same gene.
4. In homologous recombination, there is a base pairing of corresponding regions of the DNA and align chromosome which stabilizes duplexes between homologous DNA regions.

Example 1: In eukaryotes, it occurs during meiosis where homologous chromosome pairs are separated and one member of each pair is distributed to each of the two daughter cells.

Molecular basis of homologous recombination

Double stranded DNA separation

↓ endonuclease

A nick is created in one of each strand

↓

Both two strands may get separated

↓

Helix destablizing protein combines at the site where the nick occurs and aids in the opening of DNA double helix

↓

The free z' end of OH end acts as a producer for DNA synthesis (DNA poly) that produce a single strand of DNA

↓

The newly synthesized strand of DNA bears the corresponding region of the homologous chromo, stabilizing heteroduplex

↓

A heteroduplex forms when two strands of DNA that are complementary over only pair of their lengths joined together

↓

The homologous regions from a duplex and a non-complementary segment remain single stranded (heteroduplex). Both duplex with non-duplex regions are present.

↓

Endonuclease cleaves the non-duplex regions

↓

Ligase forms the z' end of DNA strand

↓

The formation of the heteroduplex is catalysed by an enzyme (*rec*-genes)–
Recombinant gene

1. Recombination does not occur in the cuts like in fraction genes.
2. The result of this enzymatic action is the formation of a bridge between the two homologous DNA strands.
3. The joining of chromosome at homologous region stabilizes chi form

↓

then chromosomes coated, so that the two DNA strands no longer cross each DNA but are still held together by covalent linkages.

↓

An endonuclease cleaves the DNA strands breaking the heteroduplex and establishing two independent chromosomes.

Example 2: In *E. coli*, homologous recombination takes place by 3 recombinant genes *rec* A, *rec* B, *rec* C

1. The *rec* A protein catalyses an ATP ribbon assimilation reaction in which the single stranded DNA hydrolyses with the duplex DNA molecules.
2. The *rec* A protein has a strong affinity for ss DNA.
3. The *rec* B and *rec* C are used in unwinding the double stranded (ds) DNA (breaking the chains into smaller fragments).

Non-homologous Recombination

1. Exchanges of DNA occurs between segments of DNA different in nucleotide sequences,
2. It takes place when the strands of homologous between the regions of DNA that are exchanged are limited,
3. This is also called as non-reciprocal recombination,
   dNA exchange occurs only at a given location within the genome,
4. it helps in the joining of DNA molecules from different sources.

Example 3

1. Viral DNA can be incorporated into the bacterial DNA.
2. *rec* genes are not involved in this mechanism.

19 GENE EXPRESSION IN YEAST

INTRODUCTION TO GENE EXPRESSION

All genes have to be expressed in order to function. The first step in expression is transcription of the gene into a complementary RNA strand. For some genes—for example, those coding for transfer RNA and ribosomal RNA molecules—the transcript itself is the functionally important molecule. For other genes, the transcript is translated into a protein molecule.

To understand how a gene is expressed, the RNA transcription must be studied. In particular, the molecular biologist will want to know whether the transcript is a faithful copy of the gene, or whether segments of the gene are missing from the transcription (Fig. 19.1(b)). These missing pieces are called introns which considerable interest centres on their structure and performed all possible functions. In addition to introns, the exact locations of the start and end points of transcription are important.

The signals that determine the start and finish of the transcription process are only partly understood, but their positions must be located, if the expression of a cloned gene is to be studied.

Transcript analysis method can be used to determine if a cloned gene contains introns and to map the positions of the start and end points for transcription.

REGULATED SYSTEMS

Although, it may be possible to operate continuous—flow culture systems for some heterologous proteins, the strong selective pressure for faster growing variants encountered in these systems make it likely that for many heterologous proteins batch cultures will be the method of choice for production. Clearly, regulated expression is desirable so that selective pressure can be minimized. The ideal regulation would be such that biomass accumulation occurred in the absence of heterologous protein production, with expression being switched on towards the end of the growth phase.

A number of regulated systems have been described that require an alteration in medium composition. Regulation of acid phosphatase has been investigated for a number of years, and it is known that a number of genes including PHO2, PHO80, PHO81, PHO85 and a gene cluster PHO82, PHO4 (Toh-e *et al.*, 1973, Toh-e and Oshima 1974, Ueda *et al.*, 1975) are involved in the expression of repressible acid phosphatase (PHO5) in response to inorganic phosphate levels in the medium.

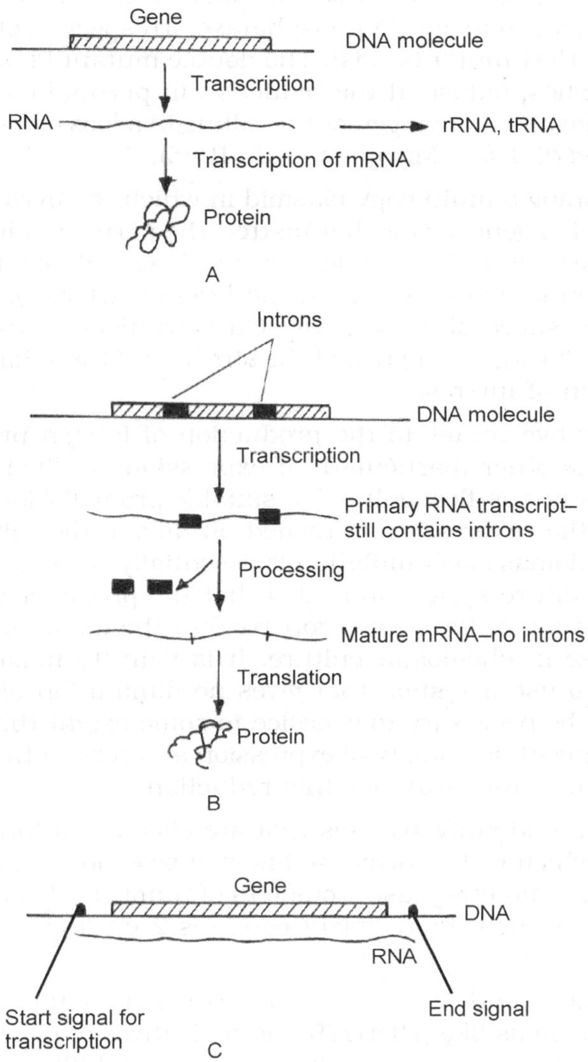

Fig. 19.1 Some fundamentals of gene expression. A. Genes are expressed by transcription and translation. B. Some genes contain introns. C. RNA transcripts include reglons either side of the gene.

Induction of acids phosphatase in medium depleted of inorganic phosphate is regulated in the transcription level (*Bostian et al.*, 1980, 1983). A number of laboratories have used the acid phosphatase promoter to achieve regulated expression of phosphotase heterologous proteins in yeast (Hinnen *et al.*, 1983, Miyanhara *et al.*, 1983, Kramer *et al.*, 1984, Rosenberg *et al.*, 1984, Lemonte *et al.*, 1985). The acid phosphatase system is not completely on–off. Basal levels, whether of acid phosphatase or heterologous proteins are produced in the presence of inorganic phosphate, but the levels can be induced up to 200-fold by removal of inorganic phosphate from the medium.

Changes in the composition of the medium are not convenient in large-scale fermenters, and this consideration led Kramer *et al.*, (1984) to combine a molecular and classical genetical approach to achieve temperature regulated expression of 1FN-alpha. The PhO80 gene product is a repressor of acid phosphatase; PHO4 mutants leads to constitutive synthesis of acid phosphatase in high and low inorganic phosphate, PHO4 encodes a positive regulates of acid phosphates, do not produce acid phosphatase, irrespective of medium conditions. Temperature-sensitive PHO4 mutants exist. The double mutant PHO4, PHO80, results in the production of acid phosphatase at the reduction in plasmid copy number or loss of plasmid; alternatively plasmid rearrangement resulting in a loss of foreign gene expression that can occur (*Jabbar et al.* 1985, Miyajima *et al.*, 1985).

If yeast cells containing a multi copy plasmid in which the pyruvate kinase promoter is used to express a foreign gene, or for that matter, the pyruvate kinase structural gene; are constantly restreaked and the streaks are used as an inoculum for cultures in which protein expression is monitored, then the level of heterologous and homologous expression declines with successive inoculam. On Petri dishes, strong growing papillae are observed against the background lawn of the streak and these have markedly reduced expression of the protein of interest.

There are basically two routes to the production of foreign proteins in yeast. One is batch culture and the other (particularly if expression is allied to protein secretion into the medium) is continuous-flow culture. A suitable promoter for batch culture is one that is regulated such that expression is turned off during the initial stages of growth and switched on when biomass accumulation is essentially complete. The use of a single stage continuous flow culture system indicates that the promoter is active throughout. Clearly, the glycolytic promoter have this property, but the instability of expression with time precludes their use in chemostat culture. It is vital (to maximize the benefits of continuous operation) to use a system that gives no diminution of yield for generation after generation. It may be necessary to sacrifice to some extent the levels of expression as a means to the attainment of stability of expression and rely on the greater productivity of continuous systems to compensate for this reduction.

We have attempted to identify systems that are effective at foreign gene expression in several ways, one of which will be discussed now. A selection system has been devised that can be used to isolate directly yeast promoters of number. Plasmid like pJDB207 will only transform strains provided the recipient contains 2 µm to provide REP1 and REP2 function in *trans* (*Beggs* 1981).

Plasmid copy number and stability are interrelated in the case of 2 µm based vectors. For example, plasmids like pJDB219 and derivatives of pJDB29 lacking bacterial sequences are capable of very high copy number and stability in *cir°* host yeast cell (*Futher* and *Cox* 1984, Walmsley *et al.*, 1983). This is presumably because (*i*) they suffer no competition for replication from native 2-µm plasmid and (*ii*) the particular yeast *leu* 2 gene they contain has a deletion in its promoter region, so it is not efficiently expressed (*Erhart* and *Hollenberg* 1983). The plasmid has to be present at high copy number to produce sufficient gene product for survival on leucine-minus medium.

It has been shown that 2-µm plasmid sequences act as ARS elements in *Saccharomyces pombe* (*Beach* and *Nurse* 1981), but do not stabilize plasmids. Indeed 2 mm has not been shown to be the basis of a stable plasmid in species of yeast or fungus other than *Saccharomyces cerevisiae*. Other yeast species do have their own native plasmids. For instance, *Zygosaccharomyces baillii* (Toh-e *et al.*, 1982), *Zygosaccharomyces rouxii* (Toh-e *et al.*, 1984) *Zygosaccharomyces bispora* (Toh-e and Utatsu 1985), have plasmids which in gross terms show resemblance to the structure of 2 µm but show no precise homology. The yeast *Kluyveromyces* lactis has two plasmid one of which is associated with the production of a secreted toxin (Gunge *et al.*, 1981). The presence of an accessible DNA plasmid in this yeast may permit the development of a host-vector system that provides efficient heterologous protein expression.

The choice of plasmid system for heterologous expression is clearly an important one; plasmid copy number and stability must be taken into account. For high level production and short fermenter runs 2-μm based vectors offer high copy number and reasonable stability. However, for long term or continuous fermentation plasmid stability will be all important and it may be advisable to use an integrating vector. In our experience a major influence on stability of heterologous protein expression is the nature of the heterologous protein itself and its effect on the yeast cell, which can be exceedingly deleterious yeast plasmid systems with a degree of copy number control similar to those which are known in *Escherichia* coli (*Twigg* and *Sherratt* 1980) would be useful for the conditional expression of proteins that are deleterious to cell health. Clearly a high copy number plasmid that is stable without selection is desirable; the native *Saccharomyces cerevisiae* 2-μm plasmid is just such a plasmid. However, manipulating its sequence and the inclusion of heterologous sequences appears to produce some degree of instability.

There is a paucity of good information on sequences involved in plasmid copy number control and stability.

VECTORS FOR YEAST

The yeast *Saccharomyces cerevisiae* is one of the most important organism in biotechnology. As well as its role in brewing and bread making, yeast has been used as a host organism for the production of important pharmaceuticals from cloned genes. Development of cloning vectors for yeast has been stimulated greatly by the discovery of a plasmid *i.e.*, present in most strains of *S. cerevisiae.* The 2 mm circle as it is called one the only a very limited number of plasmid found in eukaryotic cells.

SELECTABLE MARKERS FOR THE 2 μM PLASMID

The 2 mm circle is infact an excellent basis for a cloning vector. It is 6 kb in size, which is ideal for a vector, and exist in yeast cell at a copy number of between 70 and 200. Replication makes use of a plasmid origin, several enzymes provided by the host cell, and the proteins coded by the REP1 and REP2 genes carried by the plasmid.

However, all of it is not perfectly straightforward in using the 2 μm plasmid as a cloning vector. Firstly, there is the question of a selectable marker. One or two of the most recent yeast cloning vectors carry genes conferring resistance to inhibitors such as *methotrexate* and copper, but most of the popular yeast vectors make use of a radically different type of selection system. In practice a normal yeast gene is used, generally one that codes for an enzyme involved in amino acids biosynthesis. An example is the gene LEU2, which codes for β-isopropyl malate dehydrogenase, one of the enzymes involved in the conversion of pyruvic acid to leucine.

In order to use LEU2 as a selectable marker, a special kind of host organism is needed. The host must be an auxotrophic mutant that has a non-functional LEU2 gene. Such a *LEU2*-yeast is unable to synthesize leucine and can survive only if this amino acid is supplied as a nutrient in the growth medium.

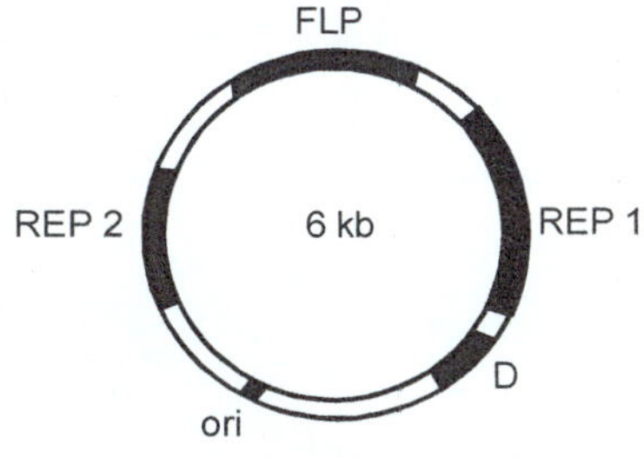

Fig. 19.2

Selection is possible because transformants contain a plasmid borne copy of the LEU2 gene, and so are able to grow in the absence of the amino acid. In a cloning experiment, cells are plated out on to minimal medium (which contains no added amino acids). Only transformed cells are able to survive and form colonies.

Vectors Based on the 2 μm Circle—Yeast Episomal Plasmids

Vectors derived from the 2 μm circle are called yeast episomal plasmids or YEps. Some YEps contain the entire 2 μm plasmid, others include just the 2 μm origin of replication. An example, of the latter type is YEp13.

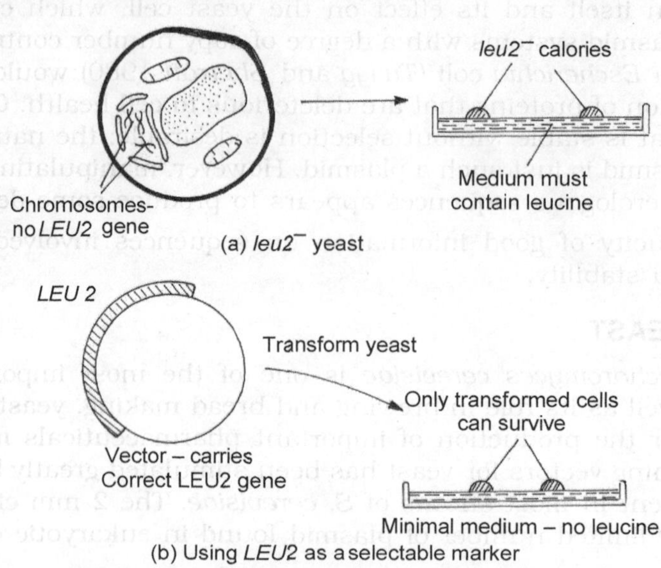

Fig. 19.3. Using the LEU2 gene as a selectable marker in a yeast cloning experiment.

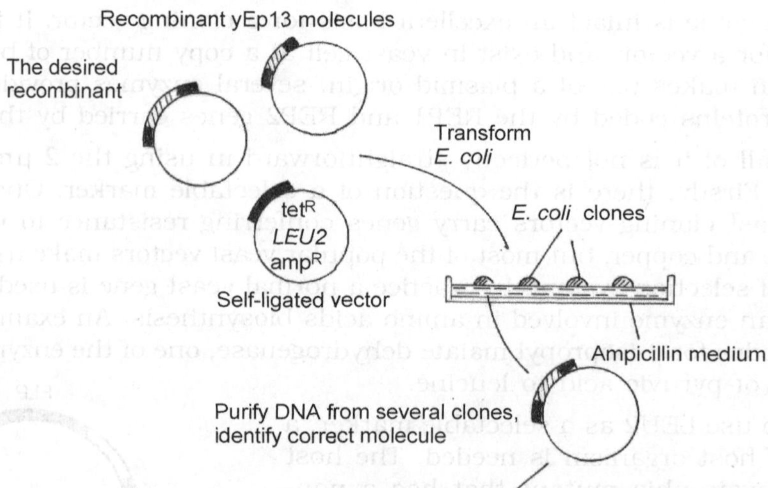

Fig. 19.4

YEp13 illustrates several general features of yeast cloning vectors firstly, it is a shuttle vector. As well as the 2 µm origin of replication and the selectable LEU2 gene, YEp13 also includes the entire pBR322 sequence, and can therefore replicate and be selected for in both yeast and *E. coli.* There are several lines of reasoning behind the use of shuttle vectors. One is that it may be difficult to recover the recombinant DNA molecule from a transformed yeast colony. This is not such a problem with YEps, which are present in yeast cells primarily as plasmids, but with other yeast vectors which may integrate into one of the yeast chromosomes purification may be impossible. This is a disadvantage as in many cloning experiments purification of recombinant DNA is essential in order for correct construct to be identified by, for example DNA sequencing. The standard procedure when cloning in yeast is therefore to perform the initial cloning experiment with *E. coli* and to select recombinants in this organism. Recombinant plasmids can then be purified, characterized and the correct molecule introduced into yeast.

Other Types of Yeast Cloning Vectors

In addition to YEps, there are several other types of cloning vectors for use with *S. cerevisiae.* Two important ones are following:

1. Yeast integrative plasmids (YIps) which is basically bacterial plasmids carrying a yeast gene. Example is YIp5, which is pBR322 with an inserted URA3 gene. This gene codes for crotidine—5'-phosphate decarboxylase, an enzyme that catalyse one of the steps in biosynthesis pathway for pyrimidine nucleotides and is used as a selectable marker in exactly same way as LEU2. A YIp can not replicate as a plasmid as it does not contain any parts of 2 µm circle, but depends for its survival on integration into yeast chromosomal DNA.
2. Yeast replication plasmids (YRps), which are capable to multiply as independent plasmids because they carry a chromosomal DNA sequence that includes on origin of replication. Replication origins are known to be located very close to several yeast genes, including one or two which can be used as selectable markers. YRpt is an example of this plasmid. It is made up of pBR322 plus the yeast gene TRpi. This gene, which is involved in tryptophan biosynthesis, is located adjacent to a chromosomal origin of replication. The yeast DNA fragment present in YRp7 contains both TRpi and the origin.

Three factors come into play when deciding which type of yeast vectors is most suitable for a particular cloning experiment. The first of these is transformation frequency, a measure of the number of transformants that can be obtained per µg of plasmid DNA. A high transformation frequency is necessary if a large number of recombinants are needed, or if the starting DNA is in short supply. YEps have the highest transformation frequency, providing between 10000 and 100000 transformed cells per µg. YRps are also quite productive, giving between 1000 and 10000 transformants/µg, but a YIp yields less than 1000 transformants /µg, and only 1–10 unless special procedures are used. The low transformation frequency of a YIp reflects the fact that the rather rare chromosomal integration events is necessary before the vector can be retained in a yeast cell.

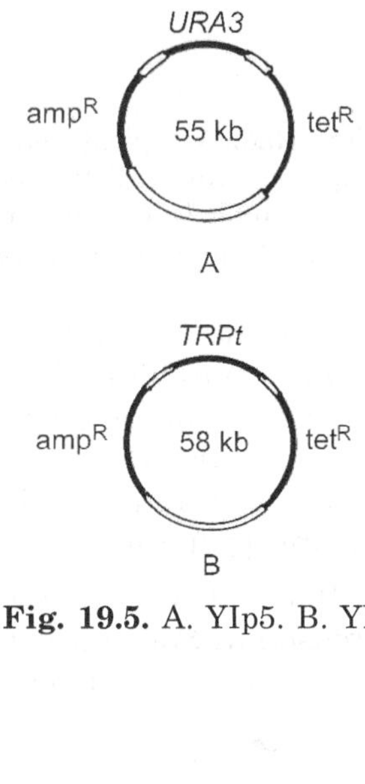

Fig. 19.5. A. YIp5. B. YRp7

YIp5 and YRp5 also have the highest copy numbers, 20–50 and 5 –100 respectively; in contrast a YIp is usually present at just one copy per cell. These figures are important if the objective is to obtain protein from the cloned gene, as the more copies there are of the gene, the greater and the expected yield of the protein product.

Vectors for other Yeasts

Cloning vectors for other species of yeast and fungi are needed for basic studies of the molecular biology of these organisms and to extend the possible uses of yeasts and fungi in biotechnology. Episomal plasmids based on the *S. cerevisiae* 2 μm circle are able to replicate in a few other types of yeast, but the range of species is not broad enough for 2 mm vectors to be of general value. In any case, the requirement of biotechnology are better served by integrative plasmids, equivalent to YIp5, as these provide stable recombinants that can be grown for long periods in bioreactors. Efficient integrative vectors are now available for a number of species, including yeasts such as *Pichia pastoris* and *Kluyveromyces lactis,* and the filamentous fungi *Aspergillus nidulans* and *Neurospora crassa.*

Expression Vectors

A typical yeast expression vector is shown in Fig. 11.6. This utilizes the alcohol dehydrogenase (ADH1) promoter and terminator which are separated by a unique *Hind* III site for the insertion of a structural gene encoding a foreign protein whose expression is desired. The ADH1 promoter was the first yeast promoter used to express a human protein, *viz* leukocyte interferon (Hitzeman *et al.*, 1981).

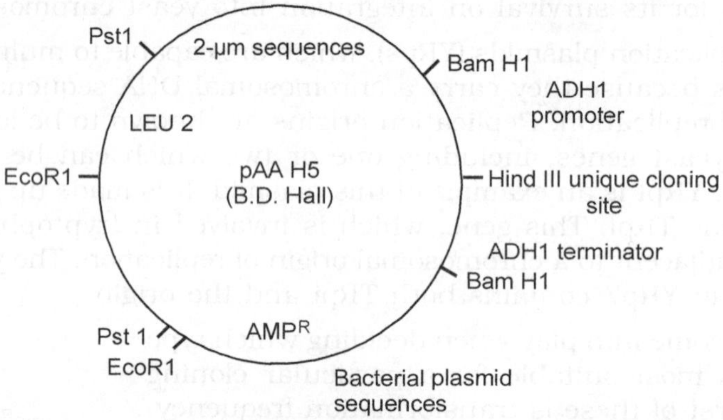

Fig. 19.6. A typical yeast expression vector. The plasmid incorporates several features: 2-μm sequences and LEU2 selection for replication in yeast, bacterial plasmid sequences and ampicillin selection for replication in bacteria, the ADH1 promoter and terminator sequences with a unique Hind III cloning site between them for the insertion of a heterologous gene of interest.

A number of glucolytic genes have been cloned and their promoters used for foreign protein production. The interest in glycolytic promoters occurred because glycolytic enzymes are amongst the best represented proteins in a yeast cell; each can represent between 1 and 5% of total cell protein, even though they are often encoded by single copy genes. It was reasonably assumed that high-level expression was due to the association of strong promoters with these genes.

It has been shown quite convincingly that heterologous eukaryotic genes can be expressed in *Saccharomyces cerevisiae* by placing them under the control of various yeast gene 'promoters' (Break *et al.*, 1984, Hitzeman *et al.*, 1983, Kramer *et al.*, 1984,

Smith *et. al.*, 1985). The yields of the heterologous proteins are determined by the 'strength' of the promoter. Thus, the yeast gene for phosphoglycerate kinase (PGK) has a promoter that can direct the expression of heterologous genes with greater than 500-fold increase in efficiency over the TRPI promoter. Mellor *et al.*, (1985) have used the PGK promoter fragment to direct the expression of human inteferon-α-2 (IFNα-2) on a high copy number plasmid in yeast. Yields of 1–3% of the total yeast protein as IFN α–2 were obtained. The highest yields of any heterologous product in yeast have been around 5% of total cell protein using an efficient promoter and a 2-μm based plasmid vector. Theoretically, however, a higher yield should be possible since the 2-μm based plasmid are usually at a copy number varying from 20 to 100 $cell^{-1}$.

Table 19.1. Glycolytic Promoters used for Heterologous Protein Expression in *Saccharomyces cerevisiae*

Promoter	*Protein*	*Reference*
GAPDH	Thaumatin	Edens *et al., (1984)*
GAPDH	α-IFN	Biter and Egan *et al., (1984)*
GAPDH	HBsAg	Bitter and Egan *et al., (1984)*
GAPDH	hEGF	Urdea *et al., (1983)*
PYK	α-IFN	Goodey *et al., (unpupl.)*
PYK	β-IFN	Goodey *et al., (unpupl.)*
PYK	γ-IFN	Goodey *et al., (unpupl.)*
TPI	PGDF	Kelly *et al., (1985)*
PGK	α-IFN	Tuite *et al., (1982)* Hitzeman *et al., (1981)*
PGK	HBsAg	Hitzeman *et al., (1983)*
PGK	γ-IFN	Derynck (*1983*)
PGK	Chymosin	Mellor *et al., (1983)*
PGK	Mouse immunoglobin	Wood *et al., (1985)*
PGK	Wheat amylase	Rothstein *et al.*, (1984)

In cloning genes of industrial significance, it is useful and economical to have the gene product(s) secreted into the medium. The secretion of proteins by *Saccharomyces cerevisiae* occurs by a mechanism very similar to that observed in mammalian cells. Schekman and collaborators (Schekman and Novick 1982, Novick *et al.*, 1981) have shown that the yeast secretory pathway involves a series of membrane bound structures that mediate the transfer of exported proteins from their site of synthesis at the endoplasmic reticulum to their site of release at the plasma membrane. As with other organisms, the secreted proteins are synthesized as large precursors including 'signal' sequences which are subsequently cleaved by a membrane bound protease to yield the mature gene products.

There have been published reports on the secretion of heterologous proteins from yeast using the signal sequences of the mammalian proteins being secreted (Hitzeman *et al.*, 1983). However, such signal sequences do into generally permit more than about 20% of the protein synthesized to be secreted. Moreover, it has been reported that with such signal sequences proteolytic cleavage seems to be imprecise. Because of such findings considerable efforts have been devoted to developing plasmids carrying signal sequences from yeast genes coding for secretory proteins such as the enzymes invertase, acid phosphatase, melibiase, glucoamylase and proteins like α-factor pheromone and killer toxin. The structural genes involved in the synthesis of these proteins have been cloned and extensively characterized.

The gene SUC2 responsible for synthesis of invertase (which is mostly secreted into the periplasmic space) has been cloned by *Carlson* and *Botstein* (1982). The SUC2 gene codes for two differentially regulated mRNAs and two forms of the sucrose–hydrolysing enzyme invertase. The secreted, glycosylated form of the enzyme is required for the extracellular hydrolysis of sucrose, and is regulated by glucose. The intracellular non glycosylated invertase has no known function and is produced constitutively in the cell. The SUC2 gene has been cloned and sequenced (*Taussig* and *Carlson* 1983). The two forms of invertases are encoded by two differently regulated mRNAs which differ only at the 5'-ends (Carlson *et al.,* 1983). A 1.9-kb RNA, which is glucose repressible, encodes a signal peptide containing precursor to the secreted invertase. However the constitutively synthesized 1.8-kb RNA does not encode a complete signal sequence, and hence the translation product remains intracellular (*Carlson* and *Botstein* 1982).

Since the SUC2 gene carries a strong promoter as well as a signal sequence for secretion, it has been found to be useful in the construction of hybrid DNA sequences by fusion heterologous genes of interest with the promoter or signal sequence region, or both of the SUC2 gene. Thus, secretion of calf prochymosin (the inactive precursor of chymosin or renin) has been achieved (Smith promoter region, the coding region for the amino terminal secretion signal and codons for the first 11 amino acids residues of mature invertase with a prochymosin encoding DNA fragment. The fused genes were placed on a plasmid vector and transformed into an appropriate recipient yeast strain for expression. Experiments involving fusion of other heterologous genes with the SUC2 promoter have also been reported (Emr *et al.,* 1983). Since depletion of glucose in the medium would activate the SUC2 promoter, the use of SUC2 promoter signal sequence system for expression and secretion of heterologous proteins such as glucoamylase is very appealing.

Transformation

The most commonly used procedure for transformation of yeast has been the one that was used for the first reported transformation of yeast (Hinnen *et al.,* 1978). Recipient yeast cells are first converted into spheroplast by removing the cell wall enzymatically. The enzymes mostly gluconases are obtained commercially, generally as crude enzyme mixtures. Several different types of enzymes or enzyme systems have been reported in the literature. Hinnen *et al.,* (1978) used *Glusulase* (Endo Laboratory, USA) which is derived from snail gut. A similar enzyme mixture Helicase (Industrie Priologique Francaise, France) was used by *Beggs* (1978), while the enzyme lyticase isolated from *Oerskovia xanthineolytica* by *Scott* and *Schekman* (1980) has been found to be very effective for preparing clean yeast spheroplasts. Kirin Breweries (Japan) and Miles Laboratories (USA) have been the major suppliers of the enzyme mixture known as Zymolyase, isolated from *Arthrobacter luteus*, while Novo has been marketing the enzyme Novozyme 234 which has been found to be effective at low concentrations. With all enzyme treatments pretreatment of cells with thiol reagents such as β-mercaptoethanol and dithiothreitol or sulphur amino acids (*e.g.,* cysteine) is required to obtain high spheroplast-formation frequencies (*Hinnen* and *Meyhack* 1982, Dowhanick *et al.,* 1984). For effective spheroplast formation it is essential that yeast cells be in mid exponential phase of growth. With some enzymes, such as Zymolyase, enzymatic treatment has to be times accurately to avoid over digestion and hence a drop in spheroplast regeneration frequency. This is particularly the case with Zymolyase. Cell wall digestion is generally monitored by check in the osmotic stability of the spheroplasts in distilled water and observing lysis under the microscope.

Osmotically sensitive spheroplast are suspended in a sorbitol buffer (1.0-1.2 mol l^{-1}), treated with Ca^{2+}, and plasmid DNA is added to the mixture. After a short incubation period the polymer polyethylene glycol (20% of PEG 4000) is added. The function of PEG is not clear, but it is believed to be involved in facilitating entry of DNA into the yeast. After an incubation period of about 20 min the PEG is usually removed by centrifugation and washing. The spheroplasts are gently mixed in regeneration agar (selective medium + sorbitol + 3% agar) and plated on regeneration plates. Several dilution are plated, as are controls (no plasmid and no sorbiton– to measure the frequency of spheroplast formation). An incubation period of 3–5 days is generally required for transformants to come up. Frequencies of transformation are strain dependent and can vary from 1×10^2 to 1×10^5 transformants μg^{-1} DNA (*Gingold* 1984, *Struhl* 1983).

A technique similar to that used for *Escherichia coli* transformation was recently described by Ito *et al.*, (1983), and has gained considerable popularity. The method is simple and does not require spheroplasting of the yeast cells. The cells are subjected to treatment with alkali cations (*e.g.*, 0.1 mol l^{-1} lithium acetate) together with PEG, after being mixed with DNA. After an incubation period of 30 min they are subjected to heat shock at 42°C for 5 min followed by cooling and plating on selective media. Transformation frequencies are generally low ($1\text{-}10^3$ μg^{-1} DNA) and strain dependent. While spheroplast transformation can be performed with as little as 1 µg of DNA, whole-cell transformations generally require about 20 µg of DNA, often with heterologous carrier DNA present. Nevertheless, the technique is simple and rapid, and transformants usually come up in less than 3 days. A modification of the technique using only $CaCl_2$ and no lithium acetate or PEG has also been reported, but a much smaller transformation frequency is obtained with such a modification (Limura *et al.*, 1983). Another modification involving treatment of intact cells with only PEG has also been reported, where transformation frequencies of up to 140 μg^{-1} DNA were obtained (Yamakawa *et al.*, 1985).

Suzuki *et al.*, (1983) reported successful transformation of yeast strains with linear DNA plasmids which were circularized *in vivo* by the host yeast ligase system. The chimeric plasmid pJDB219 carrying the yeast LEU2 gene and the 2 µm origin of replication was initially restricted with Hind III, producing four differently sized fragments. The 3.2 kb fragment was found to contain the LEU2 gene as well as the 2 µm origin of replication, but no bacterial DNA (Toh-e *et al.*, 1980). This fragment (pSLel) could transform a leu2 strain to leucine prototrophy at a high frequency and was recovered from the transformants as a circular plasmid, thus confirming *in vivo* ligation. An equimolar mixture of Hind III cut pSLe1 and Hind III linearized YIp5 (a chimeric plasmid carrying the yeast URA3 gene but no origin of replication) was used to transform a yeast strain to leucine and uracil prototrophy. A frequency of transformation of $1\text{-}5 \times 10^3$ transformants μ^{-1} DNA was obtained. Four different size plasmids were seen when the transformants were analysed for plasmid DNA contents. However, each of the plasmids could transform a yeast auxotroph to leucine and uracil prototrophy at a high frequency (10^3 – 10^4 transformants μg^{-1} DNA). All of the four plasmids were found to be circular, again confirming *in vivo* ligation. Such *in vivo* ligation was also observed when the plasmids were restricted with an enzyme (*e.g., HpaI*) that resulted in flush ends rather than cohesive ends. In fact Suzuki *et al.*, (1983) claimed that *in vivo* ligation even occurred when one plasmid was restricted with *Hind* III and the other with *EcoRI*.

20 GENE EXPRESSION IN PROTOZOAN PARASITES

REGULATION OF GENE EXPRESSION IN LEISHMANIA

As Leishmania cycle between the distinct environments provided by the insect vector and mammalian host, they differentiate into morphologically and biochemically distinct stages adapted for survival in the appropriate host. The remarkable ability of Leishmania to thrive in the vastly different environments encountered during its lifecycle is clearly the sign of a sophisticated genetic programme of work. The mechanisms used by these parasites to accomplish stage-specific gene regulation are central to the pathogenicity of Leishmania. One of the laboratory's research interests is to understand at the molecular level how gene expression is regulated during the life cycle of Leishmania.

Leishmania transcription units are polycistronic, however, unlike operons the genes in polycistronic transcription units are not coordinately regulated. Instead, polycistronic pre-mRNAs are converted into monocistronic mRNAs whose steady state levels are individually regulated by post transcriptional mechanisms. Since a key morphological distinction between promastigote and amastigote is the radical change in flagellum size and architecture. We screened for and isolated genes encoding stage specific flagellar proteins. Among the isolated genes were PFR-1 and PFR-2 which encoding major constituents of the paraflagellar rod. Steady state levels of the PFR-1 and PFR-2 mRNAs are tightly controlled by post transcriptional mechanism during the parasite lifecycle in correspondence with the appearance and loss of the paraflagellar rod structure. As such the PFR genes are paradigmatic for regulated gene expression in Leishmania. We are currently investigating the mechanism of the post-transcriptional regulation.

Structure and Function of the Leishmania Paraflagellar Rod

The paraflagellar rod (PFR) is essential for the normal motility of *Leishmania promastigotes* and is essential for survival of the parasites in the insect vector (Santrich *et al.*, in preparation), yet the molecular structure, assembly and function of this unique cytoskeletal structure are not well understood. Although a variety of cytoskeletal structures have been described in protozoans, few exhibit the architectural complexity of the PFR. The PFR extends the length of the axoneme after it emerges from the flagellar pocket and is build up from discrete filaments organized in lattice like arrays. The major protein components of the PFR do not reveal significant homology to other cytoskeletal proteins. This lack of homology together with the elaborate and unusual structure of the PFR suggests a novel type of cytoskeletal filament and organization. The functional importance of the PFR for parasite motility, the potential for utilization of PFR components as an anti-Leishmania vaccine, and the uniqueness of the PFR provides the impetus for developing a better understanding of the structure of the PFR.

21 Gene Organization and Expression in Mitochondria and Chloroplast

INTRODUCTION TO MITOCHONDRIA

Mitochondria are conspicuous, hollow, sac like cell organelles found in all eukaryotic cells except mature red blood corpuscles (RBCs) of mammals. It was firstly observed by *Kallikar. Altman* named it as bioblasts and *Benda* called it Mitochondria.

Structurally, it consists of a fluid filled cavity surroundings not by, a single but by two trilaminar unit membranes—an outer limiting membrane and thicker inner membrane. Outer membrane is smooth and straight. Inner membrane is infolded into cavity forming number of simple/branched plate like septa or tubular ridges, both called cristae. The part of inner membrane covering the cristae bears numerous, regularly spaced, minute, club shaped particles projecting into mitochondrial cavity. These particles are called elementary or f_0f_1-particles. Each particle has three distinct parts—a base piece, embedded in the membrane, a projecting stalk or pedicel and a knob like head at the tip of stalk.

Mitochondria are "energy converting organelles" or powerhouse of cells.

ORGANIZATION OF EXTRANUCLEAR GENOMES

Mitochondria

Mitochondria found in the cytoplasm of all animal and plant cells, are the principal source of energy and also called as powerhouse of cell. The presence of DNA in mitochondria has been shown by a variety of techniques, including electron microscopy, autoradiography. They contain the enzyme of the Kreb's cycle, carry out oxidative phosphorylation and are involved in fatty acid biosynthesis.

Structure

Mitochondrial DNA is a double stranded, supercoiled, circular chromosome devoid of histones and therefore not organized into nucleosomes. The size of the chromosome varies from organism to organism but is constant among mitochondria within any particular species. Interestingly, the chromosomes of higher animal mitochondria are significantly smaller than those of fungi and plants. The mitochondrial chromosome in humans, for example, is about 14000 base pairs, while that from *Drosophila* is 18000 base pairs, that from *Neurospora crassa* is 60000 base pairs, and that from yeast is 75000 base pairs. The mitochondrial genomes of plants are even larger ranging from 250000 to 2 million base pairs (80 to 800 μm), depending on the species. Maize, for example, has mitochondrial genome of 600000 base pairs (200 μm).

Mitochondrial genomes are much more variable in terms of size and topology. The genome sizes of known mitochondrial DNAs vary over a range of several orders of magnitude from ten million to over a billion relative molecular weight. Since the information encoded by the mitochondrial genome appears to be very much the same in all organisms, the size heterogeneity must reflect either the presence of non-coding DNA, or the presence of multiple copies of coding sequence.

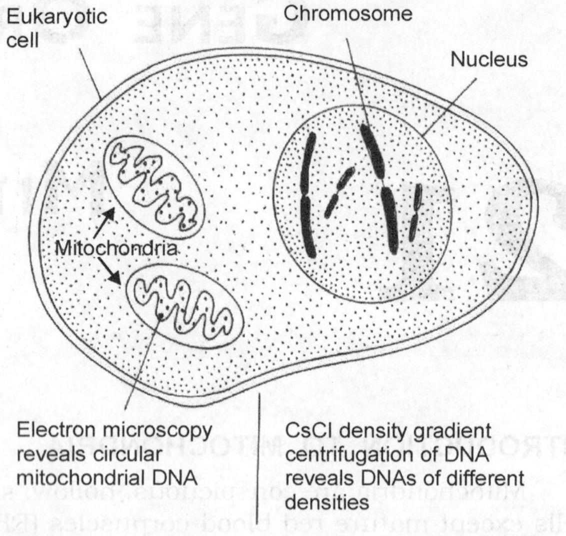

Fig. 21.1. Nuclear and mitochondrial DNA. Mitochondrial DNA can be detected under the electron microscope as small circles and can often be separated from nuclear DNA by CsCl density gradient centrifugation, since the two DNAs differ in buoyant density.

Replication

Replication of mtDNA used mitochondrial DNA polymerases that function totally independently of nuclear DNA polymerases. Moreover, the replications process occurs throughout the cell cycle without any preference for the S phase of the cell cycle when nuclear DNA replicates. Some differences appear in the details of mtDNA replication among eukaryotes, and the displacement loop (D loop) model (deduced from observing animal mitochondria *in vivo*) is useful as a general scheme.

In most animals the two strands of mtDNA have different densities, and they are called the H (heavy) and L (light) strands. The D loop model of replication shows relative asynchrony in DNA replication for the two complementary H and L strands. In the D loop model the synthesis of the new H strand is started in one origin (the L strand origin) and forms a D loop structure (which can be seen by electron microscopy). As the new H strand extends, initiation of synthesis of the new L strand at a second origin (the H strand origin) takes place. Both strands are completed by continuous replication. Lastly, the circular DNAs of each are converted to a supercoiled form with approximately a hundred superhelical twists.

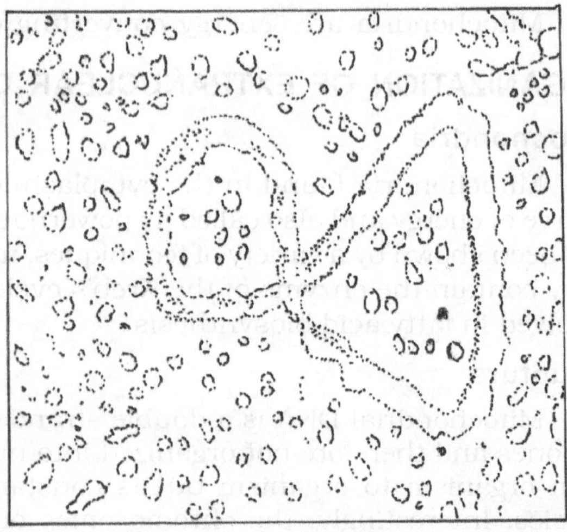

Fig. 21.2. EM of mitochondrial DNA.

As in replication of nuclear DNA, RNA primers are synthesized for initiation. Apparently, the excision of these primers is not precise since interspersed ribonucleotides are found in the mtDNA. The evidence is that mtDNA can be cleaved by treatment with alkali or RNase, both of which are known to cleave at phosphodiester linkages in which a ribose sugar is involved. About ten of these ribonucleotides are found in each mtDNA, with most of them clustered around the replication origins.

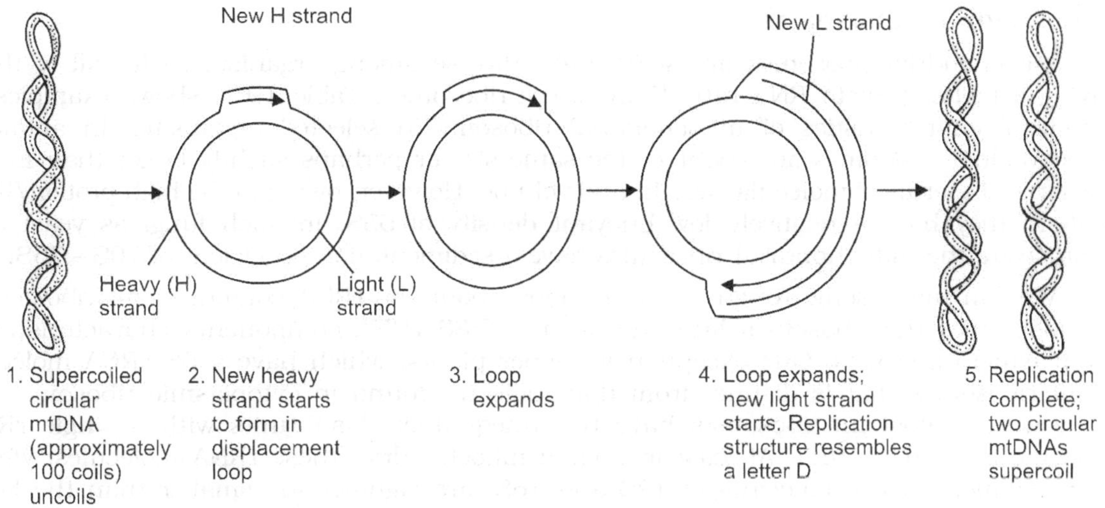

Fig. 21.3. Model for mitochondrial DNA replication that involves the formation of a D loop structure.

Gene Organization of mtDNA

Mitochondrial DNA contains information for a number of mitochondrial components such as tRNAs, rRNAs, and proteins. Mitochondria contain ribosomes that are responsible for all protein synthesis activity that occurs within mitochondria. Messenger RNAs synthesized within the mitochondria remain in the organelle and are translated by mitochondrial ribosomes. The two rRNA molecules found in mitochondria ribosomes are coded for in the mtDNA, while most, if not all, of the ribosomal proteins of mitochondrial ribosomes are coded for in the nuclear genome. In general, mtDNA codes of 2 rRNAs, 22–23 tRNAs, and 10–12 proteins that are part of the inner mitochondrial membrane and are hydrophobic to some degrees. The mtDNA molecule has been studied in a variety of organisms, and since it is relatively small, a great deal of progress has been made in determining the exact organization of its genes.

One approach to mapping mtDNA physically has been through denaturation. The DNA is placed in a medium that causes stretches of AT base pairs to denature while GC-rich regions remain hydrogen-bonded. This selective denaturation leads to the formation of denaturation bubbles that can be visualized by electron microscopy. This physical map shows AT-rich regions of the DNA that can be used as landmarks in other studies.

The structural genes are also found in both strands. Their positions were identified in two ways. One method involved searching the DNA sequence for possible start and stop signals in the same reading frame (and open reading frame or OFR), they were separated by significant lengths and that could code for amino acids. This search was accomplished by using computer programs designed for this purpose.

The second method (used primarily by *G. Attardi and his group*) was to align the 5' and 3' end proximal sequences of mitochondrial mRNAs with the DNA sequence. The mitochondrial mRNAs were isolated and were found to have a poly (A) tail at the 3' end but no 5' cap. As in the case of cytoplasmic mRNAs (*i.e.*, the mRNAs coded for by nuclear genes but found in the cytoplasm), the poly (A) tail is a great aid in isolating these molecules.

Mitochondrial Ribosomes

Mitochondrial ribosomes are structurally diverse among organisms, with all of them having a higher protein RNA ratio than *E. coli* ribosomes. Table 19.1, shows examples of structural characteristics of mitochondrial ribosome in selected organisms. In animals, mitochondrial ribosomes are about of the same size or perhaps slightly larger than *E. coli* ribosomes in terms of molecular weight and volume. However, owing to the high protein/RNA content, they have a relatively low buoyant density of 55S. In such fungi as yeast and *Neurospora*, the mitochondrial ribosomes have a sedimentation coefficient of 70S – 75S.

We can make some rough generalizations about the rRNA content of mt ribosomes. In most cases the ribosomes lack the 5S and 5.8S rRNA components characteristic of cytoplasmic ribosomes. One exception is higher plants, which have a 5S rRNA molecule in mt ribosomes that is distinct from that molecule found in cytoplasmic ribosomes. In other words, most mt-ribosomes have two unequal-sized subunits with a single rRNA molecule in each subunit. At least in animal mitochondria, these rRNA molecules, which have sedimentation coefficients of 12S and 16S, are significantly smaller than the large rRNAs of *E. coli* ribosomes, which have sedimentation coefficients of 16S and 23S.

Table 21.1. Relative Size (in S Values) of Mitochondrial (mt) Ribosomes and Cytoplasmic (cyto) Ribosomes in three Organisms.

Component	*Human (HeLa)*		*Yeast*		*Neurospora*	
	mt	*cyto*	*mt*	*cyto*	*mt*	*cyto*
RIBOSOMES	60S	74S	75S	80S	73S	77S
Large subunit	45S	60S	53S	60S	50S	60S
Small subunit	35S	40S	35S	40S	37S	37S
RIBOSOMAL RNA						
Large ribosomal subunit	16S	28S	21S	26S	25S	26S
Small ribosomal subunit	12S	18S	14S	18S	19S	17S

The number of ribosomal proteins found in mt ribosomes is not as well defined, and estimated numbers vary greatly depending on the method of electrophoresis used to separate the isolated proteins. Yeast and *Neurospora* appear to have 60–70 proteins in the mt ribosomes (significantly more than 54 found in *E. coli* ribosomes), while animal mitochondrial ribosomes have between 70 and 100 ribosomal proteins.

The proteins of mitochondrial ribosomes appear to be totally distinct from the proteins found in cytoplasmic ribosomes, and with one possible exception each in *Neurospora* and yeast, they are encoded by nuclear genes. Hence, they are synthesized on cytoplasmic ribosomes and subsequently transported into the mitochondria. This process points again to the remarkable abilities of cells to direct the product of protein synthesis to the specific sites where they are needed. We know that cytoplasmic ribosome in eukaryotes contain about 65–70 ribosomal proteins, and lower eukaryotes appear to have 60 – 70 ribosomal proteins in mitochondrial ribosomes. Thus, 125 – 140 ribosomal proteins are made on cytoplasmic ribosomes. About half of them subsequently migrate into the nucleus to be used in the assembly of cytoplasmic ribosomes, while the remainder enter the mitochondria, where they are used in the assembly of mitochondrial ribosomes along with the transcripts of the two rRNA genes located on the mitochondrial chromosome. But we have a lot to learn about this migration between cellular compartment.

Mitochondrial Protein Synthesis

Mitochondria have their own protein synthetic machinery, and the process of protein synthesis is, in some ways, analogous to the process of protein synthesis in bacteria. A species of $tRNA^{fMet}$ is present in mitochondria, and it appears to function in the initiation of protein synthesis. The initiation factors (IFs) are also very similar to those in bacteria. In a number of cases mitochondrial and bacterial components are interchangeable in an in *vitro*, protein-synthesising system. For example, *Neurospora crassa* ribosomes are able to recognize, bind and translocate *E.coli* fMet-tRNA fMet in response to AUG, and *E. coli* IFs can substitute for mtIFs in catalyzing the initiation of protein synthesis. Moreover, the tRNA fMet species from Neurospora mitochondria must have a high degree of structural similarity with its *E. coli* counterpart, since once charged with methionine, that amino acid can be formulated by using enzymes isolated from *E. coli.* Another piece of evidence for the similarity between *E. coli* and mt-ribosomes is that the latter are sensitive to most inhibitors of bacterial ribosome function, including *streptomycin, spectinomycin, neomycin,* and *chloramphenicol.* In some cases the degree of sensitivity to these agents, which are mostly antibiotics, is quite different from the situation in *E. coli.* The inhibitors may not work on mt-ribosomes by the classical mechanisms worked out in studies of bacterial ribosomes. Also, mt-ribosomes are generally insensitive to antibiotics or other agents to which cytoplasmic ribosomes are sensitive, such as cycloheximide. By the selective use of antibiotics we can investigate the site of synthesis of proteins found in mitochondria. For example, those mt-proteins that are synthesized in the presence of cycloheximide, and inhibitor of cytoplasmic ribosomes, must be made on mt-ribosomes. Conversely, those mt-proteins made in the presence of chloramphenicol and inhibitor of mt-ribosomes, must be made on cytoplasmic ribosomes. Using this approach, scientists have discovered that of the seven subunits of the mitochondrial enzyme, cytochrome oxidase in yeast mitochondria, four are made in the cytoplasm and three are made in the mitochondria.

Table 21.2. Properties of Nuclear, Mitochondrial, and Chloroplast DNAs

	Nuclear DNA		*Mitochondrial DNA*		*Chloroplast DNA*	
	% GC	*Genome Size (bp for all chromosomes)*	*% GC*	*Genome Size (bp per chromosome)*[a]	*% GC*	*Genome Size (bp per (chromosome)*[b]
ANIMALS						
Human	41	2.75×10^9	41	14×10^3	–	–
Chick	42	1.2×10^9	49	15×10^3	–	–
Sea urchin	39	7.2×10^8	45	13×10^3	–	–
D. melanogaster	42	1.75×10^8	22	18×10^3	–	–
FUNGI						
N. crassa	53	4.3×10^7	42	60×10^3	–	–
S. cerevisiae	44	1.75×10^7	18	75×10^3	–	–
PLANTS						
C. reinhardi	64	1×10^8	71	2×10^5	36	30×10^3
Corn	49	6.6×10^9	48	2.2×10^5	39	13×10^3
Tobacco	38	1.1×10^9	51	2×10^5	47	15×10^3
Sweet pea	35	9.4×10^9	38	9×10^4	46	12×10^3

[a] Each mitochondrion is believed to contain about five copies of the mitochondrial chromosome.
[b] Each chloroplast is believed to contain about twenty copies of the chloroplast chromosome.

Source : Data mainly compiled from "Chloroplast DNA: *Physical and Genetic Studies*" by Ruth Sager and Gladys Schlanger and "*Mitochondrial DNA*" by Margit M.K. Nass, 1976. *In handbook of genetics,* Robert G. King, ed. v. 5, *Molecular genetics,* P. 371 and P. 477. New York : Plenum Press.

INTRODUCTION TO CHLOROPLAST

Chloroplast is a photosynthesis plastid containing chlorophyll and other photosynthetic pigments; it is found in all green plants and algae such as chlorophyll and carotenoids and are the sites of synthesis and storage of food reserves. Chloroplast contain chlorophyll and use light energy to convert CO_2 and water to carbohydrate and O_2.

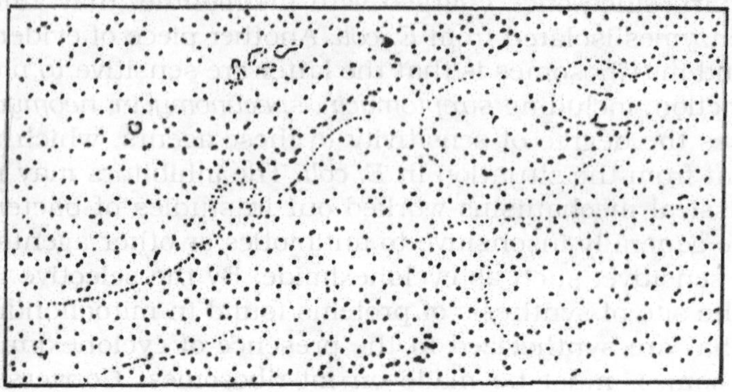

A

Satellite cpDNA band in CsCl gradient

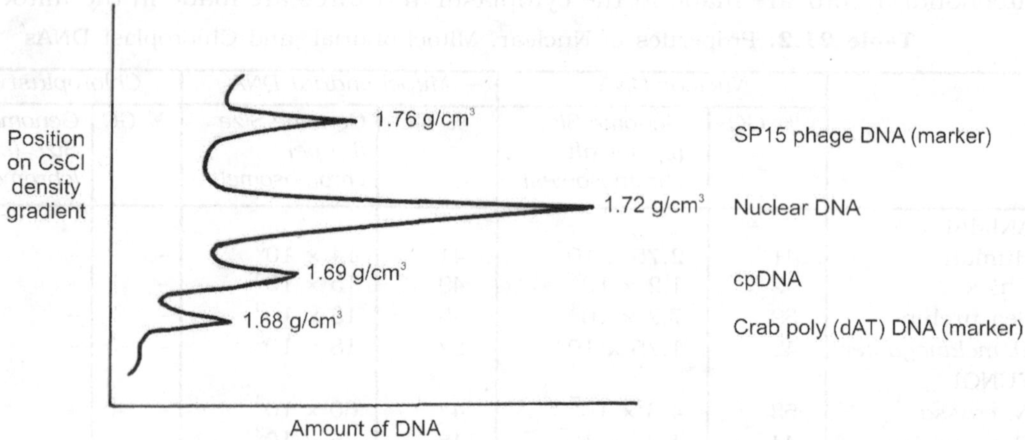

B

Fig. 21.4. Chloroplast (cp) DNA: A. Electron micrograph of a chloroplast DNA molecule. B. Example of the results of a CsCl equilibrium density gradient centrifugation experiment done with coral DNA extracted from the unicellular alga *Chlamydomonas*. The large amount of nuclear DNA results in a broad peak that obscures the mtDNA peak.

In different plant cells shape of chloroplast changes as filamentous, a void, spheroid etc. They are vascular and have a colourless centre. The chloroplast generally measures 2 – 3 µm in thickness and 5 – 10 µm in diameter. The number of chloroplast varies from cell to cell and from species to species. Algae usually have a single huge chloroplast. Cells of higher plants have 20 – 40 chloroplasts.

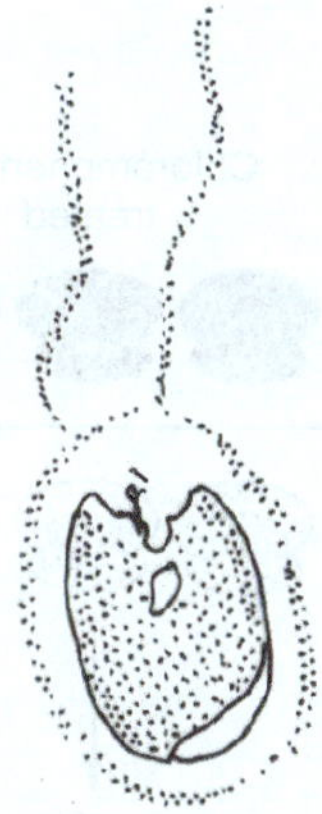

Fig. 21.5. *Chlamydomonas.*

Ultrastructure of chloroplast comprises of following three main components.

1. **Envelope:** Entire chloroplast is bounded by envelope, made of double unit membranes. Isolated membranes of envelope and chloroplast lack chlorophyll pigments but have a yellow colour due to presence of small amounts of carotenoids.
2. **Stroma:** Matrix or stroma fills the most of volume of chloroplast. It is a kind of gel fluid that surrounds the thylakoids (grana). It contains 50% proteins of soluble type. It is the place where CO_2 fixation occurs and where synthesis of sugars, starch, fatty acids and some proteins takes place.
3. **Thylakoids:** These are sac like structures, consisting of flattened and closed vesicles arranged as a membranous network. The outer surface of it is in contact with stroma. Inner surface encloses in intrathylakoid space. These may be stacked like a neat pile of coins, forming grana. There are 40 to 80 grana in matrix of chloroplast. The number of thylakoids per granum may vary from 1 to 50 or more.

CHLOROPLAST GENOME

The chloroplast genome is remarkably constant over wide taxonomic distances among the eukaryotic phototrophs (plant and algae). It is characteristically a covalent closed circular molecule of double stranded DNA. Most chloroplast genome are quite small; a relative molecular weight of about 100 million is common.

Structure

In all cases the DNA is double-stranded, circular, devoid of structural proteins, and supercoiled. In many cases, the GC content of cpDNA differs greatly from that of the nuclear and mitochondrial DNA, which allows us to isolate cpDNA by using CsCl equilibrium density gradient centrifugation. Two discrete bands of DNA are seen (apart from the two marker bands) and indicate the presence of molecular species with different base-pair compositions. Since the cpDNA has a GC content of 36 percent compared with 64 percent for nuclear DNA and 71 percent for mtDNA, the cpDNA produces a satellite band that can be isolated and studied further. The relatively large amount of nuclear DNA results in a fairly broad peak that does not permit resolution of the mtDNA peak.

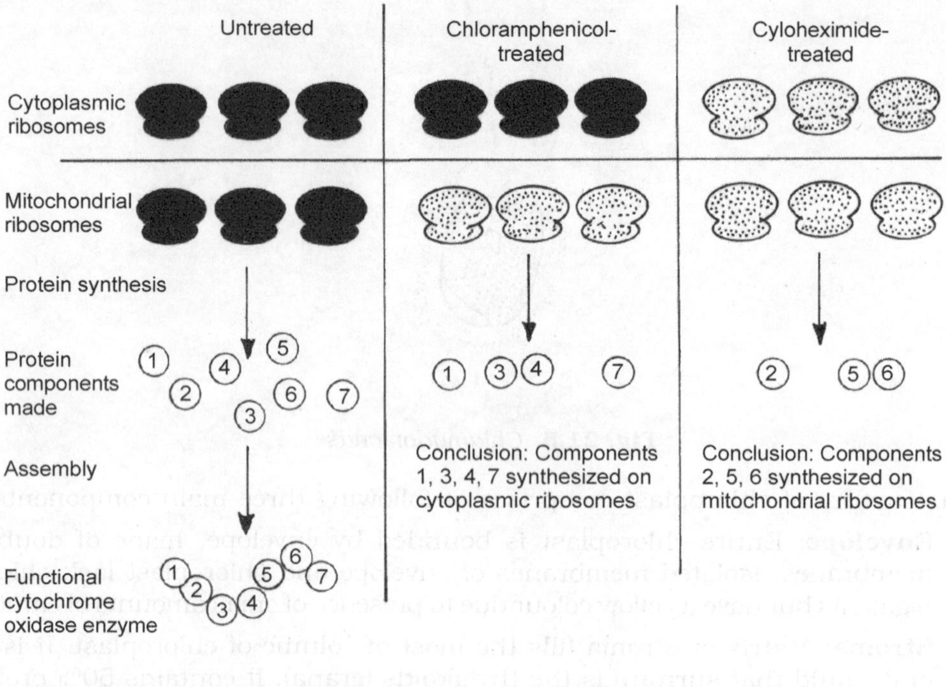

Fig. 21.6. Synthesis of the multi-subunit protein cytochrome oxidase on cytoplasmic and mitochondrial ribosomes.

The number of copies of cpDNA per chloroplast varies from species to species. In all cases there are multiple copies per chloroplast, and these copies are found in nucleoid regions that are also present in multiple copies. For example, leaf cells of the garden beet have 4–8 nucleoids per chloroplast, and about 40 chloroplasts per cell; giving almost 6000 cpDNA molecules per cell. In Chlamydomonas the one chloroplast in a cell generally contains between 500 and 1500 cpDNA molecules.

DNA of Chloroplast

Recently the chloroplasts of the algae and higher plants are found to contain DNA molecules. First of all *Ris and Plants* (1962) reported DNA molecules in the chloroplast of the *Chlamydomonas*. Later on DNA molecule has been reported from the chloroplasts of other algae and higher plants. In general, chloroplasts have a double helical DNA circle with an average length of 45 μm (about 135000 base pairs). The replication of chloroplast DNA has been followed with ^{3}H thymidine. Maps of the location of genes (genetic maps) have been made in several chloroplast DNAs by the help of restriction enzymes. The gene for the large subunit of carboxydismutase enzyme has been fully sequenced and is found to contain 1425 nucleotides.

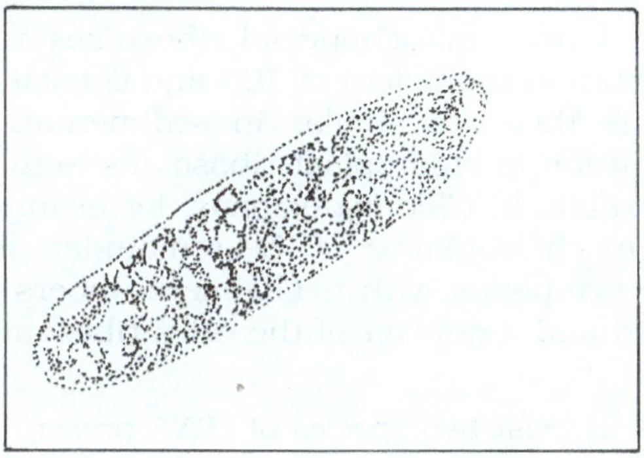

Fig. 21.7. Euglena.

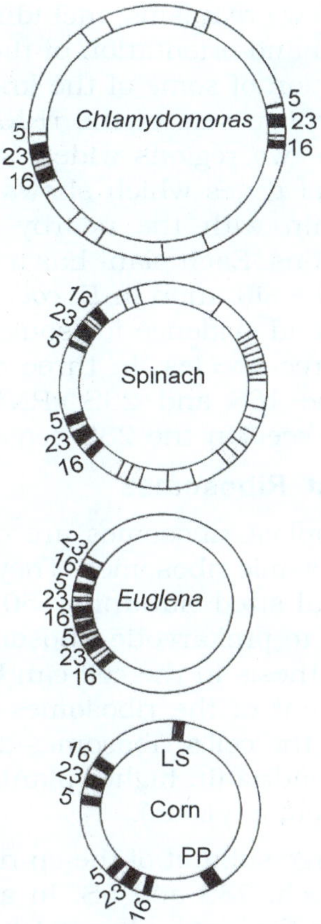

Fig. 21.8. Maps of rRNA genes in the chloroplast genome of *Chlamydomonas, Euglena,* maize and spinach. The colour bars show the positions of tRNA genes. In maize LS and PP are two structural genes.

Replication

The replication of cpDNA is a semiconservative process that probably is similar to the process for mitochondria. It is clear that chloroplast-specific enzymes are used in this replication process. These enzymes are distinct from nuclear replication enzymes and mitochondrial replication enzymes, and most of them are coded for by nuclear genes and transported into the chloroplast. Chloroplasts themselves grow and divide in essentially the same way as mitochondria do.

Gene Organization of cpDNA. The genes encoded by cpDNA have been analyzed by using antibiotics selectively to define chloroplast protein products and by using hybridization of chloroplast rRNAs and tRNAs. From this analysis we come to know that the chloroplast genome contains genes for the cp rRNAs and perhaps for some ribosomal proteins, for a number of tRNAs, and for some cp enzymes and membrane proteins that are needed for electron transport in photosynthesis. However, relatively few cp structural genes have been correlated directly with any chloroplast phenotype. This situation is in stark contrast to our state of knowledge of mitochondrial genomes for a number of organisms. All of the mRNAs from chloroplast genes are translated by chloroplast ribosomes.

One of the chloroplast proteins that has been well characterized is the part of ribulose bis-phosphate decarboxylase, the first enzyme used in the pathway for the fixation of carbon dioxide in the photosynthetic process.

The organization of the rRNA genes in chloroplast genomes has been studied in detail in a few organisms, including *Chlamydomonas, Euglena*, corn, and spinach, shows the location and orientation of the rRNA genes in the cp genome of these plants, as well as the location of some of the known tRNA genes. With the exception of *Euglena*, these plants have two sets of the rRNA genes in each genome and these sets are generally localized in two regions widely separated from one another on the DNA. *Euglena* has three sets of genes which shows the detailed organization of cpRNA genes in the four plants, again with the nearby tRNA genes. The details vary, but there are some generalizations. Each plant has a transcriptional order of 16S-23S-5S, which is extremely similar to the situation in *E. coli*. In addition, all the rRNA genes are in the same strand of cpDNA, and evidence for some systems suggests that a large precursor is processed into the three species. In three of the four plants in the figure, a tRNA gene is found between the 16S and 23S rRNA genes. Intervening sequences are present in some species (as seen in the 23S sequence in *Chlamydomonas*).

Chloroplast Ribosomes

Chloroplast ribosomes are completely distinct from mitochondrial ribosomes and from cytoplasmic ribosomes. They have a sedimentation coefficient of 70S and consist of two unequal-sized subunits, 50S and 30S—thus they are similar in sedimentation coefficients to prokaryotic ribosomes. The contribution of chloroplast ribosomes to total protein synthesis in the cell can be quite considerable. In *Chlamydomonas*, for example 30–35 percent of the ribosomes of the cell are in chloroplasts; of the remainder, the majority of the cell's ribosomes are found in the cytoplasm with negligible numbers in the mitochondria (in higher plants between a third and a quarter of the cell's ribosomes are in the chloroplasts).

The large subunit of the cp-ribosome contains at least two species of rRNA present in one copy each: 23S and 5S. In addition, some plants have other species of small rRNA molecules in the large subunit; for example, 4.5S rRNA is in spinach, and 7S and 3S rRNAs are in *Chlamydomonas*. The small subunit of the cp-ribosome contains one copy of a 16S rRNA. The two large rRNAs, 23S and 16S, are similar in size to the 23S and 16S rRNAs found in *E. coli*. The presence of two small rRNA in the large subunit is akin to the situation in cytoplasmic ribosomes, which contain one copy each of a 5S and a 5.8S rRNA, but is different from the situation in prokaryotic ribosomes, which have only a single small 5S rRNA, and from the situation in mitochondrial ribosomes, which have no small rRNA species.

Chloroplast Protein Synthesis

Protein synthesis in chloroplasts is similar to the process in prokaryotes. Formylmethionyl tRNA is used to initiate all proteins, and the formylation reaction is catalyzed by a transformylase localized in the chloroplast. The chloroplast uses IFs and EFs (elongation factors) that are distinct from those of the cytoplasmic protein synthesis system. In the algae Chlorella the cp elongation factors, EFG and EF-Tu, are not made in the presence of the antibiotic chloramphenicol and are therefore apparently synthesized in the chloroplast itself.

As we saw for mitochondrial ribosomes, cp ribosomes are resistant to cycloheximide, an inhibitor of cytoplasmic ribosomes, but are sensitive to virtually all inhibitors known to block prokaryotic protein synthesis. The ribosomes themselves are found both free and membrane bound in the chloroplast. The free ribosomes appear to make the large subunit of the ribulose bis-phosphate decarboxylase enzyme, while the membrane-bound ribosomes are presumed to make the hydrophobic proteins that are important in photosynthesis. These hydrophobic proteins are located in the membranes of the chloroplast.

Chloroplast proteins may be synthesized by three avenues:

1. By an exclusive chloroplasmic mechanism,
2. By a mechanism involving nuclear genes and chloroplastic ribosomes and
3. By nuclear genes and cytoplasmic ribosomes.

EXERCISE

(Based on Unit–3)

Q.1 What is transcription? Give differences between prokaryotic and eukaryotic transcription.

Q.2 Write short notes on:

(*i*) Structure of prokaryote gene.

(*ii*) Gene expression.

(*iii*) Catabolic repression.

Q.3 What is translation in prokaryotes ?

Q.4 Give a detailed account on structure of eukaryotic genes, eukaryotic transcription and transcription factors.

Q.5 What is translation and gene expression in eukaryotes ?

Q.6 Write an essay on one gene or one enzyme hypothesis.

Q.7 What do you understand by DNA recombination ?

Q.8 What is a molecular mechanism in prokaryotic and eukaryotic organisms ?

Q.9 Write an essay on gene expression in yeast.

Q.10 Write in detail about gene expression in protozoan parasites.

Q.11 What is gene organization and expression in mitochondria ?

Q.12 Write an essay on gene organization and expression in mitochondria and chloroplasts.

Q.13 Write short notes on:

(*i*) Gene expression in yeast.

(*ii*) Prokaryotic transcription.

(*iii*) Lac operon and tryptophan operon.

UNIT-4

22 BASIC MICROBIAL GENETICS, CONJUGATION, TRANSDUCTION AND TRANSFORMATION

INTRODUCTION

Genetic material can be transferred between bacteria by three processes. In each case: (1) Transfer is unidirectional; (2) There is no true zygote; (3) The cells are essentially always haploid. Thus, the genetic transfer processes are different from those in eukaryotes.

TRANSFORMATION

It is one of the process of genetic exchange in which there is a transfer of genetic information by means of extracellular pieces of DNA in bacteria. In DNA, fragments are derived from donor bacteria and may be taken up by another living bacterium, the recipient. If the donor and the recipient bacteria differ genetically, then genetic recombinants are produced by cross-over events involving the DNA fragments and the recipient's chromosome. The genetic recombinants generated by the transformation process are called transformation.

F. Griffth first observed it in 1928. When bacteria lyse, they release considerable amounts of DNA into the surrounding environment. These fragments may be relatively large and contain several genes. If a fragment contacts a competent cell, one able to take up DNA and be transformed, it can be bound to the cell and taken inside. The transformation frequency of very competent cells is around 10^{-3} for most general when an excess of DNA is used. That is, about one cell in every thousand will take up and integrate the gene. Competency is a complex phenomenon and is dependent on several conditions. Bacteria need to be in certain stage of growth; for example, *S. pneumoniae* become competent during the exponential phase when the population reaches about 10^7 to 10^8 cells per ml. When a population becomes competent, bacteria such as *pneumococci* secrete a small protein called competence factor that stimulates the production of 8 to 10 new proteins required for transformation. Natural transformation has been discovered so far only in certain gram-positive and gram-negative genera: *Streptococcus. Bacillus, Thermoactinomyces, Haemophilus. Neisseria, Morqxella, Acinetobacter, Azotobacter,* and *Pseudomonas.* Other genera also may be capable of transformation. Gene transfer by this process occurs in soil and marine environments and may be an important route of genetic exchange in nature.

Cells that are in a state in which they can be transformed by DNA in their environment are said to be competent. The mechanism of transformation protein intensively studied in *S. pneumoniae.* A competent cell binds a double stranded DNA fragment if the fragment is moderately large; the process is random, and donor fragments compete with each other. The DNA then is cleaved by endonucleases to double stranded fragments about 5 to 15 kilobases in size. DNA also required energy expenditure. One strand is hydrolyzed by an envelope associated exonuclease during uptake; the other strand associates with proteins and moves through the plasma membrane.

The mechanism of transformation exhibit by *Streptococcus pneumoniae* was considered to be typical of all naturally transformable Gram-positive bacteria and that exhibit by *Haemophillus influenze* to be typical of all naturally transformable Gram-negative bacteria.

The Haemophillus system of transformation differ from the streptococcus system in two additional ways.

1. Only homologous DNA is bound to end taken into cell with any significant efficiency.
2. Transforming DNA enters the recipient in double-stranded form, remaining that way up to the time of its integration into the endogenose.

Plasmids enter such cells as intact double stranded molecules. Double-stranded linear molecules also enter but in many cases these are rapidly degraded by intracellular nucleases. In the case of *E.coli,* artificial transformation by linear DNA is possible if a mutant strain that lacks two DNA-cleaving nucleases is used as a recepients.

Artificial transformation is carried out in the laboratory by a variety of techniques, including treatment of the cells with calcium chloride, which renders their membranes more permeable to DNA. This approach succeeds even with species that are not naturally competent, such as *E. coli.* Relatively high concentrations of DNA, higher than would normally be present in nature are used to increase transformation frequency. When linear DNA fragments are to be used in transformation, *E. coli* usually is rendered deficient in one or more exonuclease activities to protect the transforming fragments.

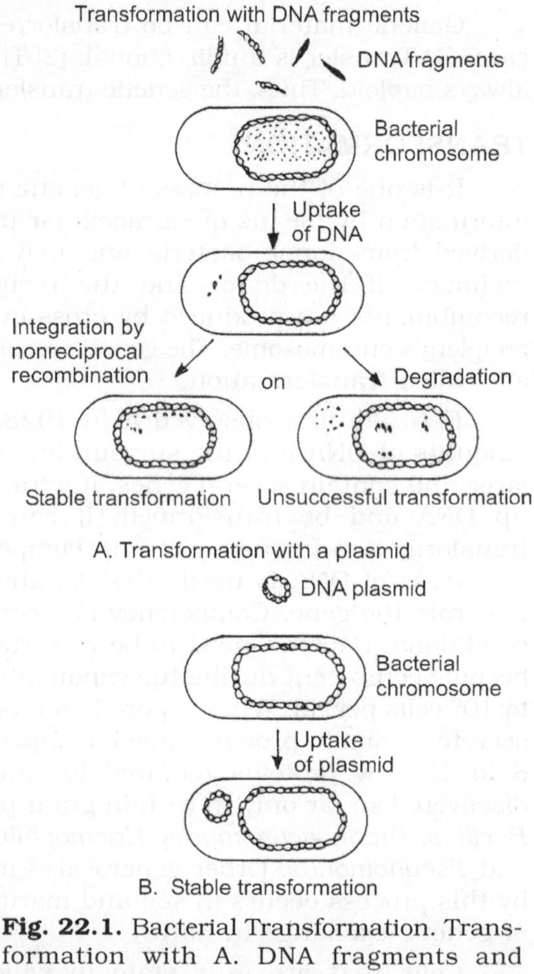

Fig. 22.1. Bacterial Transformation. Transformation with A. DNA fragments and B. plasmids. Transformation with a plasmid often is induced artificially in the laboratory. The transforming DNA is in red and integration is at a homologous region of the genome.

TRANSDUCTION

Transduction refers to the transfer of a bacterial gene from one bacterium to another by a phage; a phage carrying host as well as its own genes is called transducing phage.

Phages that occassionally produce transducing particles and therefore are capable of effecting transductional genetic exchange are relatively commonly encountered in the microbial world.

Bacterial viruses or bacteriophages participate in third mode of bacterial gene transfer. The morphology and life cycle of bacteriophages is not discussed in detail. Nevertheless, it is nececcsary to briefly describe the life cycle here as background for consideration of the bacteriophage's role in gene transfer.

After infecting the host cell, a bacteriophage (phage for short) often takes control and forces the host to make many copies of the virus. Eventually the host bacterium bursts or lyses and releases new phages. This reproductive cycle is called a lytic cycle because it ends in lysis of the host. The cycle has four phases (Fig. 22.2). First, the virus particle attaches to a specific receptor site on the bacterial surface. The genetic material, which is often double-stranded DNA, then enters the cell. After absorption and penetration, the virus chromosome forces the bacterium to make virus nucleic acids and proteins. The third stage begins after the synthesis of virus components. Phages are assembled from these components. The assembly process may be complex but in all cases phage nucleic acid is packed within the viruses protein coat. Finally, the mature viruses are released by cell lysis. Bacterial viruses that reproduce using a lytic cycle often are called virulent bacteriophages because they destroy the host cell. Many DNA phages, such as the lambda phage, are also capable of a different relationship with their host (See Fig. 22.3B). After adsorption and penetration, the viral genome does not take control of its host and destroy it while producing new phages. Instead the genome remains within the host cell and is reproduced along with the bacterial chromosome.

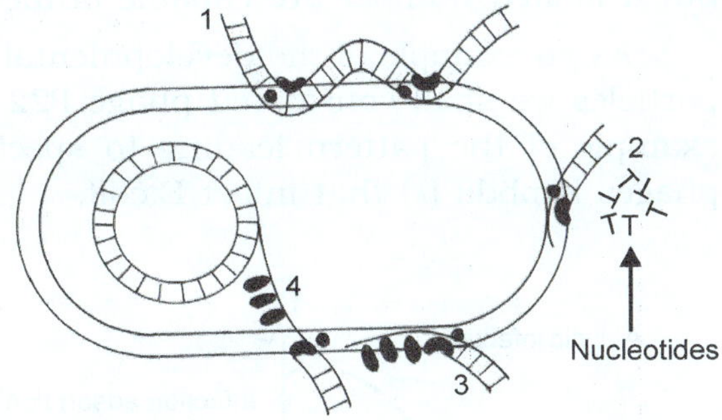

Fig. 22.2. The mechanism of transformation. (1) A long double-stranded DNA molecule binds to the surface with the aid of a DNA-binding protein (⬬) and is nicked by a nuclease (⬬). (2) One strand is degraded by the nuclease. (3) The undegraded strand associates with a competence-specific protein (⬬). (4) The single strand enters the cell and is integrated into the host chromosome in place of the homologous region of the host DNA.

A clone of infected cells arises and may grow for long periods while appearing perfectly normal. Each of these infected bacteria can produce phages and lyse under appropriate environmental conditions. This relationship between the phage and its host is called lysogeny. Bacteria that can produce phage particles under some conditions are said to be lysogens or lysogenic, and phages able to establish this relationship are temperate phages. The latent form of the virus genome that remains within the host without destroying, it is called the prophage. The prophage usually is integrated into the bacterial genome (Fig. 22.3B). Sometimes phage reproduction is triggered in a lysogenized culture by exposure to UV radiation or other factors. The lysogens are then destroyed and new phages released. This phenomenon is called induction.

Generalized Transduction

Two types of transducing particles and therefore two type of transduction exist. One of these is termed generalized or nonspecialized transduction because it mediate the exchange of any bacterial gene. The other is termed restricted or specialized transduction because it mediates the exchange of only limited number of specific genes. Most of the transducing phages are capable of mediating only one of these two types of transduction but a limited number are capable of mediating both.

As an example of the developmental pattern leading to the generalized transducing particles we shall considered phage P22 that infects *Salmonellla typhimurium*. Another example of the pattern leading to specialized transducing particle we shall consider phages lambda (λ) that infect E.coli.

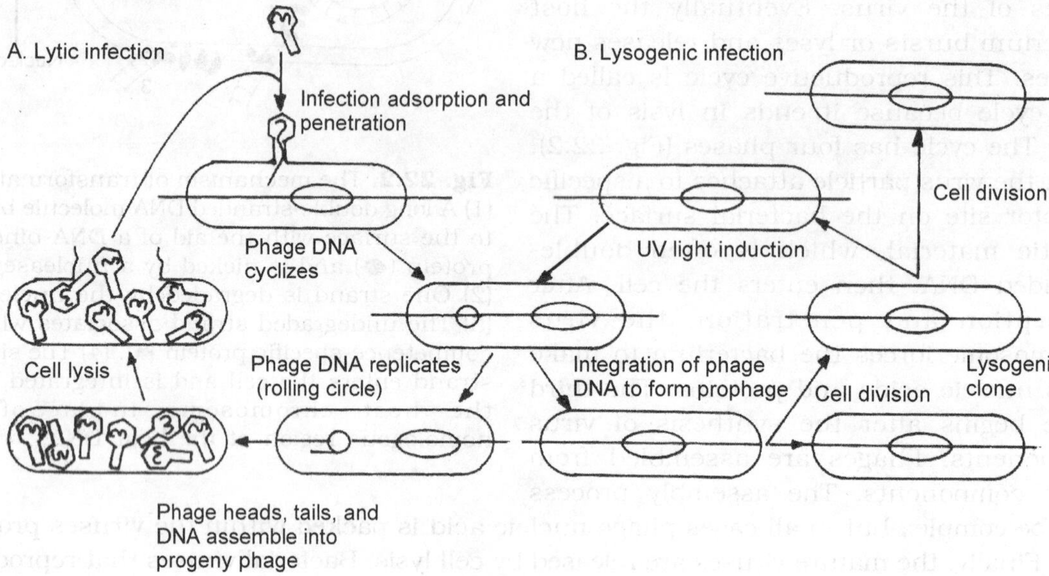

Fig. 22.3. Lytic versus Lysogems infection by Phage Lambda. A. Lytic infection. B. Lysogenic infection.

The quantity of bacterial DNA carried depends primarily on the size of the capsid. The P22 phage of *Salmonella typhimurium* usually carries about 1% of the bacterial genome : the P1 phage of *E. coli* and a variety of gram-negative bacteria carries about 2.0 to 2.5% of the genome. The resulting virus particle often injects the DNA into another bacterial cell but does not initiate a lytic cycle. This phage is known as a generalized transducing particle or phage and is simply a carrier of genetic information from the original bacterium to another cell. As in transformation, once the DNA has been injected, it must be incorporated into the recipient cell's chromosome to preserve the transferred genes. The DNA remains double stranded during transfer, and both strands are integrated into the endogenote's genome. About 70 to 90% of the transferred DNA is not integrated but often is able to survive and express itself.

The phage all sites and bacterial all sites are similar and can complex with each other, although they are not identical. The all site for lambda is next to the gal and bio genes on the *E. coli* chromosome: consequently specialized transducing lambda phages most often carry these bacterial genes. The lysate, or product of cell lysis, resulting from the induction of lysogenized *E. coli* contains normal phage and a few defective transducing particles. These particles are called lambda gal because they carry the galactose utilization genes (Fig. 22.4).

Because these lysates contain only a few transducing particles, they often are called low-frequency transduction lysates (LFT lysates). Whereas the normal phage has a complete all site, defective transducing particles have a nonfunctional hybrid integration site that is part bacterial and part phage in origin. Integration of the defective phage chromosome does not readily take place. Transducing phages also may have lost some genes essential for reproduction. Stable transductants can arise from recombination between the phage and the bacterial chromosome because of cross-overs on both sides of the gal site.

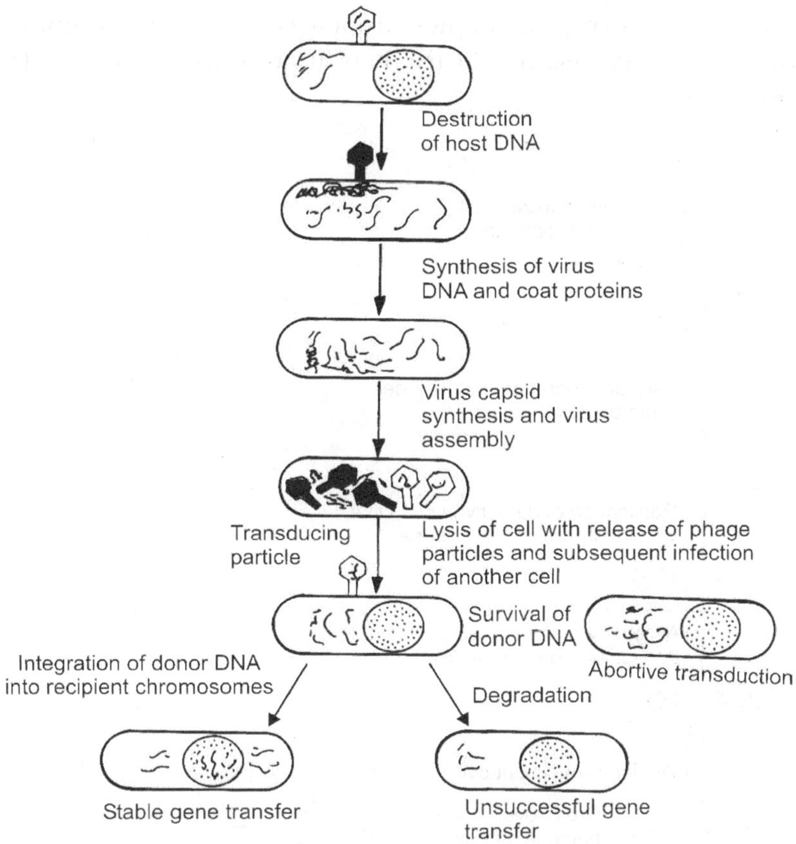

Fig. 22.4. Generalized transduction by bacteriophage.

The normal phage will integrate yielding two bacterial phage hybrid all sites where the defective lambda *dgal* phage can be inserted. It also supplies the genes missing in the defective phage. The normal phage in this instance is termed the helper phage because it aids integration and reproduction of the defective phage. These transductants are unstable because the prophages can be induced to excise by agents such as UV radiation. Excision, however, produces a lysate containing a fairly equal mixture of defective lambda *dgal* phage and normal helper phage. Because it is very effective in transduction, the lysate is called a high frequency transduction lysate (HFT lysate). Reinfection of bacteria with this mixture will result in the generation of considerably more transductants. LFT lysate and those produced by generalized transduction have one transducting particle in 10^5 or 10^6 phages. HFT lysates contain transducting particles with a frequency of about 0.1 to 0.5.

Bacterial Conjugation

In 1946, *Joshua Seclerberg* and *E.L. Tatum* performed an elegant experiment of genetic exchange. They mixed two auxotrophic strains, incubated the culture for several hours in nutrient medium and then plated in one minimal medium. They used double and triple auxotrophs on the assumption that two or three reversion would not often occur Simultaneously.

For example, one strain required biotin (Bio$^-$), phenylalanine (Phe$^-$), and cysteine (Cys$^-$) for growth and another needed threonine (Thr$^-$), leucine (Leu$^-$), and thiamine (Thi$^-$). Recombinant prototrophic copies appeared on the minimal medium after incubation. Thus, the chromosomes of the two auxotrophs were able to associate and undergo recombination.

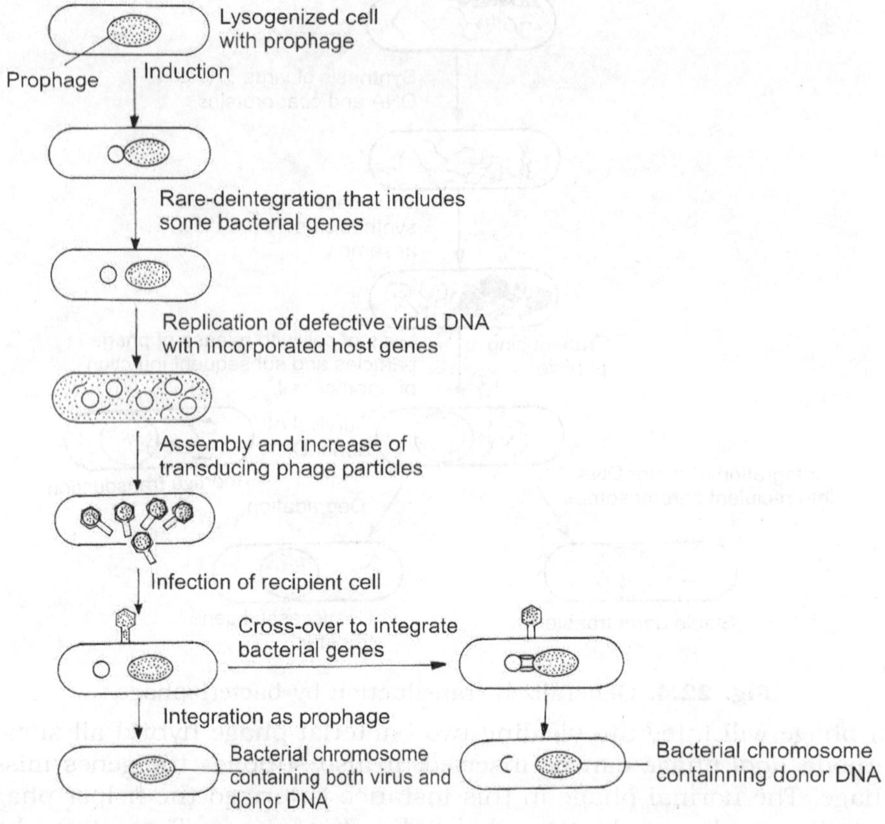

Fig. 22.5. Specialized transduction by a temperate bacteriophage. Recombination can produce two types of transductants.

F$^+$ × F$^-$ Mating

In 1952, *William Hayes* demonstrated that the gene transfer observed by Lederberg and Tatum was polar. That is there were definite donor (F$^+$) and recipient (F$^-$) strains, and gene transfer was nonreciprocal. He also found that in F$^+$ × F$^-$ mating the progeny were only rarely changed with regard to auxotrophy (that is, bacterial genes were not often transferred), but F^- strains frequently became F^+.

The F^+ strain contains an extra chromosomal F factor carrying the genes for pilus formation and plasmid transfer. During $F^+ \times F^-$ mating or conjugation, the F factor is replicated by the rolling mechanism, and a copy moves to the recipient (Fig. 22.8). The entering strand is copied to produce double stranded DNA. Because bacterial chromosome genes are rarely transferred with the independent F factor, the recombination frequency is low. It is still not completely clear how the plasmid moves between bacteria. The sex pilus or F pilus joins the donor and recipient and may contract to draw them together. The channel for DNA transfer could be either the hollow F pilus or a special conjugation bridge formed upon contact.

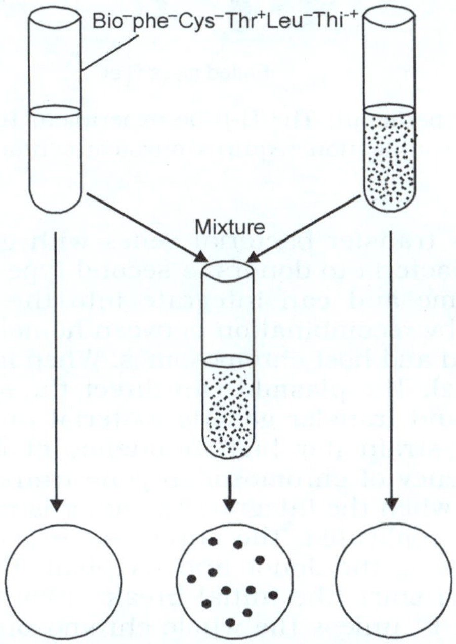

Fig. 22.6. Evidence for bacterial conjugation. *Lederberg* and *Tatum's* demonstration of genetic recombination using triple auxotrophs.

Although most research on plasmids and conjugation has been done using *E. coli* and other gram negative bacteria, self trans-missible plasmids are present in gram-positive bacterial genera such as *Bacillus, Streptococcus, Enterococcus, Staphylococcus,* and *Streptomyces*. Much less is known about these systems. It appears that fewer transfer genes are involved, possibly because a sex pilus does not seem to be required for plasmid transfer. For example *Enterococcus faecalis* recipient cells release short peptide chemical sign as that activate transfer genes in donor cells containing the proper plasmid. Donor and recipient cells directly adhere to one another through special plasmid encoded proteins released by the activated donor cell. Plasmid transfer then occurs.

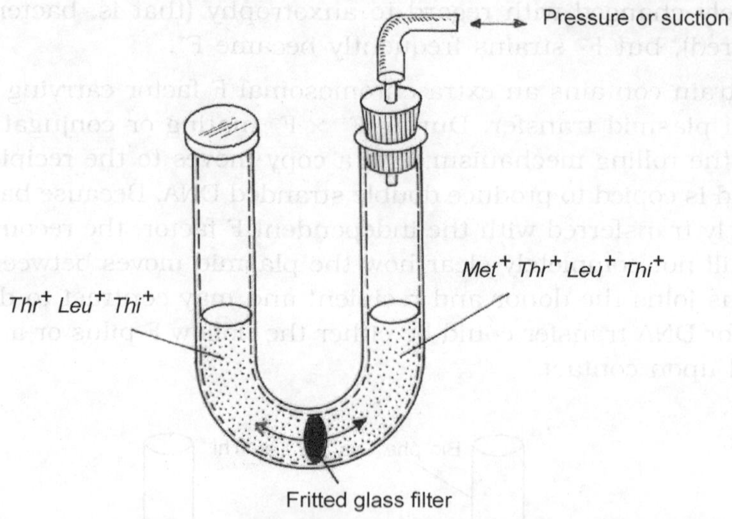

Fig. 22.7. The U-tube experiment. The U-tube experiment used to show that genetic recombination by conjugation requires physical contact among bacteria.

Hfr Conjugation

Certain donor strains transfer bacterial genes with great efficiency and do not usually change recipient bacteria to donors, a second type of conjugation must exist. The F factor is an episome and can integrate into the bacterial chromosome at several different locations by recombination between homologous insertion sequences present on both the plasmid and host chromosomes. When integrated the F plasmodia's traoperon is still functional. The plasmid can direct the synthesis of pili, carry out rolling circle replication, and transfer genetic material to an F^- recipient cell. Such a donor is called an Hfr strain (for high frequency of recombination) because it exhibits a very high efficiency of chromosomal gene transfer in comparison with F^+ cells. DNA transfer begins when the integrated F factor is nicked at its site of transfer origin (Fig. 22.9). As it is replicated, the chromosome moves through the pilus or conjugation bridge connecting the donor and recipient. Because only part of the F factor is transferred at the start (the initial break is within the F plasmid), the F^- recipient does not become F^+ unless the whole chromosome is transferred. Transfer is standardized at 100 minutes in *E. coli* and the connection usually breaks before this process is finished. Thus, a complete F factor usually is not transferred, and the recipient remains F^-.

F′ Conjugation

The F plasmid is an episome, it can leave the bacterial chromosome. Sometimes during this process the plasmid makes an error in excision and picks up a portion of the chromosomal material to form an F' plasmid. It is not unusual to observe the inclusion of one or more genes in excised F plasmids. The F' cell retains all of its genes, although some of them are on the plasmid and still mates only with an F^- recipient. $F' \times F^-$ conjugation is virtually identical with $F^+ \times F^-$ mating. Once again, the plasmid is transferred, but usually bacterial genes on the chromosome are not (Fig. 22.9). Bacterial genes on the F' plasmid are transferred with it and need not be incorporated into the recipient chromosome to be expressed. The recipient becomes F' and is a partially diploid merozygote since it has two sets of the genes carried by the plasmid. In this way specific bacterial genes may spread rapidly throughout a bacterial population. Such transfer of bacterial genes is often called sexduction.

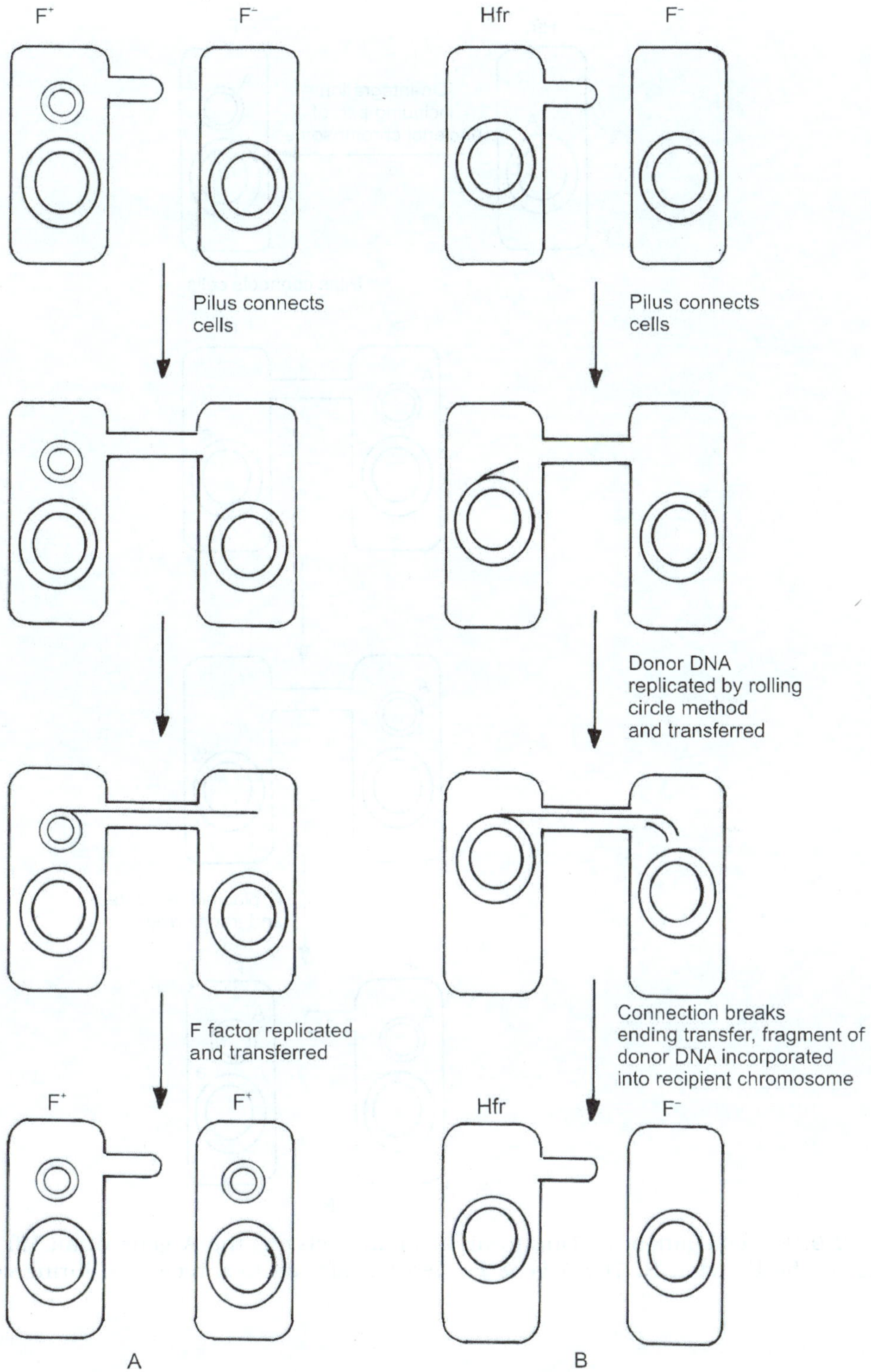

Fig. 22.8. The mechanism of bacterial conjugation. A. F$^+$ × F mating. B. Hfr × F mating.

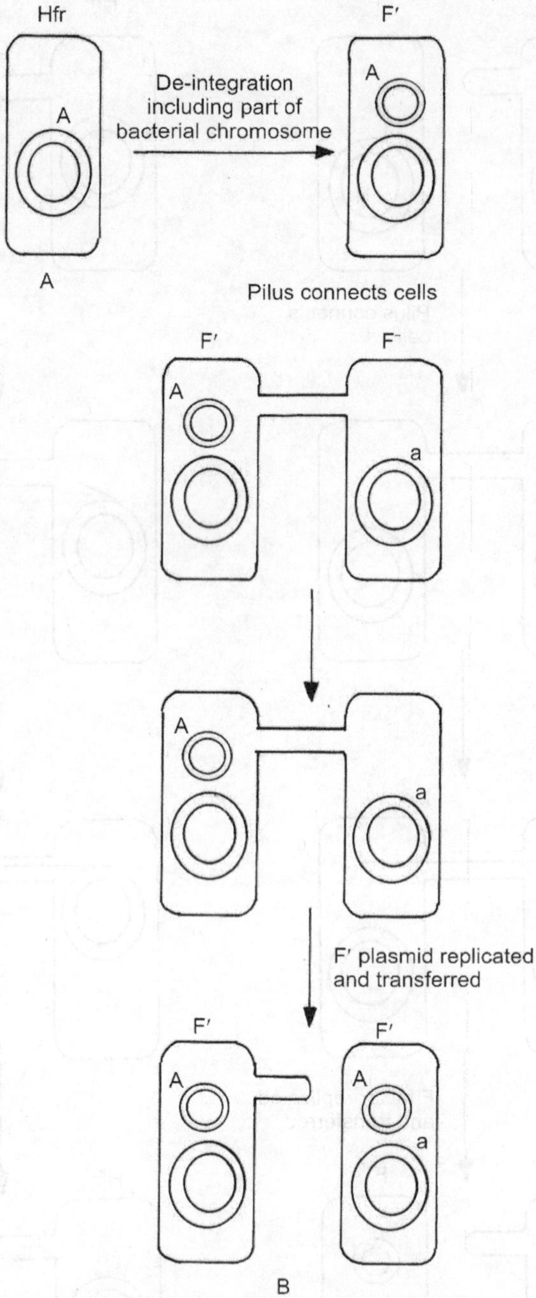

Fig. 22.9. F′ Conjugation. A. Due to an error in excision, the A gene of an Hfr cell is picked up by the F factor. B. The A gene is then transferred to a recipient during conjugation.

23 Insertion Elements and Transposons

PROKARYOTIC TRANSPOSABLE GENETIC ELEMENTS

There are four types of transposable genetic elements in prokaryotes—insertion sequence (IS) elements, transposons (In), plasmids and certain temperature bacteriophages. In this chapter, we will discuss about its insertion elements and transposons.

INSERTION SEQUENCE ELEMENTS OR IS ELEMENTS

An IS element is the simplest transposable genetic element found in prokaryotes. An IS element contains genes required for inserting the DNA segment into a chromosome and for mobilizing the element to different locations.

Discovery of IS Elements

The IS elements were first identified in *E. coli* as a result of their effects on the expression of a set of three genes whose products are needed to metabolize the sugar galatose when that sugar is supplied as carbon source. A certain set of mutations affecting the expression of these genes did not have properties typical of the classes of gene mutants. After analysis, it revealed that the mutant phenotypes resulted from the insertion of an approximately 800 bp DNA segment into a gene. This particular DNA segment is now called insertion sequences 1 or IS1.

At present geneticists think that IS1 is one of a family of genetic elements capable of integrating into a chromosome at location with which it has no homology. This event is an example of a transposition event. Integration can occur independently of normal DNA recombination activity. If the integration of an IS element places it within a gene, then that gene is usually inactivated. For example, the 1327-bp IS2 element decreases gene expression if it integrates into the chromosome in one orientation but it increases gene expression if it integrates in the opposite orientation.

An IS element also has the ability to excise itself from the chromosome. If the excision event is perfect, then a wild type revertant results. Occasionally, however, excision is imprecise, so some or all parts of the genes surrounding the IS element may be deleted and transposed with the element. Due to insertion, inversion also takes place.

Table 23.1. Prokaryote Insertion Sequence

Insertion Sequence	*Normal Occurrence in E. coli*	*Length (bp)*	*Inverted Repeating Segments (bp)*	*Comments*
IS1	5–8 copies on chromosomes	768	18/23	full sequence known for two different examples
IS2	5 on chromosomes, 1 on F factor	1327	32/41	full sequence known
IS3	5 on chromosomes, 2 on F factor	1400	32/38	sequence of 700 bp known
IS4	1 or 2 copies on chromosomes	1400	16/18	sequence of ends known
IS5	Unknown	1250	Short	no sequence information
Gamma-delta	1 on F factor, one or more on chromosome	5700	35	sequence of ends known

TRANSPOSONS

Another complex type of prokaryotic transposable genetic element is the transposon (Tn). It is a mobile DNA segment that, like IS elements, contain genes for the insertion of the DNA segment into the chromosome and for the mobilization of the element to other location on the chromosome; but unlike IS elements, it also contains genes of identifiable function. Tn's often end in long (800–1500 bp) direct or inverted reports, which are IS elements themselves. In fact, many Tn's are mobile in the genome because they have IS elements (called modular) at their ends. The properties of a number of transposons are summarized in Table 23.2.

Table 23.2. Prokaryotic Transposons

Transposone	*Gene .... (s)*	*Length (bp)*	*Inverted Repeat (bp)*	*Comments*
Tn3	Resistance to ampicillin	4957	38	Full sequence of Tn3 known
Tn4	Resistance to ampicillin, streptomycin	20500	short	Contains Tn3
Tn5	Resistance to kanomycin	5700	1500	Sequence of ends known
Tn6	Resistance to kanomycin	4200	Not detectable with electron microscopy	—
Tn7	Resistance to trimethoprim, streptomycin	14000	Not detectable	—
Tn9	Resistance to chloramphenicol	2638	18/123	Full sequence known
Tn10	Resistance to tetracycline	9300	1400	Sequence of ends known
Tn1681	Heat stable enterotoxin	2088	708	—

24 MITOCHONDRIAL AND CHLOROPLAST GENETIC SYSTEM

INTRODUCTION

Mitochondria (Gr. *mito* - thread, *chondrion* = granule) are filamentous or granular cytoplasmic organelles of all aerobic cells of higher animals and plants and also of certain microorganism including *algae*, *protozoa* and *fungi*. They are found in all eukaryotic cells except RBCs of mammals. The mitochondrion was first discovered by **Kolliker** in 1850 as granular structure in the striated muscles.

MITOCHONDRIAL GENETIC SYSTEM

Mitochondrial DNA

Mitochondrial DNA (mtDNA) is relatively small, simple, double stranded and an exception for the DNA of some algae and protozoans, it is circular. The size of mitochondrial genome is very much larger in plants than in animals. Thus, mtDNA varies in length from about 5 μm in most animal species to 30 μm or so higher in plants. It is localized in matrix and is probably attached to the inner membrane at the point where DNA duplication starts.

Mitochondrial Ribosomes

Mitochondria contain ribosomes (called mitoribosomes) and polyribosomes. In yeast and *Neurospora*, ribosomes have been described to a 70S class, similar to that of bacteria. In mammalian cells, however mitoribosomes are smaller and have total sedimentation coefficient of 55S with subunits of 35S and 25S (Altardi *et al.*, 1971). In mitochondria, ribosomes appear to be tightly associated with the inner membrane.

Mitochondrial Protein Synthesis

As already described, mitochondria can synthesize about 12 different proteins, which are incorporated into the inner mitochondrial membrane. Mitoribosomes are made up of the following proteins: three largest subunit of cytochrome oxidase, one protein subunit of the cytochrome b-c complex, four subunit of ATPase and a few hydrophobic proteins. The mitochondrial protein synthesis is inhibited by chloromphenicol, while synthesis in cytosol (cytoplasmic matrix) is not affected by this drug.

Impart Mechanism of Mitochondrial Protein: Most mitochondrial proteins are coded by nuclear genes and are synthesized for free ribosome in the cytosol. Its impart of their polypeptide involve similar mechanism both in mitochondria and chloroplast. The transport processes involved have been most extensively studied in mitochondria specially in year 1988, *Attardi* and *Schatz*. A protein translocated into the mitochondrial matrix space by passing through sites of adhesion between the outer and inner membrane called contact site. Only protein that contains a specific signal peptide are translocated into mitochondria and chloroplast. Transport to the inner mitochondrial membrane can occur as a second step if a hydrophobic peptide is also present in the impaired proteins. This second signal peptide is unmarked when the first signal peptide is cleared.

CHLOROPLAST INTRODUCTION

Chloroplasts were described as early as 17th century by *Nehemiah Grew* and *Antonie van Leeuwenhoek*. Chloroplasts are the most common plastids of many plant cells and they also perform the photosynthetic activity of greatest biological importance. By the process of photosynthesis the chloroplast synthesize the carbohydrate which contain energy in the form of chemical energy. This chemical energy is utilized by all living beings to perform various life activities.

Chloroplast Genetic System

Like the mitochondria the chloroplast have their own DNA, RNA and protein synthesis machinery and are simultaneous in nature.

Chloroplast DNA: Firstly *Rizs* and *Plant* (1962) noticed DNA molecule in the chloroplast in *Chlamydomonas*. In general, chloroplast have double helical DNA circle with an average length of 45 μm (about 135000). The replication of chloroplast DNA H–Thymidine. The gene for the last subunit of carboxydimutase enzyme has been fully sequenced and is found to contain 1425 nucleotides.

Ribosomes of Chloroplast

The chloroplast contain the ribosomes, which are smaller than the cytoplasmic ribosomes. The ribosomes of the chloroplast are of 70S type and resemble with the bacterial ribosomes. These ribosomes consist of 2 ribosomal RNAs, 23S rRNA and 16S rRNA. *Bager* and *Hamilton*, (1967) also separated polyribosomes or polysomes from the chloroplast.

Protein Synthesis

DNA`of chloroplast codes for certain proteins of thylakoid membranes. The synthesis of other chloroplast component as chlorophyll, carotenoids, lipids and photosynthetic and starch synthesizing enzymes, is controlled by nuclear gene. This mode of synthesis of proteins of chloroplast indicated towards their semiautonomas or symbiotic nature. Synthesis of carboxydimutase (CDNA) presents a good case of cooperative action of two genetic systems (*i.e.*, chloroplast and nuclear genetic system). CDNA comprises of 16 sub-units, 8 subunits of high molecular weight (55000 dalton), and 8 subunits of much smaller molecular weight (14000 daltons). The large subunit coded by genes present in chloroplastic DNA, while the subunit is produced by nuclear genes. The small unit is P 20 synthesized as precursor weighing 20000 daltons a free ribosomes, it their enters post transitionally into the stroma to be cleaved to attain its final size. Finally chloroplast protein may be synthesized by three avenues:

1. By the exclusive chloroplastic mechanism.
2. By mechanism involving nuclear gene and chloroplastic ribosomes.
3. By nuclear genes and cytoplasmic ribosome.

Protein transport into chloroplast resembles transport into mitochondria. In many respects, both occur post transitionally, both require energy and both are utilized by hydrophilic amino terminal signal peptide that are removed after use. However, there is at least one important difference that while mitochondria exploit the electrochemical gradient across their inner membrane to help drive the transport, chloroplast appears to employ. Only ATP hydrolysis to impart across their double membrane outer envelope.

25 Extra Chromosomal Inheritance

INTRODUCTION

Genes of the nuclear chromosomes play different roles in inheritance, development, mutation and cellular metabolism of the organism in which they occur. Though the genes of nuclear chromosomes have a significant role in inheritance of almost all traits from generations to generations, but they altogether cannot be considered as the rose vehicle of inheritance, because certain experimental evidences suggest the occurrence of certain entranuclear genes or DNA molecules in the cytoplasm of many prokaryotic and eukaryotic cells.

For example, bacterial cells such as *E. coli* possess a single main chromosome in the nucleoid and often extra DNA elements called plasmids in the cytoplasm. These cytoplasmic extranuclear genes or DNA molecules of plasmids, mitochondria, endosymbionts, chloroplasts have a characteristic pattern of inheritance which does not resemble with that of genes of nuclear chromosomes and is known by different terms such as non-chromosomal, natural, non-mendelian, uniparental, cytoplasmic, extra chromosomal and extranuclear inheritance.

EVIDENCES FOR CYTOPLASMIC FACTORS

Traits with extranuclear basis are identified by the accumulated evidence from a number of diagnostic criteria which are as follows:

1. Because the female gamete contributes almost all of the cytoplasm to the zygote and male gamete (sperm or pollen) contributes only a nucleus, and inheritance pattern that differs between reciprocal crosses suggests a cytoplasmic involvement. This is clear the basis for uniparental or maternal inheritance where the progeny always resemble one parent, most commonly the female parent (*e.g.*, shell coiling and *Limnaea peregra*).
2. Differences in reciprocal crosses which cannot be attributed to sex-linkage or some other chromosomal basis tend to implicate extranuclear factors (*e.g.*, chloroplast inheritance in *Mirabilis jalapa*).
3. The uniparental inheritance of a trait which cannot be attributed to unequal cytoplasmic contributions from parental gametes may, however, involved as cytoplasmic factors (*e.g., steptomycin* resistance in *Chlamydomonas*).

1. Whenever trait fails to demonstrate the classical segregation patterns and deviates from standard ratios, the conclusion is again a cytoplasm based type of inheritance (*e.g.*, mitochondrial inheritance in yeast).
2. When the trait fails to show linkage to any known nuclear linkage groups and assort independently from nuclear genes, a cytoplasmic mode of inheritance is suggested.
3. Many types of mutants that fit the above criteria will show segregation during mitotic division. This is very common in variegated plants that carry more than one type of plastids (chloroplasts) per cell.

The cytoplasmic inheritance, therefore, will be understood to be based on cytoplasmically located, independent, self replicating nucleic acids, which differ from chromosomal genes by their location within the cell, and have their own nucleotide sequences. The smallest heritable extra chromosomal unit is called a plasmagene. All of the **plasmagenes** of a cell constitute the **plasmon**.

EXTRANUCLEAR INHERITANCE IN EUKARYOTES

Many cases of extranuclear inheritance, also called cytoplasmic inheritance have been reported in eukaryotes. The sum total of self-reproducing, hereditary material of the cytoplasm is referred to as the plasmox, in contrast to the genome, the hereditary content of the chromosomes. Many geneticists have studied various cases of extranuclear inheritance in different eukaryotes. Certain most important examples of this type of inheritance in eukaryotes are as the following:

Maternal Inheritance

In certain cases, it has been observed that certain characteristic phenotypic traits of F_1, F_2 or F_3 progeny are not the expression of their own genes, but rather those of the maternal parents. Such phenotypic expressions of maternal genes (genotype) may be short-lived or may persist throughout the life-span of the individual. The substances which produce the maternal effects in the progeny are found to be transcriptional products (*i.e.*, mRNA, rRNA and tRNA) or maternal genes which have been manufactured during oogenesis and which exist in the ooplasm of unfertilized eggs in the form of inactive protein coated and late translating mRNA molecules (informosomes) or inactivated rRNA and tRNA. These transcriptional products of maternal genes produce their phenotypic effects during early cleavage and blastulation when there occurs little or no transcription since maternal and paternal genes of zygote remain engaged in mitotic replication or duplication of DNA. There may be other reasons of maternal effect, which are still little understood. The maternal inheritance has been studied in some of the following cases:

Coiling in Snail Shells

In many species of water snails (*e.g.*, *Limnaea*), coiling of the shell is either to the right (dextral) or, in some cases, to the left (sinistral). In a few cases (*L. fergera*), both dextral and sinistral individuals occur. However, dextral coiling is dominant over sinistrality; the trait of coiling not determined by genes of the individual but by those of the mother.

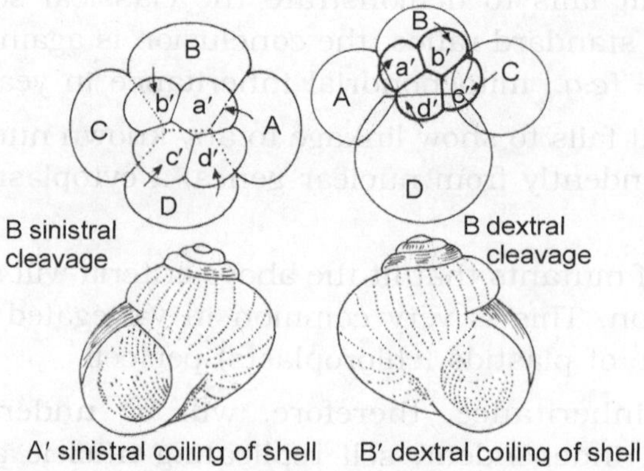

Fig. 25.1. Correlation of sinistral (A) and dextral (B) cleavage with sinistral (A′) and dextral (B′) coiling of the shell in gastropods.

The eggs of a homozygous sinistral individual (SS) are sinistral fertilized by the sperm of a dextral individual (S^+S^+), the eggs cleavage sinistrally and all the snails of this F_1 generation show a sinistral coiling of the shell. Thus, the gene of sperm (S^+) do not manifest themselves, although the genotype of the F_1 generation is S^+S. If a second generation (F_2) is bred from such F_1 sinistral individuals, it is an dextral, instead of showing segregation as would be expected in normal Mendelian inheritance. In fact, segregation does take place in the F_2 generation so far as the genes are concerned, but the new genic combinations fails to manifest themselves, since the coiling is determined by the genotype of the mother. The genotype of F_1 mother being S^+S, the gene for dextrality dominates and is responsible for the exclusively dextral coiling of the second generation. Only the F_3 generation had the genotypes — $1S^+S^+$; $2S^+S$, 1SS, ¼ of them, on the average, produce eggs developing into sinistral individuals (Fig. 25.1).

It is easy to understand that the results of a reciprocal cross that is, of the fertilization of the eggs of a homozygous dextral individual (S^+S^+) by the sperm of a sinistral individual (SS) — will lead to a somewhat different type of pedigree: the F_1 generation will be dextral (with genotypes S^+S) and the F_2 generation again all dextral (with genotypic ratio of $1S^+S^+$: $2S^+S$: 1SS). The F_3 generation will show segregation among broods, just as in the cross examined first.

Extranuclear Inheritance by Cellular Organelles

Cellular organelles such as chloroplasts and mitochondria both are known to have DNA and the protein-synthesizing mechanism. According to one theory, they were once in injectious endosymbiotic prokaryotes and developed such a dependence on gene products of the host that they are no longer able to function autonomously.

Mitochondria and chloroplasts are organelles that contain their DNA and protein synthesizing apparatus. A widely held theory proposes that they were once injectious endosymbiotic prokaryotes that evolved such a dependence on the gene products of the host that they are no longer able to function autonomously.

This theory has been supported by the fact that the genetic components of these organelles are often similar to those found in prokaryotes. For example, the chloroplasts of certain algae and *Euglena* contain 70S type small ribosomes and "naked" chromosomes or DNA which is circular. Their protein synthesis begins with the amino acid N—formyl methionine, as does prokaryotic protein synthesis, and their DNA-dependent RNA polymerase is sensitive to the inhibitor, rifampicin. The genetic materials of chloroplasts and mitochondria will be transmitted to offsprings almost exclusively via the egg. Maternal inheritance due to chloroplast and mitochondria is well illustrated by the following examples:

Chloroplast Inheritance in Variegated Four O'clock Plant

The cytoplasmic or extranuclear inheritance of colour in Four o'clock *Mirabilis Jalapa* by plastids was first described by *C. Correns* in 1908. This plant (*Mirabilis jalapa*) contains three types of leaves and parts:

1. Full green leaves or branches having chloroplast.
2. White leaves and branches having no chloroplast.
3. Variegated branches with leukoplast in pale areas and chloroplast in green patches.

It is known that chloroplast is concerned with photosynthesis while leukoplasts are incapable in performing doing this function. So non-green parts receive nourishment from green parts for their survival. Correns reported that flowers on green branches produce only green offsprings, regardless of the genotype and phenotype of both parents and likewise, from the white branches white seedlings are produced. The plants developing from white or pale seedlings do not survive as they lack chlorophyll. Correns further reported that flowers from the variegated branches yield mixed progeny of green white and variegated plants in widely varying ratios. These results are summarized in Table 25.1.

Table 25.1. Chloroplast Inheritance in Variegated Four O'clock Plants

Branch of Origin of the Male Parent	*Branch of Origin of the Female Parent*	*Progeny*
Green	Green	Green
	Pale or white	Pale or white
	Variegated	Green, pale or white, variegated
Pale or white	Green	Green
	Pale or white	Pale or white
	Variegated	Green, pale or white, variegated
Variegated	Green	Green
	Pale or white	Pale or white
	Variegated	Green, pale or white, variegated.

Mitotic Segregation

Variegated branches of *Mirabilis jalapa* produce three kinds of eggs, some contain chloroplastid, some contain only white and some contain both types of plastid. In the subsequent mitotic divisions, some form of cytoplasmic segregation occur that segregate the chloroplast into pure cell lines thus, producing the variegated phenotype in the progeny individual. This process of sorting might be described as "**mitotic segregation**" which is a pure extranuclear phenomenon. In mitotic segregation since both segregation and recombination of organelle genotype takes place, it is called cytoplasmic segregation and recombination (its acronym is CSAR).

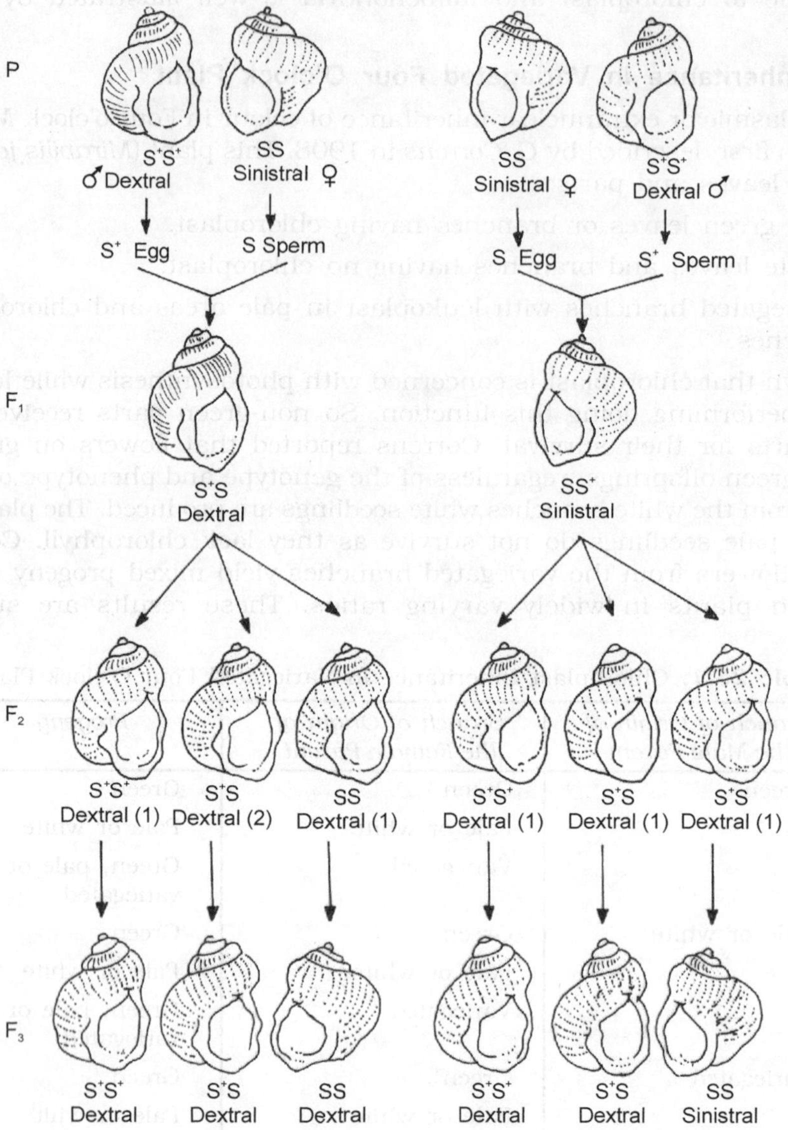

Fig. 25.2. Maternal effect in direction of coiling of the shell in *Limnaea.*

Extranuclear Inheritance by Mitochondria

On the genetics of mitochondria, an important work done in yeast was initiated by the discovery of petite mutants by *B. Ephrussi* (1953).

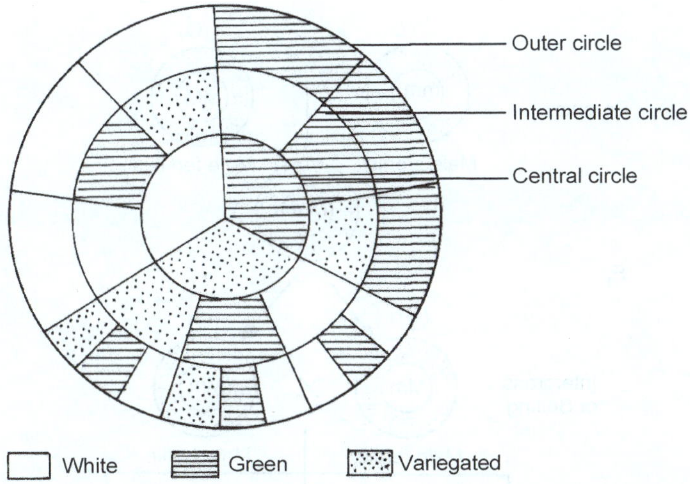

Fig. 25.3. Plastid inheritance of *Mirabilis jalapa* The *central circle* represents the type of branch that produces flowers which are pollinated. *Intermediate circle* represents branch from which pollen is used and outer *circle shows* the progeny.

Petite in yeast

Yeast, *Sacchaomyces cerevisiae*, is a single-celled ascomycetes fungus. In this life cycle, diploid and haploid adult alternates, the former reproducing by asexual meiospores called ascospores, the latter by isogametes. The petite mutants in yeast fail to grow on carbon source such as glucose and produce smaller colonies (the "littles") when grown on the same substrate. Since this difference can be observed only when such yeast cultures are kept in an oxygen-containing environment. So it is concluded that petite mutants have a defective aerobic respiratory mechanism. In other words, slow growth of petite can be attributed to yeast cells utilizing less efficient fermentation process. These petites differ from wild type, called grande and are characterized by:

(*i*) their insensitivity to inhibitors of aerobic pathways (such as cyanide),

(*ii*) absence of cytochromes a, a_3, b and a number of other changes in mitochondrial respiratory enzymes;

(*iii*) incomplete development of mitochondria; and

(*iv*) lack of stainability of petite mitochondria.

Male sterility in plants

In plants, the phenotype of male sterility is found to be controlled either by nuclear genes or plasmagenes (cytoplasm) or by both. Therefore, the trait of male sterility of plants is controlled by the following three methods:

(*a*) **Genetic male sterility.** In this type of male sterility, the sterility is controlled by a single nuclear gene which is recessive to fertility, so that the F_1 progeny would be fertile and in F_2 generation, the fertile and sterile individuals will be segregated in the typical 3:1 ratio.

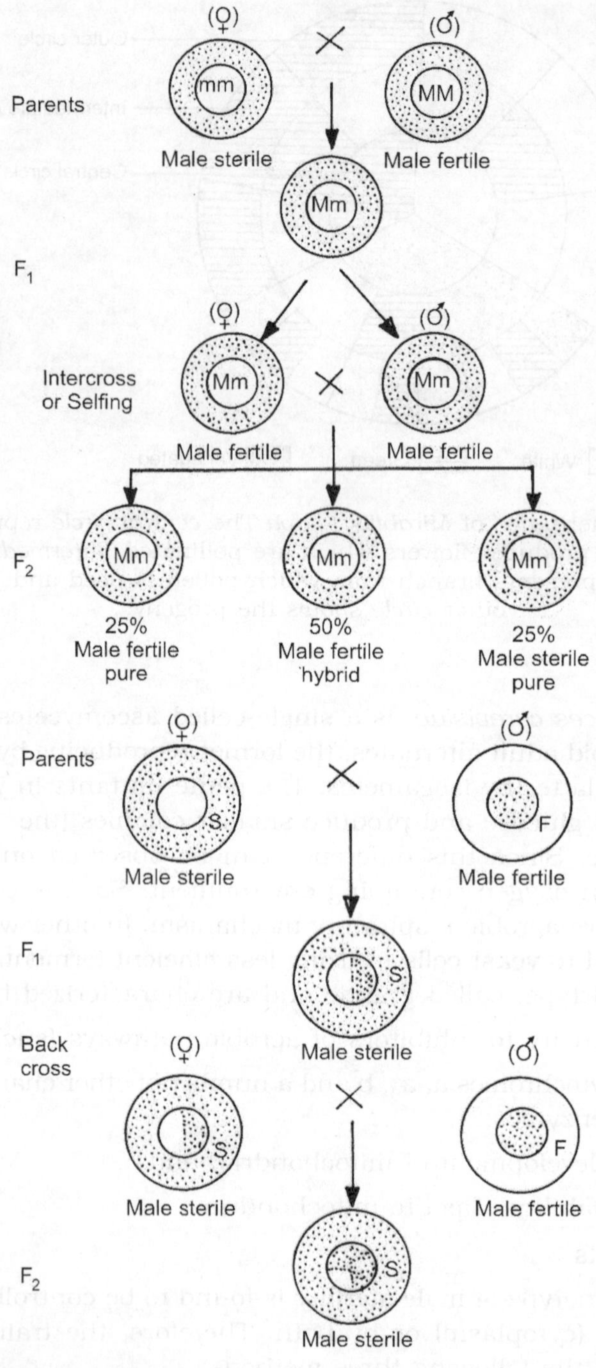

Fig. 25.4. Maternal inheritance of cytoplasmic male sterility.

(*b*) **Cytoplasmic male sterility (CMS).** In such cases as maize, when pollen grains fail to germ-inate, male sterility results. This sterility is inherited through the cytoplasm. Maize plants of this type when pollinated with pollen from any strain form F_1 hybrids that are pollen sterile.

(c) **Cytoplasmic genetic male sterility.** In certain plants, though the male sterility is fully controlled by the cytoplasm, but a restorer gene if present in the nucleus, will restore fertility. For example, if the female parent is male sterile (due to plasmagene of male sterility) then the nuclear genotype of the male parent will determine the phenotype of F_1 progeny. Thus, if male sterile female parent contains recessive nuclear genotype rr of restores gene and male parent is RR, having homozygous dominant restorer genes. Their F_1 progeny would be male fertile Rr. However, if the male parent is male fertile rr, the F_1 progeny would be male sterile rr. If the F_1 male fertile heterozygote (Rr) is test crossed with male fertile rr male, a progeny with 50 per cent male fertile and 50 per cent male sterile will be obtained.

The petite mutants can be segregational, *i.e.*, they follow Mendelian segregation and, therefore, presumably controlled by chromosomal genes. They may also be vegetative, *i.e.*, non-segregational or extra-chromosomal. The genetic basis of petite character is a cytoplasmic factor ρ^+ (rho) which may be absent or defective in petites. Thus, a vegetative petite can be neutral (ρ^0) which completely lack ρ^+ or it may be suppressive (ρ^-) having a defective ρ^+. The neutral petites are not transmitted while suppressive petites are transmitted to a fraction of vegetative diploid progeny. In various strains of yeast, the suppressiveness varies from 1–99 per cent petites. The following two lines of evidences have suggested the association of ρ^+ with mitochondrial DNA (mtDNA); (*i*) Ethidium bromide, which induces petite mutation with 100 per cent efficiency, causes degradation of mtDNA after prolonged exposure of cells. In fact, neutral petites have been found lacking in mtDNA. (*ii*) Suppressive petites contain mtDNA which is greatly altered in base composition with respect to wild mt DNA.

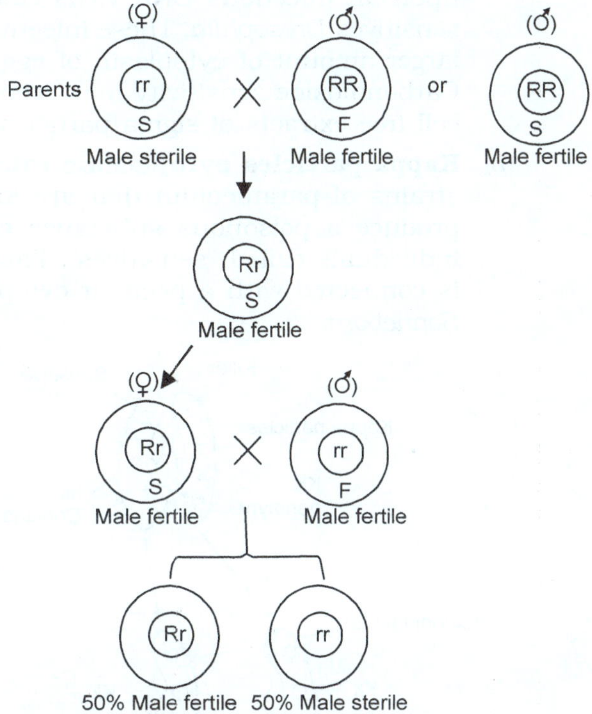

Fig. 25.5. Inheritance of cytoplasmic male sterility and effect of restorer gene in making male sterile into male fertile.

Since, in maize expression of male sterility depends on an interaction between nuclear and extrachromosomal genes male sterile lines can bear seeds only after cross-pollination. For this reason they are useful in raising hybrid seeds, especially on large scale.

Later on, in maize the following four types of cytoplasms have been recognized: the normal (N) cytoplasm and three types of male sterile cytoplasms (T, C and S). The recent studies of mitochondria in these cytoplasm revealed that the factors responsible for cytoplasmic male sterility are located in mitochondrial DNA (mtDNA) and mtDNA of N, T, C and S cytoplasms are found to be different. The cytoplasmic male sterility (CMS) of C and S type can be reversed by nuclear restore genes, however, the CMS-T cannot.

Extranuclear Inheritance by Endosymbionts

Parasites which are intra cellular such as bacterial and virus maintain symbiotic relationship with host cells. They are self reproducing and look like the cytoplasmic inclusions. Sometimes they exhibit an infection like transmission with a hereditary continuity of their own. Generally such symbionts are coined by letters of the Greek alphabets (sigma, kappa, mu, etc.). The various types of infective symbionts are as follows:

(*a*) **Sigma virus in *Drosophila*. L. Heritier and Teissier** There is a strain of *Drosophilla melanogaster* which shows a peculiar sensitivity to carbon dioxide. The wild type strain can be exposed for long periods to pure CO_2 without permanent damage. This trait (extra-sensitivity) is transmitted primarily, but not exclusively, through the maternal parent. The CO_2 sensitivity is dependent upon an infectious DNA virus called sigma, found in the cytoplasm of CO_2 sensitives *Drosophila*. These infective particles are transmitted normally via the larger amount of cytoplasm of egg, occasionally through the sperms as well. Carbondioxide sensitivity may even be induced in normal flies by injections of cell free extracts of sigma particles from CO_2 sensitive flies.

(*b*) **Kappa particles cytoplasmic inheritance in *Paramecium*.** There are some strains of paramecium that are known as killers. Such strains of *P. arrelia* produce a poisonous substance called paramecin, which is lethal to other individuals called "sensitives". Paramecin is soluble in water, diffusible, and is connected with a peculiar cytoplasmic material called "Kappa particle" by *Sonneborn*.

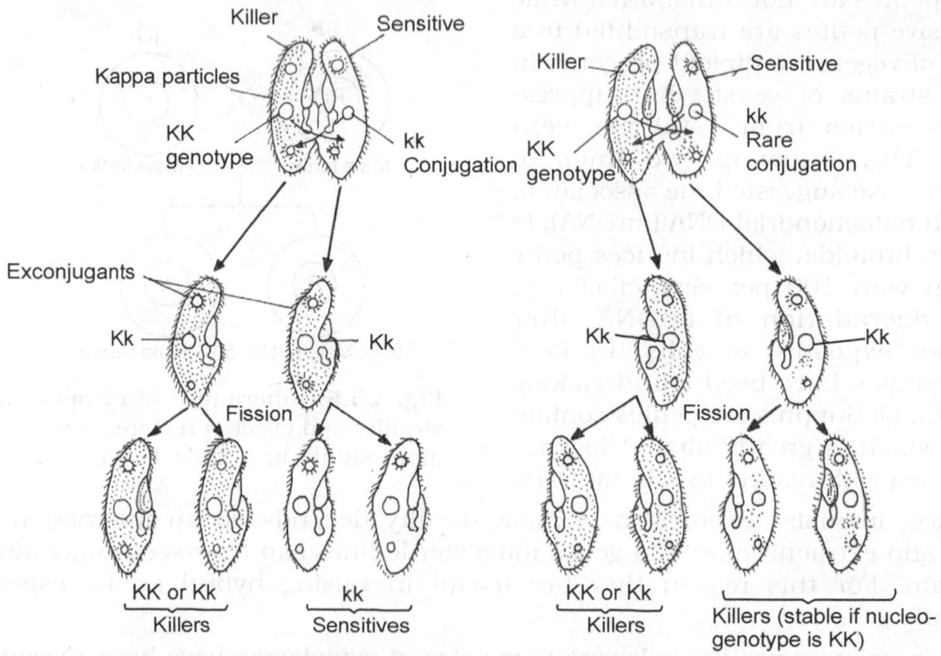

Fig. 25.6. Conjugation in *Paramecium* and the killer trait Progeny of sensitives are situations where conjugation persists for a longer period so that kappa-co is introduced into the conjugating sensitive. Kappa particles, however are the presence of K-nuclear genotype (after *Burns*, 1969 and *Gardner* 1968).

When a *Paramecium* of killer strain having the genotype KK or (K^+) conjugates with the *Paramecium* of non-killer strain having the genotype kk, the exconjugants are all heterozygous for Kk genes. The Kk genotype suggests that both conjugant should be killers. But this is not the case. If conjugation is normal, *i.e.*, lasts only for a short time, and no exchange of cytoplasm takes place between the two, both killers and non-killers (sensitive) are produced. However, rare or prolonged conjugation (*i.e.*, lasting for long time) permits mixing of cytoplasm of both conjugants and results in killers only. The killer trait is stable only in killer strain with KK genotype and is suitable in sensitive strain with kk genotype.

(*c*) **mμ particle.** Another type of killer trait known as mate killer has been reported in *Paramecium* by *R.W. Siegel* in 1952. The mate killer trait is imported by a cytoplasmic μ particle and a *paramecium* with a μ particle is called mate killer because when it conjugates with a *paramecium* without any μ particles is called mate sensitive, then it kills the latter. The μ particles exist only in those cells whose micronucleus contains at least one dominant gene of either of two pairs of unlinked chromosomal genes (M_1 and M_2). The μ particles are composed of DNA, RNA and other substances and are symbionts.

(*d*) **Milk factor in mice. Bittner** found that females of certain lines of mice are highly susceptible to mammary cancer and this trait was found to be maternally transmitted trait. Results of reciprocal crosses between these and animals of low-cancer incidence strain are allowed to be nursed by susceptible foster mothers producing a high rate of cancer in them. Apparently this is a case of infective agent transmitted in the milk. This so called milk factor resembles in many respects with a virus and has been discovered to be transmissible also by saliva and semen. The presence of milk factor also depends on nuclear genes.

Uniparental Inheritance in *Chlamydomonas reinhardi*

Like fungi, algae rarely have different sexes, but they do have mating types. In many algal and fungal species, there are two mating types that are determined by alleles at one locus. A cross can occur only if the parents are of different mating types. The mating types are physically identical but physiologically different. Such species are called heterothallic (literally "different bodied"). In *Chlamydomonas*, the mating type alleles are called mt^+ and mt^- (in *Neurospora* they are A and a; in yeast a and α).

In 1954, *Ms. Ruth Sagar* isolated a streptomycin—sensitive (sm-s) mutant of *Chlamydomonas* with a peculiar inheritance pattern. In the following crosses, sm-r and sm-s indicate *streptomycin* resistance and *streptomycin* sensitivity, respectively, and mt is the mating type gene:

$$mt^+ \text{ sm-r} \times mt^- \text{ sm–s} \rightarrow \text{progeny all sm-r}$$

$$mt^+ \text{ sm-s} \times mt^- \text{ sm–r} \rightarrow \text{progeny all sm-s}$$

Here, occurs a difference in reciprocal crosses all progeny cells show the *streptomycin* phynotype of the mt^+ parent. Like the maternal inheritance this is a case of **uniparental inheritance.** In fact, *Sagar* now refers to the mt^+ mating type as the female, using this analogy.

26 POST TRANSLATION REGULATION OF GENE EXPRESSION

INTRODUCTION

Post translational control involves the enzymic modification of a protein that alters its activity. Examples are common among eukaryotes. Some eukaryotic enzymes are synthesized as inactive proenzymes that become activated, when required, by proteases that cleave them at specific sites. The activity of many other eukaryotic proteins is altered by the action of protein kinases that catalyse the transfer of a γ phosphate group from ATP to the protein.

Examples of post transcriptional control in prokaryotes are much less common, but the best studied one is the control of the activity of glutamine synthetase, an enzyme that plays a pivotal role in the nitrogen metabolism of many bacteria. It catalyzes the reaction,

$$\text{glutamate} + \text{ATP} + NH_4^+ \xrightarrow[\text{synthetase}]{\text{glutamine}} \text{glutamine} + \text{ADP} + Ⓟ$$

which is the only one by which glutamine can be synthesized. If this amino acid is not present in the medium, the reaction must function to satisfy the requirement of protein synthesis and a number of other biosynthetic reactions in which glutamine participates as an obligatory nitrogen donor. But the majority of organic nitrogen in a cell is derived from glutamate via transmination; hence, the major flow from ammonia to organic nitrogen passes through glutamate, which can be synthesized either from glutamine by the glutamate synthase reaction,

$$\text{glutamine} + \alpha\text{-ketoglutarate} \xrightarrow[\text{synthase}]{\text{glutamate}} 2\ \text{glutamate}$$

or by the ammonia-incorporating glutamate dehydrogenase reaction

$$\alpha\text{-ketoglutarate} + \text{NADH} + NH_4^+ \xrightarrow[\text{dehydrogenase}]{\text{glutamate}} \text{glutamate} + NAD^+ + H_2O$$

Thus, the glutamine synthetase reaction, in addition to being the sole route of synthesizing glutamine, also serves as part of an alternate route by which glutamate can be synthesized, and by which the major flow of ammonia into nitrogenous constituents of the cell can occur. However, the two major ammonia-incorporation reactions differ in important respects. The glutamine synthetase reaction requires ATP and functions well at low ammonia concentrations; the glutamate dehydrogenase reaction does not require ATP but has only a relatively low affinity for ammonia. Therefore, the former must function as the major route of ammonia incorporation when the supply of ammonia is restricted, but the latter is a more economic one when the supply of ammonia is adequate.

Part of the complex set of mechanisms by which the activity of glutamine synthetase is controlled is one mediated by post transcriptional modification of glutamine synthetase (Fig. 26.1) by means of an enzyme that adds an adenyl group to it, rendering it less active when the supply of ammonia is high. Another enzyme hydrolytically removes this group when the concentration of ammonia decreases. The adenylating and deadenylating enzymes themselves are allosterically controlled.

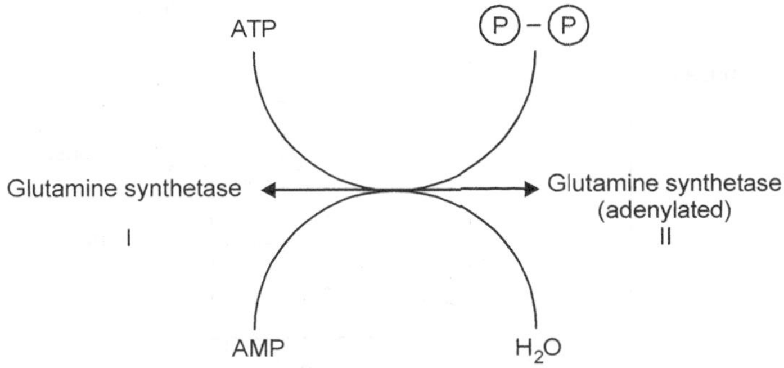

Fig. 26.1. Covalent modification of glutamine synthetase through adenylation is stimulated by high concentration of ammonia and results in a reduction of activity of the enzyme. The adenylated form of glutamine synthetase is sensitive to inhibition by the end products of glutamine metabolism, whereas the unsubstituted form is considerably less sensitive.

ALTERATION OF GENE STRUCTURE

One example of a mechanism that alters gene structure in a manner that increases the ability of bacteria to survive in natural environments has been discovered and preliminary evidence suggests that a number of other similar mechanisms probably operate in prokaryotes. The well studied example changes the type of flagella synthesized by a *Salmonella typhimurium* cell. This mechanism is not a control mechanism in the sense that the alteration occurs in response to a change in the environment (although preliminary evidence suggests that some mechanisms that change gene structure might respond to specific environmental signals). Rather the change occurs randomly, but at a frequency that assures some cells will always be present that can survive in a particular hostile environment.

Salmonella typhimurium possess the genetic capacity to produce two types of flagella from non-allelic genes, but an individual cell produces only one type or the other: either the so-called H1 type or H2 type. Cells producing the former are said to be in phase 1 and those producing the latter in phase 2. Phase variation, as the phenomenon of abrupt change from one phase to the other is termed, occurs about once in every 1000 bacterial division. It is selectively advantageous to the organism in its parasitic state because flagella are actively antigenic, stimulating the host to produce antibodies that mediate the destruction of cells that produce the corresponding antigen. But not all the infecting bacteria will be eliminated, if a single type of antibody is produced by the host. Owing to phase variation, some cells will be in the phase that produces the other flagellar type thereby rendering these cells resistant to the action of the antibody.

The genes encoding H1 and H2 type flagellins (residing at different locations on the chromosome) are transcribed for different promoters. The H2 encoding gene lies in an operon that also contains a repressor that acts at the promoter of the H1 gene. Thus H1 is expressed only if H2 is not.

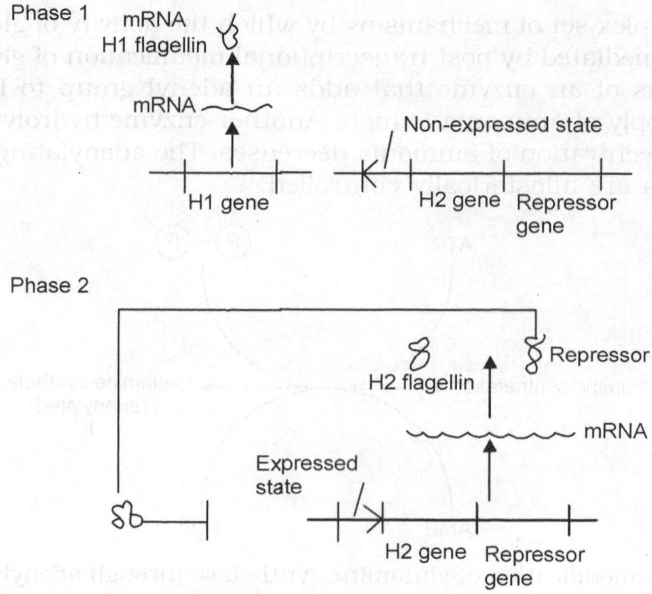

Fig. 26.2. Regulation of phase variation. In phase 1, genes of the H2 operon are not expressed, and the H1 gene is. In phase 2 the H2 operon is expressed producing H2 flagellin and a repressor that prevents expression of the H1 gene.

Expression of the H2 repressor operon is determined by the orientation of a 995-base pair segment that lies immediately adjacent to it because the promoter lies within it. In the orientation encoding the H2 phase, transcription proceeds in the direction of structural genes; in the other, it probably proceeds in the opposite direction; H2 and repressor are not synthesized, so H1 is.

Fig. 26.3. Regulation of expression of the H2 operon. In one orientation (phase 2) of the 995- base pair segment (shaded) a promoter (p) is in the proper position for transcription of the H2 and repressor genes. In the other (phase 1) they are not transcribed. In either orientation, the *hin* gene is transcribed from its own promoter, p.

Inversion of the controlling segment is mediated by a protein, Hin, encoded by a gene *hin*, that lies within the segment. *hin* acts to invert the segment by catalyzing a recombination between the 14 base pair repeated regions that occur in inverted order at the ends of the sequence.

Pattern of Regulation

Certain generalization can be made about the regulation of metabolic pathways. In unbranched biosynthetic pathways, the first enzyme of the pathway is regulated by end product inhibition and the biosynthesis of all enzymes of the pathway are subject to end product repression. In catabolic pathways involving the metabolism of carbon sources that are commonly present in the environment of the organism, flow through the pathway is regulated largely by allosteric enzymes; the enzymes of the pathway are constitutively synthesized and are subject to minimal catabolite repression. Pathway of catabolism of substrates that occurs more rarely in the environment of the organism are regulated primarily by enzyme induction and by catabolite repression, rather than by allosteric enzymes.

Enzyme Repression in Branched Biosynthetic Pathway

The repression of enzyme synthesis in branched biosynthetic pathways is, the feedback inhibition, both complex and variable in mechanism, for instance, the synthesis of carbonyl phosphate synthetase by *E. coli* is partially repressed either by arginine or by cytidine triphosphate (CTP) and is fully repressed by both together. Thus, the synthesis, of this rey allosteric enzyme is regulated independently by two end products. Neither of which affects the activity of the enzyme. In the case of isofunctional enzymes subject to independent allosteric control by different end products, the synthesis of each enzyme is frequently controlled by the end product that inhibit its activity.

27 Development and Environmental Regulation of Gene Expression

INTRODUCTION

As we know every organism originates from a single cell called zygote. The genetic constitution of the zygote only specifies the potential for the organism to develop and function. Many things can influence gene expression as the organism develops and differentiates, and one such influence is the organism's environment.

A gene is a segment of DNA, and an organism's set of genes or called genome, is found on its chromosomes. An organism develops from a zygote by a process of regulated growth and differentiation that results from the interactions of the genome with the internal cellular environment and also with the external environment. Development is not a single step process but it is a programmed series of phenotypic changes that are under temporal, spatial and quantitative control. Development is essentially irreversible under normal conditions. In the complex process of development four major processes interact with one another to constitute the complex process of development. These four processes are: (*i*) replication of the genetic material, (*ii*) growth, (*iii*) differentiation of the various cell types, and (*iv*) the aggregation of differentiated cells into defined tissues and organs. We must think of development as a series of intertwined, complex biochemical pathways whose steps are under gene control. Any of these pathways are susceptible to environmental influences if the products of the genes controlling the pathways are affected by the internal or external environmental parameters. Though genes influence development, no gene by itself totally determines a particular phenotype. In other words, a phenotype is the result of closely interwoven interactions among many factors during development. Therefore, the same genetic constitution does not necessarily mean the same phenotypes will result for all individuals with that constitution. This phenomenon is most readily studied in experimental organisms where the genotype is unequally known. The extent to which the gene manifests its effects under varying environmental conditions can then be seen.

In some cases not all individuals who are known to have a particular gene show the phenotype specified by that gene. The frequency with which a dominant or homozygous recessive gene manifests itself in individuals in a population is called the **penetrance** of the gene. For example, an organism may be genotypically A or aa but may not display the phenotype typically associated with that genotype. If 80 per cent, say, of the individuals carrying a particular gene shows the corresponding phenotype, we say that there is 80 per cent **penetrance.** Penetrance depends on both the genotype (*e.g.*, the presence of epistatic or other genes) and the environment. Penetrance is complete

(100 per cent) when all the homozygous recessives show one phenotype, when all the homozygous dominants show another phenotype, and when all the heterozygotes are alike. Many genes show complete penetrance. All the seven gene pairs in Mendel's experiments and the alleles in the human ABO blood group system are two examples. If less than 100 per cent of the carriers of a particular genotype exhibit the phenotype expected, penetrance is incomplete. Figure 27.1 illustrates the concept of penetrance.

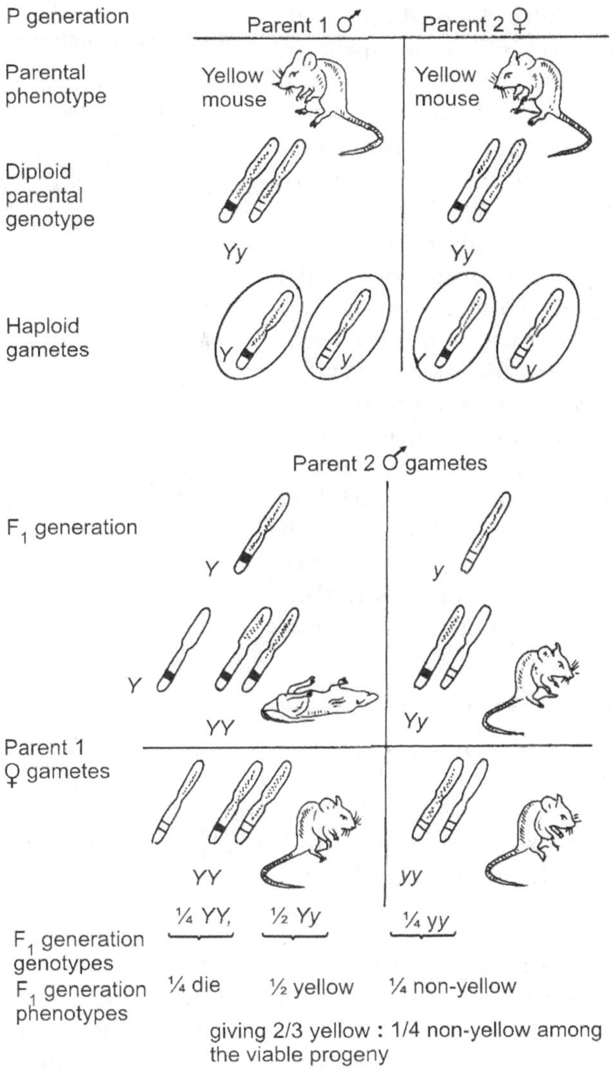

Fig. 27.1. Inheritance of lethal gene Y in mice mating of yellow and yellow gives 1/4 non-yellow (black) mice 1/2 yellow mice and 1/4 mice dead. The variability of yellow mice are heterzygous. Yy and YY are dead individuals are homogygous.

An example of incomplete penetrance is retinoblastoma (tumor of the eye) disease in humans. This disease is inherited as an autosomal dominant, and the mutant gene shows incomplete penetrance since not all the individuals who have the gene for retinoblastoma express that trait. Another human gene that shows incomplete penetrance is the autosomal dominant gene that determines polydactyly (extra fingers or toes).

At a different level we can also determine the degree to which a gene influences a phenotype. Expressivity refers to the kind of or degree to which a penetrant gene or genotype is phenotypically expressed. Expressivity may be described in either qualitative or quantitative terms (Fig. 27.2). Like penetrance, expressivity depends on both the genotype and the external environment, and it may be constant or variable. A classical example of variation in expression is found in the human condition osteogenesis imperfect. The three main features of this disease are blue sclerae (the whites of the eyes), very fragile bones, and deafness. This condition is inherited as an autosomal dominant with almost 100 per cent penetrance; that is 100 per cent of individuals with the genotype have the condition. However, the trait shows variable expressivity: A person with the gene may have anyone or any combination of the three traits. Moreover, the fragility of the bones for those who exhibit this condition is also very variable. Therefore, we see that in medical genetics it is important to recognize that a gene may vary widely in its expression, a qualification that makes the task of genetic counseling that much more difficult.

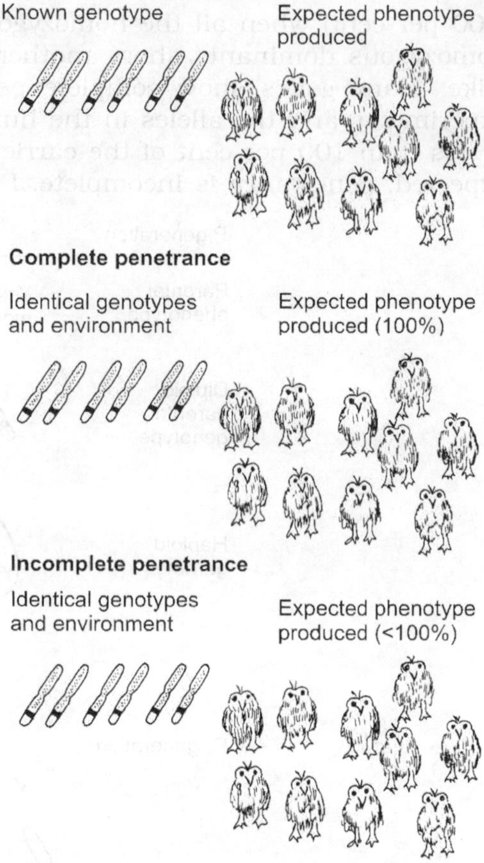

Fig. 27.2. Complete and incomplete penetrance.

Effects of the Internal Environment

Complex biochemical responses occur in cells and in the organism as a result of certain stimuli from the internal environment. These stimuli effects the biochemistry of cell or sometimes they change the cell biochemistry. Not enough is known about cell biochemistry as it relates to development and adult function, though to describe these environmental effects at the molecular level.

Age

The age of the organism is a relevant internal environmental change that can affect gene function. All genes do not continually function; instead, programmed activation and deactivation of genes occur as the organism develops and functions with time. The Huntington's disease gene is one example of a gene whose effects are not manifested until later in the organisms's existence. Somehow, the gene recognizes the age of the individual, and the phenotypic effects, devastating as they, commence. Numerous other age dependent genetic diseases occur in humans; pattern baldness (*i.e.,* the bald spot creeping forward from the crown of the head) appears in males between 20 and 30 years, and Duchenne severe muscular dystrophy appears in children between 2 and 5 years. While such interactions with age can be described, and while in some cases we know the defective gene product involved, in most cases the nature of the age dependency is not understood.

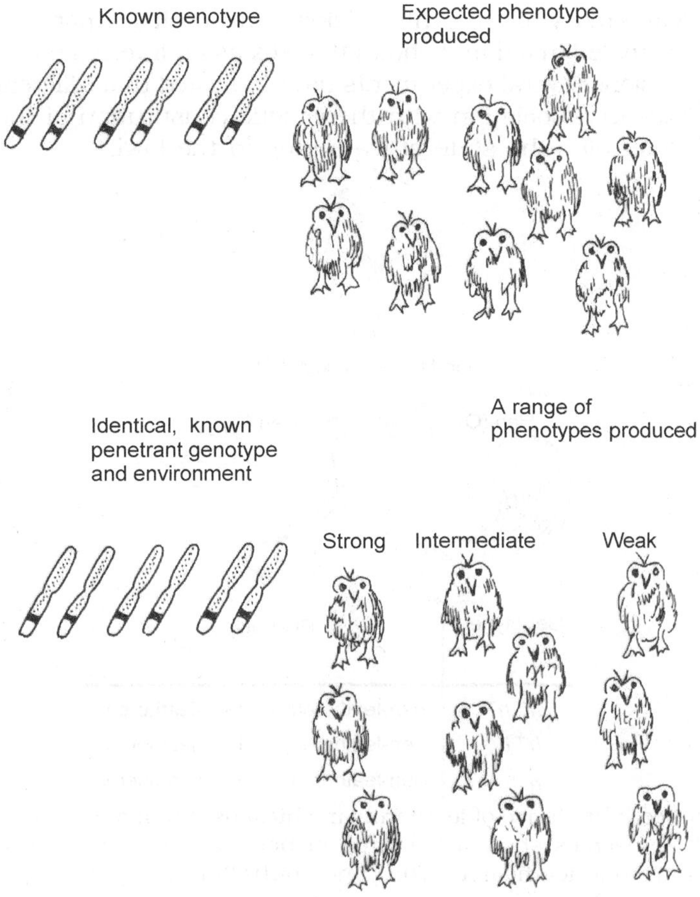

Fig. 27.3. Expressivity of gene mild intermediate and strophe.

Sex

Sex is also an important factor which influence the expression of particular genes of the individual. In the case of sex-linked genes as mentioned earlier, differences in the phenotypes of the two sexes are related to different complements of genes on the sex chromosomes. However, in some cases genes that are not located on the sex chromosomes affect a particular character that appears in one sex but not the other. Traits of this kind are called sex-limited traits.

One such trait is the feathering pattern of chickens. A common difference between male and female fowls is in the structure of many of the feathers. Cocks often have long, narrow, pointed feathers on their hackles and saddle, pointed feathers on the cape, back and wing, and long, curving, pointed sickle feathers on the tail. This pattern is called cock feathering. Hens usually show the contrasting feather characteristics of short, broad, blunt, and straight feathers. This pattern is called hen feathering. If F_1 heterozygous (h^+h) hen-feathered males and females are interbred, all F_2 females are hen-feathered, whereas F_2 males exhibit a 3 : 1 ratio of hen feathered : cock feathered. The interpretation is that the trait is controlled by an autosomal gene, with hen feathering (h^+) dominant to cock feathering (h). Assigning gene symbols, we can conclude that all three possible F_2 genotypes in the females, h^+ h^+, h + h, and hh, give rise to hen feathering of the tail, but in the F_2 males the h^+h^- and h^+h

chickens are hen-feathered while the hh chickens are cock feathered. In other words, the cock-feathered trait is male-limited and shows that sex as an internal environmental parameter can affect gene expression. Several experiments have indicated that differences in sex hormones between the two sexes in association with the genetic constitution of the skin (in which the feathers originate) determine the state of feathering in the fowl.

Genotype	Phenotype ♀	Phenotype ♂
h^+h^+	Hen-feathered	Hen-feathered
h^+h	Hen-feathered	Hen-feathered
$h\ h$	Hen-feathered	Cock-feathered

Fig. 27.4. Sex limited inheritance of feathering in chickens. From a cross between heterozygous F_1 parents the F_2 progenies show a 3:1 ratio of hen-feathered cocks-feathered male birds, while all females are hen-featheared. The gene controlling the phenotype is autosomal.

A slightly different situation is found in sex influenced traits. Such traits appear in both sexes, but either the frequency of occurrence in the two sexes is different or the relationship between genotype and phenotype is different. For example, the human trait *Spina bifida*, in which the spinal cord does not completely close as the embryo develops, is found much more frequently in females than in males.

Another straight forward example of a sex influenced trait is pattern baldness in humans. One pair of alleles of an autosomal gene is involved in this trait, b^+ and b. The bb genotype specifies pattern baldness in both males and females and the b^+b^+ genotype gives a nonbald phenotype. The difference lies in the heterozygote: in males it leads to the bald phenotype, and in females it leads to the non bald phenotype. In other words, the B allele acts as a dominant in males but as a recessive in females; its expression is influenced by the sex hormones of the individual. This pattern of inheritance and gene expression explains why pattern baldness is far more frequent among men than among women. From a large sampling of the progeny of matings between two heterozygotes, 3/4 of the daughters are nonbald and 1/4 are bald, and 3/4 of the sons are bald and 1/4 are nonbald.

EFFECTS OF THE EXTERNAL ENVIRONMENT

Like internal environment, external environment also influence the gene expression of an individual. Although many factors in the external environment influence gene expression (nutrition, light, chemicals, infectious agents etc.) but here we will discuss only two of them, temperature and chemical (nutrition).

Temperature

The rate at which a chemical reaction occurs depends in large part on temperature as is also true, within limits, for biochemical reactions taking place within organisms. Therefore it is reasonable to expect that temperature might affect, gene expression, with significant effects on development. A good example is fur colour in Himalayan rabbits. Certain genotypes of this white rabbit cause dark fur to develop at the extremities (ears, nose, and paws), where the temperature is lower. Since all the body cells develop from a single zygote, this distinct fur pattern cannot be the result of a genotype difference of the cells in those areas. It must result from the external environmental influence.

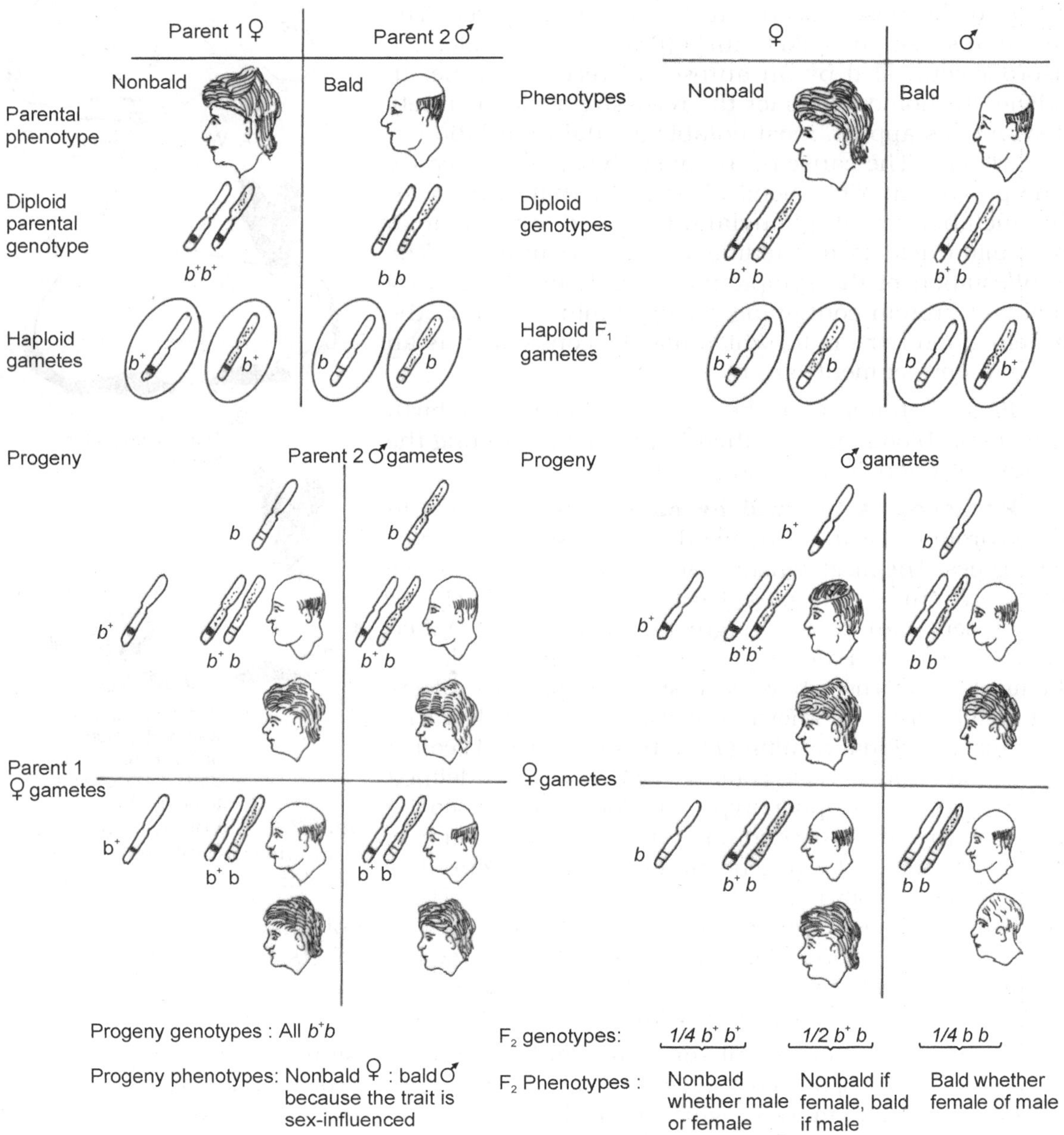

Fig. 27.5

This hypothesis can be tested by rearing Himalayan rabbits under different temperature conditions. When a rabbit is reared at a temperature above 30°C, all its fur, including the ears, nose and paws, is white (Fig. 27.6A). If a rabbit is raised at a temperature of approximately 25°C, the typical Himalayan phenotype results (Fig. 27.6B). Lastly, if a rabbit is raised at 25°C while its left rear flank is artificially cooled to a temperature below 25°C, the rabbit develops the Himalayan coat phenotype, with additional dark fur on the cooled flank area (Fig. 27.6C).

Chemicals

Phenylketonuria. Ultimately, all the interactions we have discussed occur at the chemical level. The human disease phenylketonuria (PKU), for example, is a disorder controlled by an autosomal recessive gene. In individuals homozygous for the recessive allele, a variety of symptoms appear, most notably mental retardation at an early age. The cause of the phenotypes is a defective gene product in a biochemical pathway that is used for the metabolism of phenylalanine, and amino acid. (An amino acid is a building block of a protein). The manifestation of the symptoms depends on the dietary intake of protein containing phenylalanine such as the protein of mother's milk, which may be considered as an external environment effect.

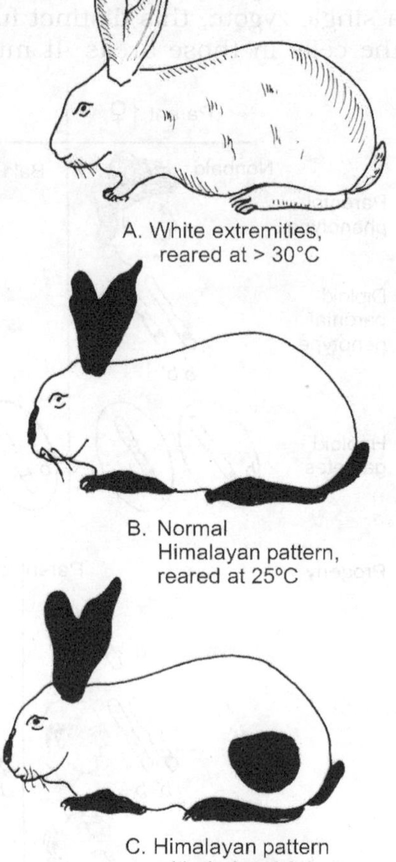

Fig. 27.6. Phynotypic appearance of Himalayan rabbits raised under different temperature conditions.

Phenylketonuria can be diagnosed soon after birth by a simple blood test and then treated by restricting the amount of phenylalanine in the diet.

Phenocopies induced by chemicals. Changes in the environment can influence that expression of one or more genes. The most sensitive area is development, since relatively small changes that take place early in development can result in great changes later. When administered to a developing embryo, certain drugs, chemicals and viruses have been shown to produce effects that mimic those of known specific mutant alleles. The abnormal individual resulting from this treatment is called a phenocopy (phenotypic copies). A phenocopy is defined as a nonhereditary, phenotypic modification (caused by special environmental conditions) that mimics a similar phenotype caused by a gene mutation. The agent that produce a phenocopy is called a phenocopying agent. There are many examples of phenocopies, and studies of phenocopies have in some instances, provided useful information about the actual molecular defects caused by the mutant counterpart. Remember that the abnormalities seen in these situations occur in a genotypically normal individual without any alteration of the genetic constitution.

Phenocopies occur in all sorts of organisms from the simple to the complex. In the bread mold *Neurospora crassa*, for example, the normal growth phenotype is a many branched, web like mass in which individual filaments extend fairly rapidly. The result is an orange-coloured, fluffy growth similar to the grey or green mold that sometimes grows on bread. In some mutants of *Neurospora*, the branching is much more frequent and growth is slower, resulting in what is called a colonial-like growth habit since it extends only slowly from the point of inoculation and looks like a button on a solid medium. The colonial growth is a result of a defect in cell wall structure, which makes it weaker. Consequently, the internal pressure of the cytoplasm of the filaments induces many more branches than normal. This same colonial phenotype can be phenocopied by adding chemicals to the growth medium that affect normal cell wall development. In wild type *Neurospora*, the addition of the unusual sugar sorbose brings about colonial growth. If a piece of the colonial culture is now placed onto a normal medium, the wild type growth habit is re-established. At the biochemical level cell was abnormaly similar to those found in some colonial mutants can be detected in the sorbose-induced phenocopies.

There are also many examples of phenocopies in humans. Cataracts, deafness, and defects of the heart are sometimes produced when an individual is homozygous for rare recessive alleles and may also result if the mother is infected with rubella (German measles) virus during the first 12 weeks of pregnancy. Another human trait for which there is a phenocopy is phocomelia, a suppression of the development of the long bones of the limbs caused by a rare dominant allele with variable expressivity. Between 1959 and 1961, similar phenotypes were produced by the sedative thalidomide when taken by expectant mothers during the 35th to 50th day of gestation. The drug was removed from the market when the devastating effects were discovered.

In conclusion, at least for the more complex organisms, phenocopies are produced when an organism exposed to a phenocopying agent during a sharply defined period of its development, and the same effect can often be produced by very different agents.

28 POPULATION GENETICS

INTRODUCTION

The science of genetics can be mainly categorised into three main categories; these are: Transmission genetics (also called classical genetics), molecular genetics and popular genetics. Each of the three disciplines focuses on a different aspect of heredity. For instance. Transmission genetics is primarily concerned with genetic processes that occur within individuals, and with the help of this branch we come to know that how genes are passed from one individual to another.

Thus, our unit of study for transmission genetics is the individual. In molecular genetics, we are largely interested in the molecular nature of heredity—how genetic information is encoded within the DNA and how biochemical processes of the cell translate the genetic information into the phenotype. Consequently, in molecular genetics we focus on the cell.

Population Genetics

The field of genetics that studies heredity in groups of individuals. Population geneticists investigate the patterns of genetic variation found within groups and how these patterns change, or evolve over time. Because population genetics is concerned with the genetic basis of evolution, the field is also called evolutionary genetics. In this discipline, our perspective shifts away from the individual and the cell, and focuses instead upon a Mendelian population. A Mendelian population is not just any group of individuals, but a group of interbreeding individuals, who share a common set of genes. The genes shared by the individuals of a Mendelian population are called the gene pool. To understand the genetics of the evolutionary process, we study the gene pool of a Mendelian population, rather than the genotypes of its individual members. Except for rare mutations, individuals are born and die with the same set of genes; what changes genetically over time (evolves) is the hereditary makeup of a group of individuals, reproductively connected in a Mendelian population.

GENOTYPIC AND GENE FREQUENCIES

These two terms are the basis of the study of population genetics. The whole concept of the population genetics depends on these two terms.

Genotypic Frequencies

To study the genetic composition of Mendelian population, population geneticists must first quantitatively describe the gene pool of the group. This is done by calculating genotypic frequencies and allelic frequencies for the individuals within the population.

A frequency is a per cent, or a proportion and always ranges between 0 and 1. If 43 per cent of the people in a group have red hair, the frequency of red hair in the group is 0.43. The genotypic frequencies of a population consist of the frequencies of all the genotypes in the population. To calculate the genotypic frequencies at a specific locus, we count the number of individuals with one particular genotype and divide this number by the total number of individuals in the population. We do this for each of the genotypes at the locus, and the sum of all the genotypic frequencies should add up to 1.

Consider a locus that determines the pattern of spots in the scarlet tiger moth, *Panaxia dominula.*

In this example, three genotypes are possible and each genotype produces a different phenotype. *E. B. Ford* collected moths at one locality in England and found the following number of genotypes: 445 BB, 39 Bb and 4 bb out of a total of 495 moths:

The genotypic frequencies are therefore:

$$P = f(BB) = 445/495 = 0.898$$

$$H = f(Bb) = 39/495 = 0.078$$

$$Q = f(bb) = 4/495 = 0.008$$

$$\text{Total} = 0.984$$

Population geneticists often use the capital letter P, H, and Q to represent the frequencies (*f*) of the three genotypes at a locus. Be careful that you do not confuse these symbols with the small letters p and q which are used to represent the allelic frequencies.

Gene Frequencies

While genotypic frequencies are useful for examining the effect of certain evolutionary forces on a population, in most cases population geneticists are frequencies of alleles to describe the gene pool. The frequencies of alleles at a locus are called the allelic frequencies or more commonly the gene frequencies.

The use of gene frequencies offers more advantages over to genotypic frequencies. First there are always fewer alleles that genotypes, and so that gene pool can be described with fewer parameters when gene frequencies are used. If a locus has three alleles, six genotypic frequencies must be calculated to describe the gene pool, whereas use of the allelic frequencies requires the calculation of only three allelic frequencies. Further more, in sexually reproducing organisms genotypes breakdown to alleles when gametes are formed, and alleles, not genotypes, are passed from one generation to the next. Consequently, only alleles have continuity over time, and the gene pool evolves through changes in the frequencies of alleles.

Gene frequencies may be calculated in either of two ways. First, we can calculate the gene frequencies directly from the number of genotypes. In this method, we count the number of alleles of one type and divide it by the total number of alleles in the population:

$$\text{Gene frequency} = \frac{\text{Number of copies of a given allele in the population}}{\text{sum of all alleles in the population}}$$

As an example, imagine a population of 1000 diploid individuals with 353 AA, 494 Aa, and 153 aa individuals. Each AA individual has two A alleles, while each Aa heterozygote possesses only a single A allele. Therefore, the number of A alleles in the population is 2 × (the number of AA homozygotes), or (2× 353) + 494 = 1200. Because every diploid individual has two alleles, the total number of alleles in the population will be twice the number of individuals, or 2 × 1000. Using the equation given above, the gene frequency is 1200/2000 = 0.60.

Fig. 28.1. *Panaxia dominula*, the scarlet tiger morh. The top moths are normal homozygotes (BB), those in row 2 and 3 are heterozygotes (Bb) and the bottom moth is the rare homozygote (bb).

When two alleles are present at a locus, we can use the following formulas for calculating gene frequencies:

$$p = f(\mathrm{A}) = \frac{2\times(\text{number of AA homozygotes}) + (\text{number of Aa heterozygotes})}{2\times(\text{total number of individuals})}$$

$$q = f(\mathrm{a}) = \frac{2\times(\text{number of aa homozygotes}) + (\text{number of Aa heterozygotes})}{2\times(\text{total number of individuals})}$$

The frequencies (*f*) of two alleles are commonly symbolized as p and q. The gene frequencies for a locus should always add up to 1; therefore, once p is calculated, q can be easily obtained by subtraction

$$1 - p = q$$

Gene Frequencies with Multiple Alleles. Suppose we have three alleles—A^1, A^2 and A^3 at a locus, and we want to determine the gene frequencies. Here, we employ the same rule that we used with two alleles; we add up the number of alleles of each type and divide by the total number of alleles in the population:

$$p = f(A^1) = \frac{2 \times (A^1A^1) + (A^1A^2) + (A^1A^3)}{2 \times (\text{total number of individuals})}$$

$$q = f(A^2) = \frac{2 \times (A^2A^2) + (A^1A^2) + (A^2A^3)}{2 \times (\text{total number of individuals})}$$

$$r = f(A^3) = \frac{2 \times (A^3A^3) + (A^1A^3) + (A^2A^3)}{2 \times (\text{total number of individuals})}$$

We will illustrate the calculation of gene frequencies when more than two alleles are present with data from a study on genetic variation in milkweed beetles. *Walter Eanes* and his coworkers examined gene frequencies at a locus that codes for the enzyme phosphoglucomutase (PGM). Three alleles were found at this locus; each allele codes for a different molecular variant of the enzyme. In one population, the following numbers of genotypes were collected:

AA	=	6
AB	=	40
BB	=	82
AC	=	28
BC	=	88
CC	=	39
Total	**=**	**283**

The frequencies of the alleles are:

$$f(A) = p = \frac{(2 \times 6) + 40 + 28}{2 \times 283}$$

$$= \frac{12 + 40 + 28}{566}$$

$$= 0.141$$

$$f(B) = q = \frac{(2 \times 82) + 40 + 28}{2 \times 283}$$

$$= \frac{164 + 40 + 28}{566} = 0.409$$

$$f(C) = r = \frac{(2 \times 39) + 88 + 28}{2 \times 283}$$

$$= \frac{78 + 88 + 28}{566}$$

$$= \frac{194}{566} = 0.342$$

As seen in these calculations, we add twice the number of homozygotes that possess the allele and one time each of the heterozygotes that have the allele. We then divide by twice the number of individuals in the population, which represents the total number of alleles present. Notice that for each allelic frequency, we do not add all the heterozygotes, do not have the allele; for example, in calculating the gene frequency of A, we do not add the number of BC heterozygotes in the top part of the equation, because we are adding the number of A alleles, and BC individuals do not have an A allele. We can use the same procedure for calculating gene frequencies when four or more alleles are present.

Gene Frequencies from Genotypic Frequencies. A second method for calculating gene frequencies is from the genotypic frequencies. This calculation may be quicker if we have already determined the frequencies of the genotypes. For two alleles, the gene frequencies are:

p = f(A) = (frequency of the AA homozygote) + 1/2 × (frequency of the Aa heterozygote)

q = f(a) = (frequency of the aa homozygote) + 1/2 × (frequency of the Aa heterozygote)

The frequency of the homozygote is added to half of the heterozygote frequency because half of the heterozygotes alleles are A and half are a. If three alleles (A^1, A^2 and A^3) are present in the population, the gene frequencies are:

$$p = f(A^1) = f(A^1A^1) + 1/2 \times f(A^1A^2) + 1/2 \times f(A^1A^3)$$

$$q = f(A^2) = f(A^2A^2) + 1/2 \times f(A^1A^2) + 1/2 \times f(A^2A^3)$$

$$r = f(A^3) = f(A^3A^3) + 1/2 \times f(A^1A^3) + 1/2 \times f(A^2A^3)$$

Although calculating the gene frequencies from genotypic frequencies may be quicker than calculating them directly from the numbers of genotypes, more rounding error will occur. As a result, calculations from direct counts are usually preferred.

Gene Frequencies at an X-linked Locus. Calculation of gene frequencies at an X-linked locus is slightly more complicated, because males have only a single X-linked allele. However, we can use the same rules we used for autosomal loci. Remember that each homozygous female carries two X-linked alleles; heterozygous females and all males have only a single X-linked allele. To determine the number of alleles at an X-linked locus, we multiply the number of homozygous females by 2 and add the number of heterozygous females and the number of homozygous males. We then divide by the total number of alleles in the population. When determining the total number of alleles, we add twice the number of females (because each female has two X-linked alleles) to the number of males (who have a single allele at X-linked loci). Using this reasoning the frequencies of two alleles at an X-linked locus (X^A and X^a) are determined with the following equations:

$$p = f(X^A) = \frac{2 \times (X^AX^A \text{ females}) + (X^AX^a \text{ females}) + (X^AY \text{ males})}{2 \times (\text{number of females}) + (\text{number of males})}$$

$$q = f(X^a) = \frac{2 \times (X^aX^a \text{ females}) + (X^AX^a \text{ females}) + (X^AY \text{ males})}{2 \times (\text{number of females}) + (\text{number of males})}$$

Allelic frequencies at an X-linked locus can be determined from the genotypic frequencies by

$$p = f(X^4) = f(X^AX^A) + 1/2 \times f(X^AX^a) - f(X^AY)$$

$$q = f(X^a) = f(X^aX^a) + 1/2 \times f(X^AX^c) - f(X^aY)$$

Students should strive to understand the logic behind these calculations and not just memorize the formulas. If the basis of the calculations is fully understood, the exact equations need not be remembered and gene frequencies can be determined for any situation.

THE HARDY-WEINBERG LAW

The Hardy-Weinberg law is divided into three parts—a set of assumptions and two major results. These are as follows:

Part 1: In an infinitely large, randomly mating population, free from outside evolutionary forces (mutation, migration, natural selection);

Part 2: The frequencies of the alleles do not change over time; and

Part 3: After one generation of random mating, the genotypic frequencies will remain in the proportions p^2 (frequencies of AA), 2pq (frequency of Aa), and q^2 (frequency of aa), where p is the allelic frequency of A and q is the allelic frequency of a.

In short, the Hardy-Weinberg law explains what happens to gene and genotypic frequencies of a population as the genes are passed from generation to generation in the absence of evolutionary forces.

Assumptions of the Hardy-Weinberg Law

There are certain conditions/assumptions, which are presented by part-1 of the Hardy-Weinberg law. These assumptions must be present for the law to apply.

First, the law indicates that the population must be infinitly large. If a population is limited in size, chance deviations from expected ratios can cause changes in gene frequency. Changes in gene frequency are termed as genetic drift. The assumption of infinite size in part 1, however, is unrealistic—no population has an infinite number of individuals. Yet, population size determines the magnitude of genetic drift, and sampling error has a significant effect on gene frequencies only in "fairly small populations" which will be discussed later, when we examine genetic drift in more detail. At this point we merely want to stress that populations need not be infinitely large for the Hardy-Weinberg law to hold true.

A second condition of the Hardy-Weinberg law is that mating must be random. Random mating refers to matings between genotypes occurring in proportion to the frequencies of the genotypes in the population. More specifically, the probability of a mating between two genotypes is equal to the product of the two genotypic frequencies.

To illustrate random mating, consider the MN blood types in humans. The MN blood type is an antigen on the surface of red blood cell, similar to the ABO antigens except that incompatibility at the MN system does not cause problems during blood transfusion. The MN blood type is determined by one locus with two codominant alleles, M and N. In a population of Eskimos, the frequencies of the three MN genotypes are MM = 0.835, MN = 0.156, and NN = 0.009. If Eskimos interbreed randomly, the probability of a mating between an MM male and an MM female is equal to the frequency of MM times the frequency of MM = 0.835 × 0.835 = 0.697. Similarly, the probabilities of other possible matings are equal to the products of the genotypic frequencies when mating is random.

The requirement of random mating for the Hardy-Weinberg law is often misinterpreted. Many students assume, incorrectly, that the population must be interbreeding randomly for all traits, do not mate randomly. Humans mate preferentially for height. IQ, skin colour, socioeconomic status, and other traits. While mating may be nonrandom for some traits, most humans still mate randomly for the MN blood types: few of us even know what our MN blood type is. The principles of the Hardy-Weinberg law apply to any locus for which random mating occurs, even if mating is nonrandom for other loci.

A third condition of the Hardy-Weinberg law is that the population must be free from outside evolutionary forces. Remember in the Hardy–Weinberg law we are interested in whether heredity along changes gene frequencies and how reproduction influences genotypic frequencies. Therefore, the influence of other evolutionary factors must be excluded. Later we will discuss these outside evolutionary forces and their effect on the gene pool of a population. This condition (no evolutionary forces acting on the population) applies only to the locus in question—a population may be subject to evolutionary forces acting on some genes, while still meeting the Hardy-Weinberg assumptions at other loci.

Predictions of the Hardy-Weinberg Law

If the conditions in part 1 of the Hardy-Weinberg law are met, the population is in genetic equilibrium and two results are expected. First, the frequencies of the alleles do not change from one generation to the next, and therefore the gene pool is not evolving at this locus. Second, the genotypic frequencies will be in the proportions p^2, $2pq$, and q^2 after one generation of random mating. The genotypic frequencies also remain in these proportions as long as all the conditions required by the Hardy-Weinberg law are met. When the genotypes are in these proportions, the population is said to be in Hardy-Weinberg equilibrium. An important use of the Hardy-Weinberg law is that the genotypic frequencies can be determined from the gene frequencies when the population is in equilibrium.

To summarize, the Hardy-Weinberg law makes several predictions about the gene frequencies and the genotypic frequencies of a population when certain conditions are satisfied. The necessary conditions are that the population is large randomly mating, and free from outside evolutionary forces. When these conditions are met the Hardy-Weinberg law indicates that gene frequencies will not change, genotypic frequencies will change only in the first generation, and thereafter the genotypic frequencies will be determined by the gene frequencies, occurring in the proportions p^2, $2pq$ and q^2.

Derivation of the Hardy-Weinberg Law

The Hardy-Weinberg law states that when a population is in equilibrium, the genotypic frequencies will be in the proportions p^2, $2pq$ and q^2. To understand the basis of these frequencies at equilibrium, consider a hypothetical population in which the frequency of allele A is p and the frequency of allele a is q. In producing gametes, each genotype passes on both alleles that it possesses with equal frequency; therefore, the frequencies of A and a in the gametes are also p and q. Table 28.1 shows the combinations of gametes when mating is random. This table illustrates the relationship between the allelic frequencies and the genotypic frequencies, which forms the basis of the Hardy-Weinberg law. We see that when gametes pair randomly, the genotypes will occur in the proportions p^2 (AA), $2pq$ (Aa), and q^2 (aa). These genotypic proportions result from the expansion of the square of the allelic frequencies $(p + q)^2 = p^2 + 2pq + q^2$, and the genotypes reach these proportions after one generation of random mating.

Table 28.1

Gametes		*Sperm*	
		*p*A	*q*a
Eggs	*p*A	p^2AA	*pq* Aa
	*q*a	*pq* Aa	q^2aa
In sum, p^2 AA + 2 pq Aa + q^2 aa = 1.00			

We see that the probability of AA × Aa ($2p^3q$) and AA × Aa ($2p^3q$) matings is $4p^3q$, and we know from Mendelian principles that these crosses produce 1/2 AA and 1/2 Aa offsprings. Therefore, the probability of obtaining AA offsprings from these matings is $4p^3q \times 1/2 = 2p^3q$. The frequencies of offsprings produced by each type of mating are presented in the body of the table. At the bottom of the table, the total frequency for each genotype is obtained by addition. As we can see, after random mating the genotypic frequencies are still p^2, $2pq$ and q^2 and the gene frequencies remain at p and q. The population can thus be represented in the zygotic and gametic stages as follows:

Zygotes			Gametes
p^2 AA	+ $2pq$ Aa	+ q^2 aa	pA + qa

Each generation of zygotes produces A and a gametes in proportions p and q. The gametes unite to form Aa, Aa and aa zygotes in the proportion $p^2 + 2pq + q^2$, and the cycle is repeated indefinitely as long as the assumptions of the Hardy-Weinberg law hold. This shotproof gives the theoretical basis for the Hardy-Weinberg law.

The Hardy-Weinberg law indicates that at equilibrium, the genotypic frequencies depend upon the frequencies of the alleles. This relationship between allelic frequencies and genotypic frequencies for a locus with two alleles is represented in Table 28.2. Several aspects of this relationship should be noted: (*i*) the maximum frequency of the heterozygote is 0.5, and this maximum value occurs only when the frequencies of A and a are both 0.5; (*ii*) if gene frequencies are between 0.33 and 0.66, the heterozygote is the most numerous genotype; (*iii*) when the frequency of one allele is low, the homozygote for that allele is the rarest of the genotypes. When f (a) = q = 0.1, the frequencies of the genotypes are f(AA) = 0.81, f(Aa)' = 0.18, and f(aa) = 0.01. This point is also illustrated by the distribution of genetic diseases in humans, which are frequently rare and recessive. For a rare recessive trait, the frequency of the gene causing the trait will be much higher than the frequency of the trait itself, because most of the rare alleles are in non affected heterozygotes. Albinism, for example, is a rare recessive condition in humans. One form of albinism is tyrosinase-negative albinism. In this type of albinism, affected individuals have no tyrosinase activity, which is required for normal production of pigment. Among North American whites, the frequency of tyrosinase negative albinism is roughly in 40000 or 0.000025. Since albinism is a recessive disease, the genotype of affected individuals is aa. According to the Hardy-Weinberg law, the frequency of the AA genotypes equals q^2. If q^2 = 0.000025, then q = 0.005 and $p = 1 - q = 0.995$. The heterozygote frequency is therefore $2pq = 2 \times 0.995 \times 0.005 = 0.00995$ (almost 1 per cent). Thus, while the frequency of albinos is low (1 in 40000), individuals heterozygous for albinism are much more common (almost 1 in 100). Heterozygotes for recessive traits can be common, even the trait is rare.

The Hardy-Weinberg law also states that gene and genotypic frequencies remain constant generation after generation if the population remains large, randomly mating and free from outside evolutionary forces. This result can be demonstrated by considering a hypothetical, randomly mating population, as illustrated in Table 28.2. In this table, all possible matings are given. By definition, random mating means that the frequency of mating between two genotypes will be equal to the product of the genotypic frequencies. For example, the frequency of an AA × AA mating will be equal to p^2 (the frequency of AA) × p^2 (the frequency of AA) = p^4. The frequencies of the offsprings produced from each mating are also presented in Table 28.2.

Table 28.2. Algebraic Proof of Genetic Equilibrium in a Randomly Mating Population for Two Albinism in a Population in which $(p + q)^2 = p^2 + 2pq + q^2 = 1.00$.

Type of Mating	*Mating Frequency*	*Offspring Frequencies*		
		AA	Aa	aa
p^2 AA × p^2 AA	p^4	p^4		
p^2 AA × $2pq$ Aa }* 2 pq Aa × p^2 AA	$4p^3q$	$2p^3q$	$2p^3q$	
p^2 AA × q^2 aa q^2 aa × p^2 AA	$2p^2q^2$		$2p^2q^2$	
2pq Aa × 2pq Aa	$4p^2q^2$	p^2q^2	$2p^2q^2$	p^2q^2
2 pq Aa × q^2 aa q^2 aa × $2pq$ Aa	$4pq^3$		$2pq^3$	$2pq^3$
q^2 aa × p^2 aa	q^4			q^4
Totals	$(p^2 + 2pq + q^2)^2 = 1$	$p^2(p^2 + 2pq + q^2) = p^2$	$2pq(p^2 + 2pq + q^2) = 2pq$	$q^2(p^2 + 2pq + q^2) = q^2$

Genotype frequencies = $(p+q)^2 = p^2 + 2pq + q^2 = 1$ in each generation afterward.

Gene (allele) frequencies = p(A) + q(a) + 1 in each generation afterward.

* For example, matings between AA and Aa will occur at $p^2 \times 2pq = 2p^2q$ for AA × Aa and at $p^2 \times 2pq = 2pq$ for Aa × AA, for a total of $4p^3q$. Two progress types AA and Aa, result in equal proportions from these matings. Therefore offspring frequencies are $2p^3q$ (*i.e.*, $1/2 \times 4p^2$a) for AA and for Aa.

EXERCISE
(Based on Unit–4)

Q.1 Give an introduction of basic microbial genetics.

Q.2 What is transformation ? Give the mechanism of transformation in bacteria.

Q.3 What is transductor ? How many types of transduction is there ?

Q.4 What is conjugation ? Give mechanism of bacterial conjugation ?

Q.5 Write short notes on:

(*i*) F^1 conjugation

(*ii*) Hfr conjugation

(*iii*) $F^+ \times F^-$ mating

Q.6 What are insertion elements ?

Q.7 Give an account of:

(*a*) Transposons

(*b*) Insertion sequence elements

Q.8 Write an essay on mitochondrial and chloroplast genetic system.

Q.9 Write short notes on:

(*i*) Mitochondrial protein synthesis

(*ii*) Mitochondrial DNA

(*iii*) Protein synthesis in chloroplast

(*iv*) DNA and ribosome of chloroplast.

Q.10 Write an essay on extrachromosomal inheritance.

Q.11 Write a note on extranuclear inheritance in eukaryotes.

Q.12 Write an essay on post translational control.

Q.13 Write short notes on:

(*i*) Pattern of regulation

(*ii*) Alteration of gene structure

Q.14 What are the internal and external environmental effects on gene expression ?

Q.15 Give detailed account on development and environmental regulation of gene expression.

Q.16 What are gene frequencies ? What are advantages of these types of frequencies ?

Q.17 What is Hardy-Weinberg's law ?

Q.18 Describe the assumptions of Hardy-Weinberg's law.

Q.19 Write an essay on population genetics.

Q.20 Write short note on genotypic frequency.

UNIT-5

29 MUTATIONS—SPONTANEOUS AND INDUCED

MUTATIONS—AN INTRODUCTION

The term mutation refers both to the change in the genetic material and to the process by which the change occurs.

Mutation is the ultimate source of all genetic variation, it provides the raw material for evolution. Spontaneous mutations are those that occur without a known cause. This class of mutation may result from errors in DNA replication or they may actually be caused by mutagenic agents present in environment. Spontaneous mutation occurs infrequently, although the observed frequencies vary from gene to gene and from organism to organism.

SPONTANEOUS MUTATIONS

Spontaneous mutations arise without exposure to external agents. This class of mutations may result from errors in DNA replication. A few of the more prevalent mechanisms are described in the following paragraphs.

Generally replication errors occur when the base of a template nucleotide takes on a rare tautomeric form. Tautomerism is the relationship between two structural isomers that are in chemical equilibrium and readily change into each other. Bases typically exist in the keto form. However, they can at times take either an enol form. These tautomeric shifts change the hydrogen bonding characteristics of the bases, allowing purine for purine or pyrimidine for pyrimidine substitutions that can eventually lead to a stable alteration of the nucleotide sequence. Such substitutions are known as transition mutations and are relatively common, although most of them are repaired by various proofreading functions. In transversion mutations, a purine is substituted for a pyrimidine or a pyrimidine for a purine. These mutations are rarer due to the steric problem of pairing purines with purines and pyrimidines with pyrimidines.

Spontaneous mutations also arise from frameshifts, usually caused by the deletion of DNA segments resulting in an altered codon reading frame. These mutations generally occur where there is a short strech of the same nucleotide. In such a location, the pairing of template and new strand can be displaced by the distance of the repeated sequence leading to additions or deletions of bases in the new strand.

Spontaneous mutations originate from lesions in DNA as well as from replication errors. For example, it is possible for purine nucleotides to be depurinated — that is, to lose their base. This results in the formation of an apurinic site, which will not base pair normally and may cause a transition type mutation after the next round of replication. Cytosine can be deaminated to uracil, which is then removed to form an apyrimidinic site. Reactive forms of oxygen such as oxygen free radicals and peroxides are produced by aerobic metabolism. These may alter DNA bases and causes mutations. For example, guanine can be converted to 8-oxo-7, 8-dihydrodeoxyguanine, which often pairs with adenine rather than cytosine during replication.

Finally, spontaneous mutations can result from the insertion of DNA segments into genes. This results from the movement of insertion sequences and transposons and usually inactivates the gene. Insertion mutations are very frequent in *E. coli* and many other bacteria.

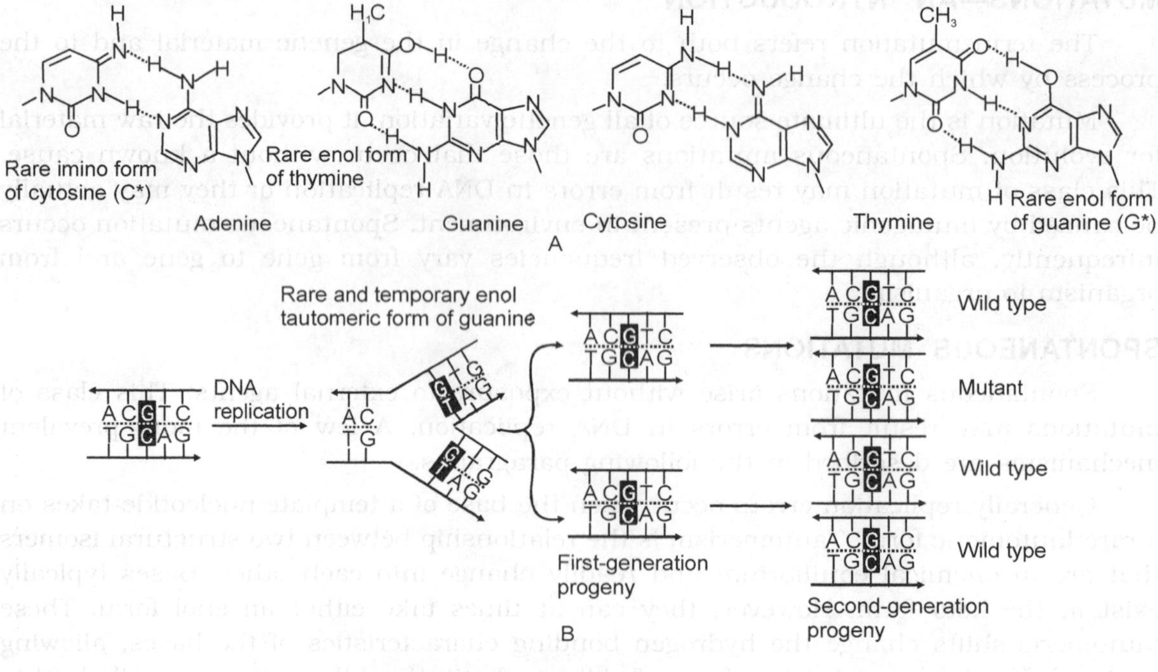

Fig. 29.1. Transition and Transversion Mutations. Mirrors in replication due to base tautomerization. A. Normally AT and GC pairs are formed when keto groups participate in hydrogen bonds. In contrast, enol tautomers produce AC and GT base pairs. B. Mutation as a consequence of tautomerization during DNA replication. The temporary enolization of guanine leads to the formation of an AT base pair in the mutant, and a GC to AT transition mutation occurs. The process requires two replication cycles. The mutation only occurs if the abnormal first-generation GT base pair is missed by repair mechanisms.

INDUCED MUTATIONS

Induced mutations are those resulting from exposure of organism to mutagenic agents such as Ionizing, Irradiation, ultraviolet light or various chemicals that react with DNA mutagens can be conveniently classified according to their mechanism of action. For common modes of mutagens action are incorporation of base analogs, specific mispairing, intercalation and by pass of replication. Virtually any agent that directly damage DNA, alters its chemistry, or interferes with repair mechanisms will induce mutations. Mutagens can be conveniently classified according to their mechanism of action. Four common modes of mutagen action are incorporation of base analogs, specific mispairing, intercalation, and bypass of replication.

Base analogs are structurally similar to normal nitrogenous bases and can be incorporated into the growing polynucleotide chain during replication. Once in place, these compounds typically exhibit base pairing properties different from the bases they replace and can eventually cause a stable mutation. A widely used base analog is 5-bromouracil (5-BU), an analog of thymine. It undergoes a tautomeric shift from the normal keto form to an enol much more frequently than does a normal base. The enol forms hydrogen bonds like cytosine and directs the incorporation of guanine rather than adenine (Fig. 29.1). The mechanism of action of other base analogs is similar to that of 5-bromouracil.

Specific mispairing is caused when a mutagen changes a base structure and therefore alters its base pairing characteristics. Some mutagens in this category are fairly selective; they preferentially react with some bases and produce a specific kind of DNA damage. An example of this type of mutagen is methyl-nitrosoguanidine, an alkylating agent that adds methyl groups to guanine, causing it to mispair with thymine. A subsequent round of replication could then result in a GC-AT transition. DNA damage also stimulates error-phone repair mechanisms. Other examples, of mutagens with this mode of action are the alkylating agents ethylmethanesulfonate and hydroxylamine. Hydroxylamine hydroxylates the C-4 nitrogen of cytosine, causing it to base pair like thymine. There are many other DNA modifying agents that can cause mispairing.

Intercalating agents distort DNA to induce single nucleotide pair insertions and deletions. These mutagens are planar and insert themselves (intercalate) between the stacked bases of the helix. This results in a mutation, possibly through the formation of a loop in DNA. Intercalating agents include asidines such as proflavin and acridine orange.

Many mutagens, and indeed many carcinogens, directly damage bases so severely that hydrogen bonding between base pairs is impaired or prevented and the damaged DNA can no longer act as a template. For instance, UV radiation generates cyclobutane type dimers, usually thymine dimers, between adjacent pyrimidines. Other examples are ionizing radiation and carcinogens such as aflatoxin B1 and other banzo (a) pyrine derivatives. Such damage to DNA would generally be lethal but may trigger a repair mechanism that restores much of the damaged genetic material, although with considerable error incorporation.

Mutagenesis by the Base Analog 5-Bromouracil. (*a*) Base pairing of the normal keto form of 5-BU is shown in the top illustration. The enol form of 5-BU (bottom illustration) base pairs with guanine rather than with adenine might be expected for a thymine analog. (*b*) If the keto form of 5 BU is incorporated in place of thymine, its occasional tautomerization to the enol form (BU_e) will produce an AT to GC transition mutation.

Retention of proper pairing is essential in the prevention of mutation. Often the damage can be repaired before a mutation is permanently established. If a complete DNA replication cycle takes place before the initial lesion is repaired, the mutation frequently becomes stable and inheritable.

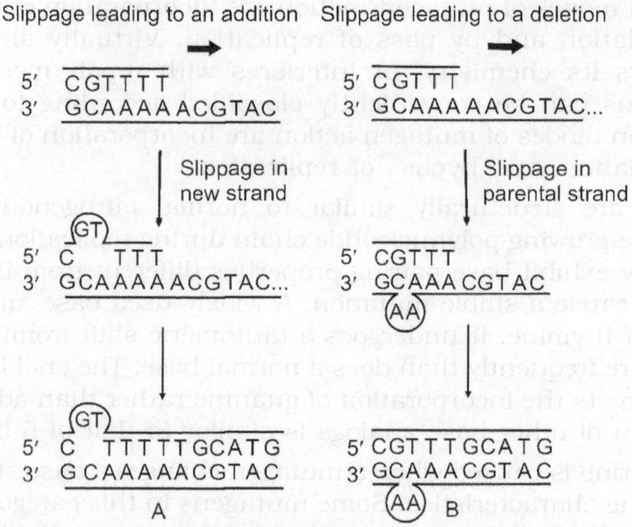

Fig. 29.2. Additions and Deletions. A hypothetical mechanism for the generation of addition and deletion during replication. The direction of replication is indicated by the large arrow. In each case there is grand slippage resulting in the formation of a small loop that is stabilized by the hydrogen bonding in the repetitive sequence, the AT stretch in this example. DNA synthesis proceeds to the right in this figure. A. If the new strand slips an addition of one T results B. Slippage of the parental strand yields a deletion (in this case a loos of to Ts).

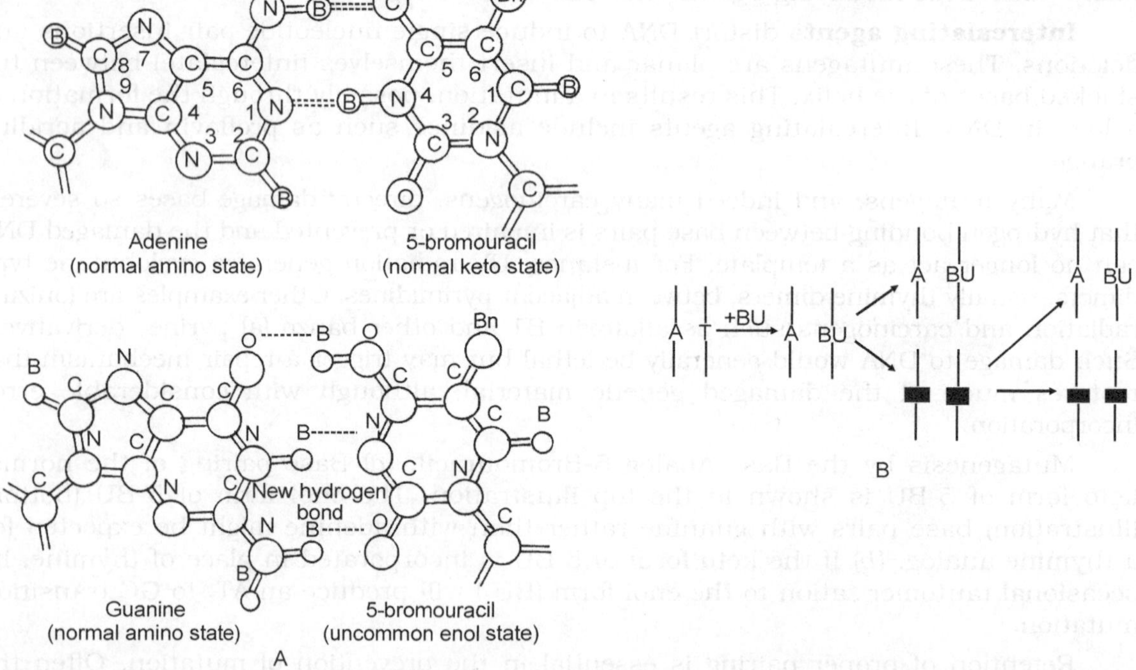

Fig. 29.3. Mutagenesis by the Base Analog-5-Bromouracil. A. Base pairing of the normal keto form of 5-BU is shown in the illustration. The enol form of 5-BU (bottom illustration) has pairs with guanine rather than with adenine straight be experts for a thymine analog. B. If the keto form as 5 BU incorporated in place of thymine, its occasional tautomerization to the end of (BU) will produce an AT to GC transition mutation.

30 CHEMICAL AND PHYSICAL MUTAGENS

INTRODUCTION

Muller's discovery of the mutagenic effect of irradiation provided a method for the induction of large numbers of mutations with a minimal amount of efforts. However, because of the large number of effort effects of radiation on living tissues, both primary and secondary studies on radiation—induced mutation provided little information about the molecular details of mutation processes. Subsequent discoveries of the mutagenic effects of a large variety of chemicals, many with very specific effects on DNA, provided the tools needed to work out many of the details of mutation processes at the molecular level. Hundreds of chemicals are known to have from slight to very large mutagenic effects.

5-Bramodeoxyuridine (5-BDU) can replace thymidine in DNA.

2-Aminopurine (2-AP) and *2,6-Diaminopurine* (2,6-DAP) are purine analogues.

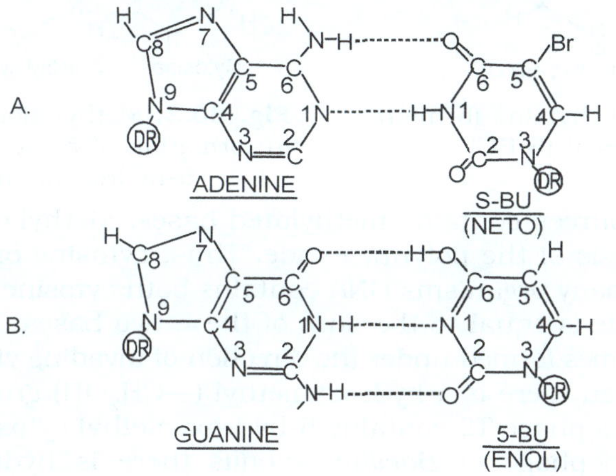

Fig. 30.1. Tautomerism of 5 bromouracil (5–BU).

A. Regular base pairing of adenine with 5–bromouracil in the normal keto form.

B. Forbidden base pairing of 5–BU (in the rare enol form) with guanine.

The first chemical mutagen discovered was mustard "gas". In the 1950s, chemical mutagens with more or less specific action were developed. Chemical mutagens can be classified according to the way in which they bring about mutations: (*i*) base analogues which are incorporated into DNA instead of normal bases, (*ii*) agents modifying purines and pyrimidines and agents labilizing bases, and (*iii*) agents producing distortions in DNA.

The agents in categories (*i*) and (*iii*) require replication for their action, while agents in category (*ii*) can modify even non-replicating DNA.

Base Analogues

A chemical substance resembling a base is called a base analogue. A base analogue may be incorporated into newly synthesized DNA instead of a normal base.

The pyrimidine analogue 5-bromouracil (5-BU) is structurally very much similar to thymine. If bacteriophages are grown in the presence of 5-BU they incorporate the substance as thymine. 5-BU does not have a lethal action because it is incorporated in place of T and functions almost normally. 5-BU can, however, undergo internal arrangement (tautomerization) from the usual keto state to the rare enol state. 5-BU now pairs with guanine instead of adenine, the natural partner of thymine. Thus, there is 5-BU-G pairing instead of T-A pairing. Because of this property, 5-BU is used in the chemotherapy of viral infections and cancer. By pairing with guanine disturbes the normal replication mechanism of micro-organisms.

5-Bromodeoxyuridine (5-BDU) can replace thymidine in DNA.

2-Aminopurine (2-AP) and 2, 6 diaminopurine (2, 6-DAP) are purine analogues.

2-Aminopurine can be read as either adenine or guanine. It normally pairs with thymine but can also form a single hydrogen bond with cytosine. It can therefore produce A-T→G-C transitions. 2-AP and 2,6-DAP are less effective as mutagens than 2-BU and 5-BDU.

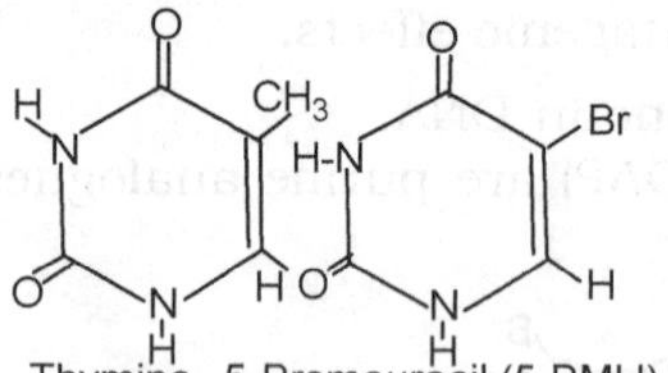

Thymine 5-Bromouracil (5-BMU)

Fig. 30.2. Structure of thymine (5–methyl uracil) and (5–bromouracil (5 BU).

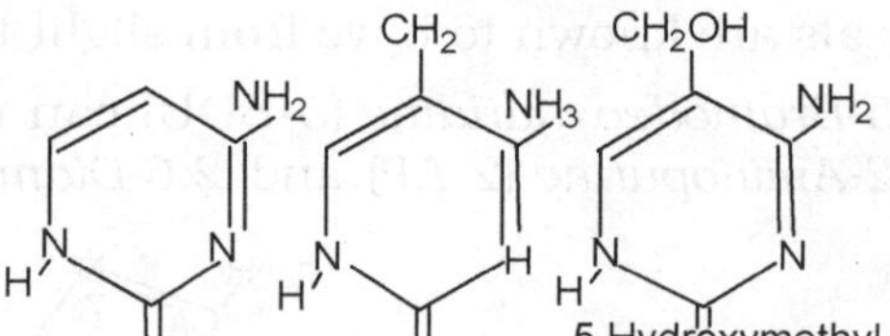

Cytosine 5 methyl cytosine 5 Hydroxymethyl cytocsine

Fig. 30.3. Methylation of cytosine resulting in formation of 5–methyl cytosine and 5–hydroxymethyl cytosine.

DNA from any sources contains methylated bases. Methylation of the bases takes place after the synthesis of the polynucleotide. Thus, cytosine on methylation becomes 5-methylcytosine. In many organisms DNA contains both cytosine and 5-methyl cytosine. The amount of guanine is equal to the sum of these two bases. Methylation appears to protect DNA from enzymes formed under the direction of invading viruses. 5-hydroxymethyl cytosine is formed when there is a hydroxymethyl (—CH_2OH) group at the fifth position of cytosine. The bacteriophage T2 contains 5-hydroxymethyl cytosine instead of cytosine. Similarly in a bacteriophage of *Bacillus subtilis* there is hydroxymethyl uracil and 5-dihydroxypentyl uracil instead of thymine. It should be noted that the methylated bases mentioned above are normal constituents of DNA and are not mutagens.

Agents Modifying Purines and Pyrimidines

Agents which labilize the bases include nitrous oxide, hydroxylamine and alkylating agents.

(*i*) **Nitrous oxide** (HNO_2) reacts with bases containing amino groups. It can change the structure of such bases by deamination (removal of the amino group). When purines or pyrimidines containing the amino group are treated with nitrous oxide, the amino group ($-NH_2$) is replaced by the hydroxyl group (—OH). The order of frequency of deamination is adenine, cytosine and guanine.

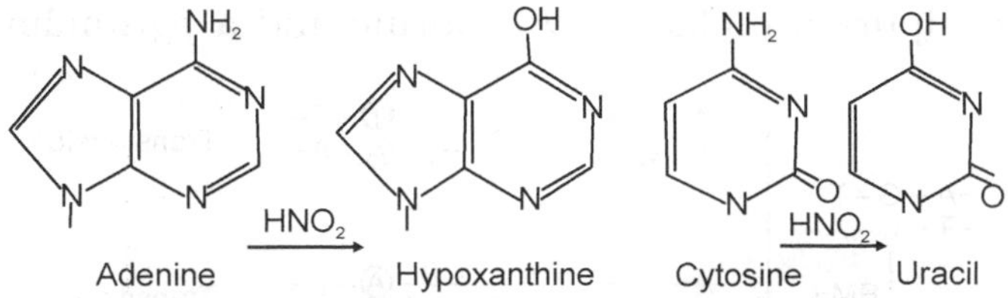

Fig. 30.4. Deamination by nitrous acid.

Deamination of *adenine* results in the formation of hypoxanthine. The pairing behaviour of hypoxanthine is like that of guanine. Therefore, hypoxanthine pairs with cytosine rather than with thymine. Thus, A-T pairing is replaced by G-C pairing.

The deamination of cytosine (at the 6-position) results in the formation of uracil. The hydrogen bonding properties of uracil are similar to those of thymine. Therefore, instead of C-G pairing there is U-A pairing.

Guanine is deaminated to xanthine. There is no change in pairing behaviour in this case, because xanthine behaves like guanine and pairs with cytosine. Instead of G-C pairing there is X-C pairing. Thus, deamination of guanine is not mutagenic.

Table 30.1. Changes in Structure and Pairing Behaviour of DNA Bases as a Result of Deamination by Nitrous Oxide

Normal Bases of DNA	*Normal Pairing*	*Bases Formed by Deamination*	*New Pairing*
Adenine	A-T	Hypoxanthine	G-C
Cytosine	C-G	Uracil	U-A
Guanine	G-C	Xanthine	X-C

The base formed after deamination of adenine and cytosine have a different pairing behaviour. As a result, changes in DNA take place in 50% of the progeny. Deamination of guanine, however, does not result in a heritable mutation, since there is no change in the pairing behaviour of the deaminated base (xanthine).

(*ii*) **Hydroxylamine** (NH_2OH) is very specific in its action. It reacts mainly with cytosine and guanine residues and brings about transition and mispairing. It deaminates cytosine to a base which pairs with adenine instead of guanine. Thus, C-G pairing is changed to A-T pairing.

(*iii*) **Alkylating agents** are the most widely used mutagenic reagents. They include:

dimethyl sulphate (DMS)

ethyl methane sulphonate (EMS) — $CH_3CH_2SO_3CH_3$

ethyl ethane sulphonate (EES) — $CH_3CH_2SO_3CH_2CH_3$.

The main chemical reaction of these agents is alkylation at the N-7 position of guanine residues or at the N-3 position of adenine residues. Alkylation increases the probability of ionization and introduces pairing error. The base-sugar linkage undergoes hydrolysis and releases the base from the DNA molecule. This creates gap in one chain.

EMS specifically removes guanine from the chain. During replication the chain without gaps will give rise to normal DNA. In the chain with gaps, however, any base (A, T, G or C) may be inserted across the gap. This may be a correct base or an incorrect one. In the next replication the gap is filled by a base which is complementary to the inserted base. Where the correct base is inserted the DNA is normal. Insertion of an incorrect base may result in a transversion (purine replaced by a pyrimidine and vice versa) or a transition (purine replaced by a purine and a pyrimidine by a pyrimidine).

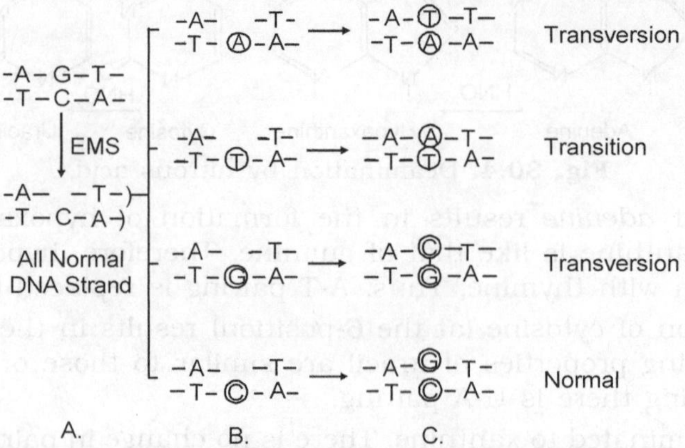

Fig. 30.5. Effect of the alkylating agent ethyl methane sulphonate (EMS) on DNA. A. EMS removes G and creates a gap in DNA. B. 1st replication, A/T/G/C inserted across the gap. C. 2nd replication. Complementary base fills the gap.

Agents Producing Distortions in DNA

Certain flourescent acridine dyes such as proflavine and acridine orange cause mutations by insertion or deletion of bases. Crick's work on acridine mutants has provided strong evidence for the genetic code. The acridines are planar (flat) molecules, like the purine bases, and can be intercalated between the bases of the DNA helix. This distorts the structure of DNA and can result in deletion or insertion of bases during recombination.

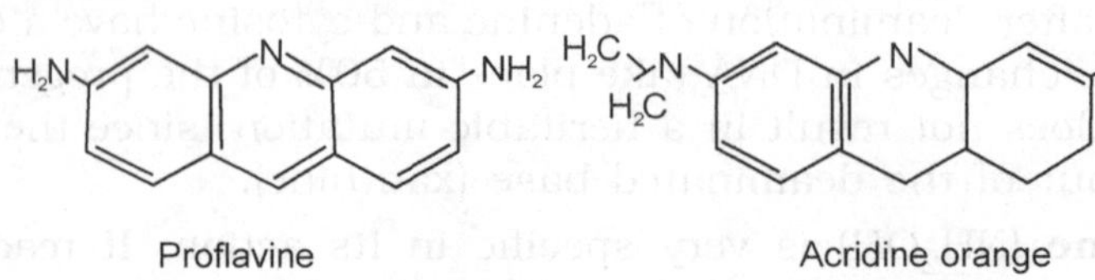

Fig. 30.6. The acrine dyes proflavine and acridine orange.

(*i*) **Intercalation resulting in insertion of base.** Intercalation of the acridine molecule between two bases of the template strand results in the lengthening of the DNA molecule. During replication a base (X') is inserted at random opposite the acridine molecule in the new chain. In the next replication a complementary base (X) will pair with the newly inserted base. Thus, the new DNA has an additional base pair.

(*ii*) **Intercalation resulting in deletion of base.** The acridine molecule may be inserted in the new chain during synthesis. This blocks the base in the template strand and does not permit any base to pair with it. The chain produced is deficient in one base, and in the next replication produces DNA with a deficient base pair.

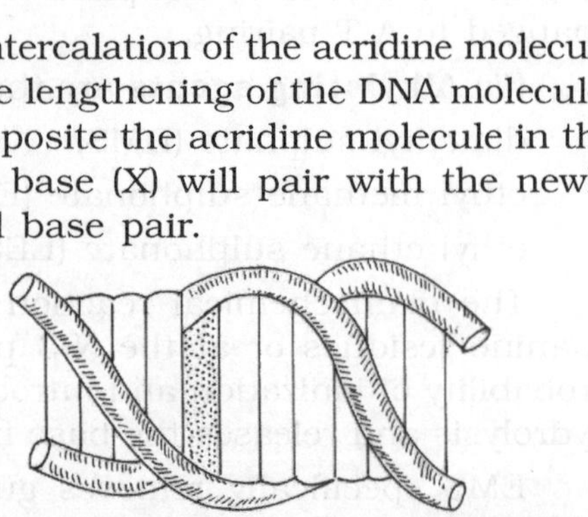

Fig. 30.7. Insertion of acridine dye molecule (block) between bases of DNA.

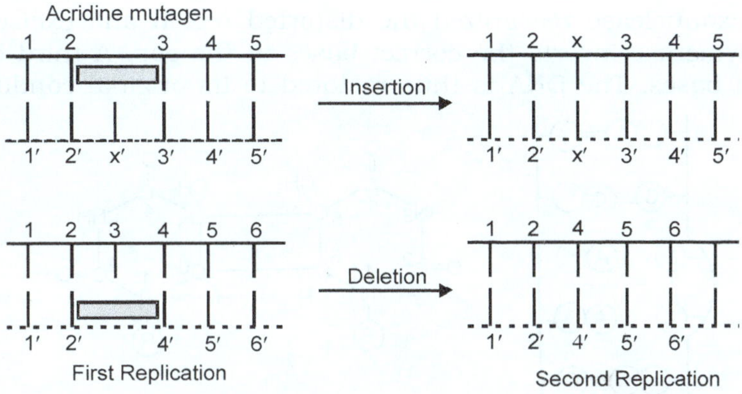

Fig. 30.8. Diagram of mutagenic action of acridine dyes.

RADIATION

Among the physical mutagens, radiation is the most important. The energy content of a radiation depends upon its wavelength. In general, the shorter the wavelength the greater the energy value of the radiation. High-energy radiations can change the atomic structure of a substance by causing the loss of an electron and the formation of an ion. Sometimes an electron pair may be moved from an inner to an outer orbital shell. This brings about excitation of the atom. In this excited state the atom is highly reactive and is called a free radical. Radiation which brings about such state is called ionizing radiation.

Alterations in nucleic acids caused by radiation are of great genetic importance. High energy ionizing radiation and ultraviolet (UV) light are important mutagenic agents. Ionizing radiation has greater penetration power than UV-radiation and produces free radicals which tend to labilize molecules. This type of radiation causes single-strand breaks in DNA and produces deletions.

Both DNA and RNA preferentially absorb UV-light, causing their nitrogen-containing bases to become highly reactive free radicals. The resulting unstability causes the conversion of one base to another (a purine to another purine or a pyrimidine to another pyrimidine). If this change occurs in mRNA only a few inactive proteins will be formed, because mRNA is soon broken down. Substitutions in DNA, however, may have a lasting effect. All the proteins coded by the DNA may be defective. Moreover, if the mutation happens to take place in germ cells the mutated DNA strands could be passed on to succeeding generations.

The primary mutagenic effect of UV-light appears to be due to the production of thymine dimers (Fig. 30.9). The 5, 6 unsaturated bonds of adjacent pyrimidines become covalently linked to form a cyclobutane ring. Irradiation of a bacterial culture and subsequent extraction of DNA yields three possible types of pyrimidine dimers in DNA:

Thymine–thymine – 50%

Thymine–cytosine – 40%

Cytosine–cytosine – 10%

Pyrimidine dimers can also be formed between adjacent strands. In RNA, pyrimidine dimers are formed between adjacent uracil and cytosine rings. Pyrimidine dimers cannot fit into the DNA double helix and cause distortion of the molecule. If the damage is not repaired, replication is blocked, leading to lethal effects.

Distortions in DNA caused by thymine dimers can be corrected by a repair mechanism. An exonuclease recognized the distorted region and excises it. A second enzyme, DNA polymerase inserts the correct bases in the gap. A third enzyme, ligase, joins the inserted bases. The DNA is thus restored to its original condition.

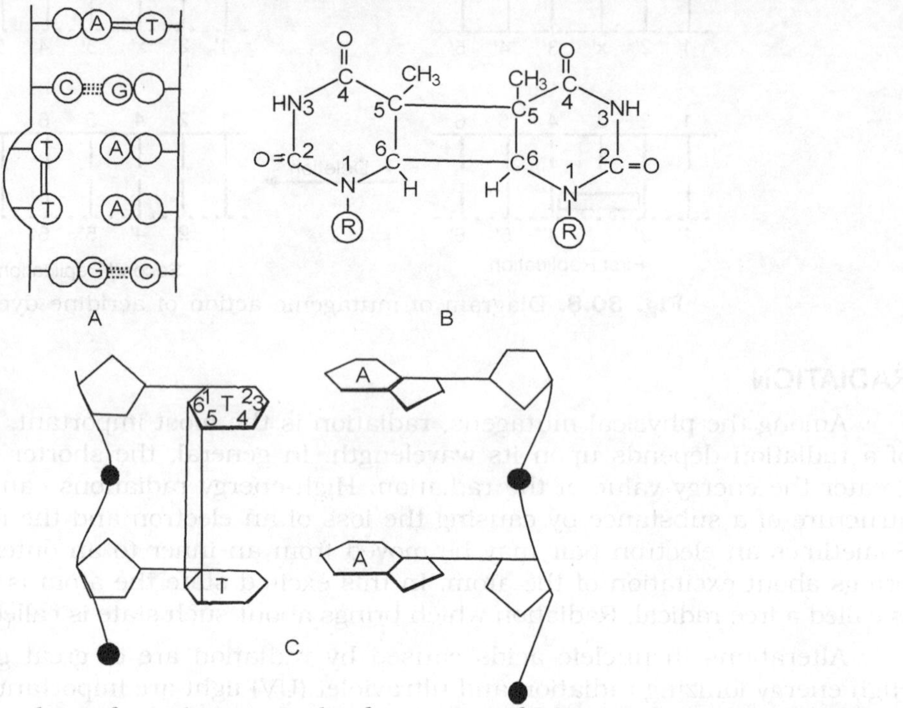

Fig. 30.9. Thymine dimer formed as a result of exposure of DNA to ultraviolet radiation.

A. Distortion of DNA by thymine dimer.

B. Molecular structure of a thymine dimer.

C. Linking of two adjacent thymine residues to form a dimer.

UV-radiation also causes addition of water molecules to pyrimidines in both DNA and RNA, resulting in the formation of photohydrates. The water molecule is added across the C5-C6 double bond.

X-rays bring about mutations by breaking the phosphate ester linkages in DNA. The breakage may takes place at one or more points. As a result a large number of bases are lost (deletion) or rearranged. In double stranded DNA breaks may occur in one or both strands. Only the later type is lethal. Sometimes two double-stranded breaks may occur in the same molecule and the two broken ends may rejoin. The part of the DNA. between the two breaks is eliminated, resulting in a deletion. The damage caused to nucleic acids by UV-light and X-rays is utilized to sterilize bacteria and viruses.

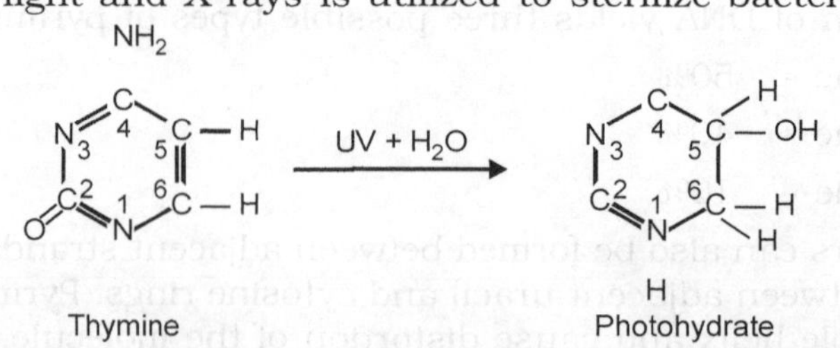

Fig. 30.10. Formation of photohydrate by ultraviolet radiation.

31 INDUCED MUTATIONS IN MICROBES AND ANIMALS FOR ECONOMIC BENEFIT OF MAN

INTRODUCTION

Microbes are one of the most important partners in our life. Their presence is felt by human beings. Hence one cannot forget the gifts which have been given by these organisms to this world. The beneficial contribution of the micro-organisms to the world are as follows:

MICRO-ORGANISMS AS SOURCE OF FOOD

It is estimated that worldwide food protein deficiency was about 12 million ton in 1985, and likely to increase to approximately 23 million tonn by 2000. SCP has developed hope to solve this problem. Remarkable increase in productivity is also one of the achievements of what come to be known as green revolution and it could be possible by sincere efforts of agricultural scientists and by microbiologists. After the Second World War, research works on the new sources of food have been done all over the world which have increased remarkable food production and productivity through crop improvement and biotechnological processes. This was the first phase of green revolution. In broad sense, biotechnology has three impacts on the development of agriculture: (*i*) with the improvement of genetic engineering techniques, the time for generating and evaluating new germplasm can be drastically reduced, (*ii*) chemical and biotechnological processes have opened new frontiers to products of improved value from agricultural raw materials and (*iii*) new strains of micro-organisms for food production. It is imperative that new food sources be found in order to that future generations of mankind be adequately fed. A food source that is nutritionally complete and that requires a minimum of land, time and cost of produce is highly desirable.

Micro-organisms are tackle the food problem by improving the fertility of soil. They are capable of rapidly transforming agricultural wastes, sludge etc., into organic matter of manurial value. The availability of essential elements to crop depends upon the microbial activities, which are also responsible for the improvement of the soil. Micro-organisms, specially bacteria and blue green algae, fix atmospheric nitrogen and enrich the nitrogen contents of soil. From a better soil, farmers can get good crops.

Yeast are also utilized as food. The large scale production of pest is referred as Microbial farming. Food yeast contains vitamins, like thiamin, riboflavin, nicotinic acid, biotin etc. Mushrooms have fascinated man since time immemorial. About 2000 species of edible mushrooms have been reported from all over the world. *Agaricus campestris* is the common field edible mushroom. A few other edible mushrooms are *Lentinus, Volvariella,*

Collybia volutipes etc. *Morchella* is another well known edible mushroom popularly called **gucchi**. The genus **Tuber** and **Lycoperdon** are also edible at young stage. Japan and China are producing several million pounds of mushrooms for food. Certain states of India like Himachal Pradesh, Tamil Nadu and Uttar Pradesh are advancing forward the cultivation of mushrooms. Chemically mushrooms are very high in protein content. It is about 51%, which is more than any other comparable fruit or vegetable. Apart from protein, mushroom are the good source of nicotinic acid, riboflavin, panthothenic acid, vitamin B, C and K, biotin and thiamine. Carbohydrate content, however, vary from 28–76%.

SINGLE CELL PROTEIN

Unicellular micro-organisms as a direct source of food for man were suggested in the early 1900s. The expression Single Cell Protein (SCP) was coined at the Massachusetts Institute of Technology around 1966 to depict the idea of micro-organisms as food sources. Although SCP is a misnomer in that proteins are not the only food constituent represented by microbial cells, it obviates the need to refer to each product generally as in **algal protein, yeast cell protein** and so on. Although SCP as a potential and real source of food for man differs from the other products.

In addition to meeting these criteria, SCP can be produced on a variety of waste materials. Among the overall advantages of SCP over plants and animal sources of proteins are the following:

1. Micro-organisms have a very short generation time and can thus provide a rapid mass increase.
2. Micro-organisms can be easily modified genetically to produce cells that bring about desirable results.
3. The protein content is high.
4. The production of SCP can be based upon raw materials readily available in large quantities.
5. SCP production can be carried out in continuous culture and thus it is independent of climatic changes.

The greater speed and efficiency of microbial protein production compared to plant and animal sources may be illustrated as follows: a 1,000-lb, steer produces about 1 lb, of new protein/day, soyabeans (prorated over a growing season) produce about 80 lb; while yeasts produce about 50 tons.

MICRO-ORGANISMS AND THEIR FERMENTATION SUBSTRATES

Single cell protein (SCP) can be obtained from certain members of algae, cyanobacteria, yeasts molds and higher fungi. These plant species are harvested and dried for use of protein source. The SCP is characterised by high percentage of protein by weight (sometimes upto 85%). For the production of SCP from plant sources, a variety of raw materials are required which serve as carbon source including solar radiation (for photoautotrophic plant species). The important are CO_2, $NaHCO_3$, sunlight (*Spirulina maxima*, protein 60%), CO_2 and sunlight (*Scenedesums acutus*, protein 55%), Bagasse (*Cellulomonas*, protein 85%), molasses (*Saccharymyces cerevisiae*, protein 50%). The SCP is being manufactured under different commercial names like Brovile, Marmite, Pruteen, etc. Pruteen is manufactured by Imperial Chemical Industries, UK from the bacterium *Mythylophilus methylotropous* which is grown on methyl derived from natural gas. The company produces about 50,000 ton of Pruteen evey year as animal feed. Similarly Rank Hovis McDougall Company, UK produce SCP under the trade name Mycoprotein using potato, wheat, cassava as carbohydrate source for protein production. A large number of algae, yeasts, molds and bacteria have been studied as SCP sources. Among the most promising genera and species are the following:

Algae: *Spirulina maxima, Chlorella sp., and Scenedesmus sp.*

Yeasts: *Candida guilliermondii, C. utilis, C. lipolytica and C. tropicalis; Debaryamyces kloeckeri; Torulopsis candida, T. methanosorbosa; Pickia sp.; Kluyveromyces fragilis; Hansenulo polymorpha; Rhodotorula sp.; and Saccharomyces sp.*

Filamentous fungi: *Agaricus sp.; Aspergillus sp.; Fusarium sp.; Pencillium sp.; Edomycopsis fibuligera; and Trichosporon cutaneum.*

Bacteria: *Bacillus sp.; Cellulomonas sp.; Acinetobacter calcoaceticus; Nocardia sp.; Methylomonas sp.; Aeromonas hydrophilia; Alcaligenes eutrophus (Hydrogenomonas eutrophe), Mycobacterium sp.; and Rhodopseudomonas sp.*

Choice of Fermentable Materials

Bacteria, yeasts and molds can be grown on a wide variety of fermentable materials including food processing wastes (such as cheese, whey and brewery, potato processing, cannery, and coffee wastes); industrial wastes (such as sulfite liquor in the paper industry and combustion gases; and cellulosic wastes (including bagasse, newsprint mill, and barley straw). In the case of cellulosic wastes, it is necessary to use organisms that can utilize cellulose, such as *Cellulomonas sp. or Trichoderma viride.* A mixed culture of a combination of *Endomycopsis fibuligera* and a *Candida sp.* such as *C. utilis* has been employed, in which the former effects hydrolysis of starches and the latter subsists upon the hydrolyzed products to produce biomass. Some other representative substrates and fermenting organisms are listed in Table 1.

Table 31.1. Some Examples of Substrate Materials that Support the Growth of Micro-organisms in the Production of SCP.

Substrates	*Micro-organisms*
CO_2 and sunlight	*Chlorella pyrenoidosa, Scenedesmus quadricauda, Spirulina maxima,*
n-Alkanes, Kerosene	*Candida intermedia, C. lipolytica, C. tropicalis, Nocardia sp.*
Methane	*Methylomonas sp. (Methenomonas), Methylococcus capsulatus sp., Trichoderma*
H_2 and CO_2	*Alcaligenes eutrophus (Hydrogenomonas eutropha)*
Gas oil	*Acinetobacter calcoaceticus (Micrococcus cerificans), Candida lipolytica*
Methanol	*Methylomonas methanica (Methanomonas methanica), Candida utilis*
Ethanol	*Acinetobacter calcoaceticus*
Sulphite liquor wastes	*Candida utilis*
Cellulose	*Cellulomonas sp. Trichoderma viride*
Starches	*Endomycopsis fibuligera*
Sugars	*Saccharomyces cerevisiae, Candida utilis, Kluyuveromyces fragilis.*

Spirulina (a cyanobacterium) offers a rich source of SCP. *S. platensis* contains about 60% crude protein and is rich in tryptophan, lysine and vitamin B_{12}. In India, NBRI, Lucknow and CFTRI, Mysore have started mass culture and utilization of this cyanobacterium as a source of SCP. *S. platensis* is cultured in sewage water containing

9 g/1 of sodium bicarbonate and 1 g/1 of sodium nitrate. The average yield of the micro-organisms varies between 6 g/m^2/day and 9 g/m^2/day on dry weight basis. Similarly A.M.M. Marugappa Chettiar Research Centre, Madras, has isolated *Spirulina fusiformis* and produced SCP after filtration from cloth filters (30 mesh/inch).

SCP is also produced from paper industry waste, because the industry faces a serious problem in disposing one of the waste products—sulphide liquor. If this pollutant is discharged in the river, it quickly depletes the water's oxygen reserves and causes death of fishes. The sulphide liquor is fed to *Paecilomyces molds*. This process is known as Pelilo. This process purifies the waste liquids from paper factories but also yields a rich harvest of microbes which are sold as animal feed.

SCP Products

While the products of fermentation of other foods in this section consist of an altered or modified substrate, the product of SCP fermentation consist solely of the microbial cells that grow upon the substrate. The cells may be used directly as a protein source in animal feed formulations, hereby freeing animal feed, such as corn, for human consumption; or they may be used as a protein source or food ingredient for human food. In the case of animal feed or food supplements, the dried cells may be used without further processing. As noted above, whole cells of *Spirulina maxima* are consumed by man in at least one part of Africa.

Forms of SCPs for Human Use

For human use, the most likely products are SCP concentrates or isolates that may be further processed into textured or functional SCP products. To produce functional protein fibres, cells are mechanically disrupted, cell walls removed by centrifugation, proteins precipitated from disrupted cells, and the resulting protein extruded from syringe like orifices into suitable menstra such as acetate buffer, $HClO_4$, acetic acid etc. The SCP fibres may now be used to form textured protein products. Baker's yeast protein is the only product of this type presently approved for human food ingredient use in the United States, but as *Litchfield* has noted, it is reasonable to assume that similar products from other food-grade yeasts such as *Candida utilisa* and *Kluyveromyces fragilis* will ultimately be approved.

Success has been achieved in rat-feeding studies with a variety of SCP products, but human feeding studies have been less successful, except in the case of certain yeast cell products. Gastrointestinal disturbances are common complaints in man following the consumption of algae and bacteria SCP, and other problems associated with the consumption of SCP have been reviewed elsewhere. When gram-negative bacteria are used as SCP sources for human use, the endotoxins must be removed or detoxified.

Microbes in Food Processing

The knowledge of bread-making by fermentation dates back very ancient times. Fermentation is a metabolic process characterized by incomplete oxidation. The process where oxygen is used as the final acceptor, is known as oxidative fermentation, *e.g.*, acetic acid, citric acid, fumaric acid and oxalic acid fermentation. Those in which organic substances, particularly aldehydes act as acceptors are referred as an oxidative fermentation. The fermentation of alcohol is an example of anoxidative fermentation. It consists of conversion of invert sugars into carbon dioxide and ethyl alcohol by certain strains of micro-organism. The micro-organisms which are responsible for anoxidative fermentation are yeast species of *Saccharomyces, Torulopsis, Kloekera, Candida* and certain species of mucor and bacteria.

Microbes in Bread Preparation

During the preparation of common bread yeast is added to a part of flour having waste and salt. A little sugar is also added. Yeasts convert soluble sugar into alcohol and carbon dioxide. This carbon dioxide causes swelling.

MICROBES IN INDUSTRY

Alcoholic Fermentation

Ethyl alcohol is considered to be the oldest product of fungi. Many varieties of fungi bring about this alcohol through fermentation of sugar. Yeasts (*Saccharomyces spp*) are most commonly used in this connection. There are two species of Saccharomyces; one is the top fermenter (*S. cervisiae*) and the second is bottom fermenter (*S. carlsbergonsis*). Alcohol is usually produced by fermentation of molasses which is an aerobic process. Micro-organisms utilize the carbohydrate present in molasses to obtain energy for growth and metabolic activities lead to the formation of alcohol. The reaction takes place via the Embden Mayerhof Paranas (EMP) pathway.

There are three main by-products of alcohol industry—carbon dioxide, higher alcohol (isoamyl alcohol, butyl alcohol, collectively named as fuel oil) and alcohol. Alcohol is often used in manufacturing of ether, acetaldehyde, acetic acid, butanol, etc. Ethyl alcohol is one of the best known solvent of hundreds of different materials from which various compounds are synthesized. As a solvent, it is used in manufacture of such widely diverse materials as dye, soaps, plastics, polishes, synthetic rubber etc.

Butyl alcohol and acetone are also manufactured by the fermentation, which is brought about by the bacterial activity (*Clostridium acetobutylicum*) on molasses. Both chemicals are of great industrial use.

Microbes in Organic Acid Production

A number of organic acids are produced by micro-organisms specially moulds and bacteria through the process of fermentation. These are as follows:

Citric acid: It is the most important product and is now produced commercially after fermentation by moulds like *Aspergillus niger*, *Citromyces sp.* and certain species of Mucor. Apart from these fungi, *Pencillium* can also produce citric acid from sugar. The acid is used in manufacturing citrates, flavouring citrates, soft drinks etc.

Gluconic acid: *Aspergillus niger* synthesizes gluconic acid from sugar. Calcium, gluconate cures cow's suffering from milk fever. It is also used in increasing calcium content of egg and in the preparation of toothpaste.

Gallic acid: *Aspergillus gallomyces* and *Penicillim glaucum* synthesize gallic acid.

Kojic acid: It is extracted from the mycelium of *Aspergillus oryzae.*

Acetic acid: Bacteria like *Mycoderma aceti* synthesize acetic acid from fermentation of sugary solution and is used in vinegar production like *Acetobacter aceti. A. xylinim* and *A. rances.*

Microbes in Enzyme Production

Various micro-organisms like bacteria, fungi and protozoa produce different kinds of enzymes. They produce extracellular enzymes to break those organic and inorganic compounds, which have higher molecular weights. Simpler substances can easily be assimilated through cell membranes of micro-organisms.

Bacillus mesentericus is used for commercial production of amylase and protease. *Bacillus macerans* and *B. polymyxa* are also used for amylase production. Bacterial protease is employed to desize cotton and degum silk.

Fungi (*Aspergillus oryzae*) also synthesize amylase. Amylase is a starch digesting enzyme and is generally used in conversion of starch to fermentable sugars in alcohol industries. The commercial scale production of amylase is carried out by yeasts. Yeasts (*Saccharomyces cervisiae*) also produces an important enzyme called invertase which is used in production of confections. It brings about a partial hydrolysis of sugar into glucose and fructose. Apart from these important enzymes fungi, such as *Penicillium glaucum* and *Aspergillus niger* also produce enzyme like pectinase and cellulose oxidase respectively.

Microbes in Fat Production

Micro-organisms, specially fungi synthesize fats. The most common fungi of this kind are: *Aspergillus nidulans, A. sydowi, A. fisheri, Penicillium piscarum, P. javinicus. Endomyces vernatis* etc.

Microbes in Retting of Fibres

Retting is the process in which plant fibres (jute, hemp, flax) are separated by the activity of micro-organisms. Plants are immersed in water for a considerable period. During the period, they absorb water and swell. The water medium becomes rich in harbouring bacteria (*Clostridium butyricum*) which hydrolyze the pectic substance of the middle lamella of fibre tissues. Consequently the fibres get separated and are then used in preparing ropes, sacks etc.

Curing of Tobacco and Tea

Flavour and taste of tobacco and tea leaves are due to the fermentation of these leaves which is brought about by certain bacteria like *Micrococcous candisans* by the process referred as curing.

MICROBES IN DAIRY INDUSTRY

Milk and Curd

Milk is a complete food but it can also transmit diseases. It is an excellent nutritive medium for the growth of certain micro-organisms like bacteria, yeasts and moulds. These microbes alter the physical and chemical properties of milk. The common souring and curding of milk is brought about by the activity of bacteria (*Streptococcus lactis*) which is to convert lactose milk sugar into lactic acid by the action of enzyme produced by them. The production of lactic acid gives milk a sour taste and in turn causes the precipitation of casein. These bacteria can be eliminated by heating the milk at 140°F continuously for 30 minutes. This heating process is called holding process. Flash pasteurization is an another method which involves heating of milk at 160–165 °F for 15 seconds and rapid cooling. This method also does not permit the growth of bacteria. The whole technique is referred to as pasteurization. This process does not affect the digestibility of milk. Pasteurization has two advantages: (a) reduction in the spread of pathogens through milk and (*b*) an improvement in preserving its quality.

Butter

Butter is composed of about 80% fat and 10–16% water. At first, cream is separated from milk and it is then pasteurized to destroy the micro-organisms present in it. Starter of the bacterial culture is then added which consists of lactic acid bacteria like *Steptococcus lactis*.

These bacteria curdle and coagulate the milk. Apart from *S. lactis* other bacteria like *Leuconostoc cirtovorum, S. paracitrovorus* are also added. These bacteria act on citric acid producing the products, such as diacetyl to impart flavour and aroma. At the required time the ripening is checked by freezing and the product is churned to form butter.

Cheese

It is an another important item of dairy product. It is concentrated constituent of milk, principally fat, casein and insoluble salts. To retain these constituents in concentrated form, milk is coagulated either by fermenting bacteria, which produce lactic acid, or by the addition of rennet. Fermentation makes the casein spongy, soft and tasty.

Ice Cream

It is made of cream and sugar and contains about 8–14% butter fat. Ice cream also contains some non-pathogenic bacteria.

MICROBES IN MEDICINE

Micro-organisms are of an enormous significance in medicine. Some important drugs synthesized by micro-organisms are as follows:

Antibiotics

An antibiotic is a substance which is produced by living organisms and has the capacity to inhibit the growth or metabolic activities of other living organisms. Green alga *Chlorella* produces chlorellin—an antibiotic, which is known to inhibit the growth of several species of bacteria and is effective against other algae. Bacteria and fungi are the well known groups which produce various kinds of antibiotics (see chapter Antibiotics in this book). Apart from these micro-organisms, certain protozoans, *e.g.*, *Paramecium aurelia* also produce antibiotic.

Ergotin

Ergot of rye (the sclerotia of *Claviceps purpurea*) has been used as a drug to induce uterine contraction in case of delayed child birth. This, like almost any other drug, can even fatal if taken frequently or in excess quantity.

Ephedrine

This drug has been synthesized by yeasts from benzaldehyde and is often used against troubles and asthma.

Vitamins

They are synthesized by micro-organisms like bacteria and fungi. In 1958, *H. A. Barker* and *H. Weissbarch* discovered that cell-free-extract of *Clostridium tetanomorphum* produce vitamin B. It has been discovered recently that *Streptomyces griseus* and *S. aurofacius* produce vitamin B_{12}. This vitamin is used as a preventive measure against nervous disorder. Fungi like *Ashbya gossipy* and *Eremothecium ashpyi* synthesize the vitamin riboflavin which is an important component of vitamin B complex group.

Glycerin

It has been noted that glycerin can be produced commercially by the action of certain yeasts, *e.g.*, *Saccharomyces ellipsoideus*, on sugar.

Microbes as Pesticides

Certain micro-organisms like bacteria, fungi, viruses, protozoa and even nematodes infect insects and kill them. In other sense they may be called microbial pesticides and they are now being used as means of pest control.

MICROBES AS TOOLS FOR BIOLOGICAL RESEARCH

Some of the micro-organisms are widely used for academic studies and research work.

Genetical, Cytological and Bio-chemical Studies

Micro-organisms such as *Chlorella, Anacystis nidulans, Chlamydomonas, Nostoc, Anabaena, Neurospora, Saccharomyces, Ascobolus, Sordaria* etc., are the important tool in aforementioned researches. Protozoa are essential for the studies of geology and palaentology.

Gibbrella fujikuroi produces an important growth hormone called gibbrellic acid which has its great importance in physiological studies. *Penicillium notatum* produces an antibiotic penicillin which also shows activity like a plant hormone. It has been shown to affect plant metabolism by promoting several hormone-regulated phenomena. It shows Gibberellin-like activity, Cytokinin-like activity and Auxin-like activity.

Animals are of great economic importance for human beings. A number of products are obtained from animals, which are very useful since time immemorial. In the present age with the advancement of biotechnology a number of techniques have been evolved for the betterment of life. One of such techniques involved the transgenic animals.

TRANSGENIC ANIMALS

Transgenic Mice and other Animals. One of the first reports of transgenic animals published in December, 1982, involved transfer of **growth hormone (GH)** gene (from rat) fused to the promoter for the mouse **metallothionein 1 (MT)** gene. Since then many transgenic animals, including those in cattle, sheep, goats, pigs, rabbits, chickens and fish have been produced (Table 2) and will be utilized in future for a variety of purposes including (*i*) their efficiency in utilizing feed; (*ii*) ability to give leaner meat, (*iii*) their ability to grow to marketable size sooner and (*iv*) their resistance to certain diseases. More than this recently efforts are being made to use transgenic animals as living bioreactors. Transgenic animals produced for this purpose will secrete valuable recombinant proteins and pharmaceuticals into their milk, blood and urine which can be used for extraction of these drugs. This new possibility of manufacturing drugs through transgenic animals is often described as 'molecular farming' or 'molecular pharming'.

Table 31.2. A List of Transgenic Animals Produced and the Promoter, Enhancer and Structural Transgenes Involved in Each Case.

	Animal	*Genes Transferred (Promoter or Enhancer/Structural Genes)*
1.	Mouse	*mMT/rGH, mMT/bGH, mMT/oGH, mMT/hGH, mMT/hGRF, mMT/hFIX*
2.	Chicken	*ALV, REV*
3.	Cow	*BPV, lactoferring*
4.	Fish	*hGH, mMT/hGH, mMT/bGal, cd-crystallin SV/hygro, AFP*
5.	Pig	*mMT/hGH, mMT/bGH, hMT/pGH, MLV/rGH, bPRL/bGH*
6.	Rabbit	*mMT/hGH, hMT/hGH, rbEu/rb, c-myc*
7.	Sheep	*mMT/hGH, mMT/TK, mMT/bGH, mMT/hGRF, oBLG/hFIX, oBLG/a. lAT, oMT/oGH*
8.	Goats	*A varient of tPA gene (LAtPA)*

Note : The list of genes; promoter/or enhancer is separated from structural gene by a slash (/). The source of the gene is given by a lower case preceding the gene: *b* = bovine (or); *c* = chicken: *h* = human; *m* = mouse; o = ovine; *p* = porcine; *r* = rat; *rb* = rabbit.

Gene abbreviations – *ALV* = avian leukosis virus; *alAT* = Al anti trypsin; *BPV* = bovine papilloma virus; *Eu* = immunoglobulin heavy chain; *FIX* = factor IX; *GH* = growth hormone; *GRF* = growth releasing factor, β*Gal* = β galactosidase; *hygo* = hygromycin; *BLG* = β – lactoglobulin; *MT* = metallothionein; *MLV* = moloney murine leukemia virus; c-myc = a cellular proto oncogene; *REV* = reticuloendotheliosis; *PRL* = prolactin; *SV* = *SV* = *SV 40 TK* = thymidine kinase; *AFP* = antifreeze protein; *tPA* = human tissue–type plasminogen activator.

Although, initially many experiments leading to the production of transgenic animals did not give commercially attractive results, success has been obtained in some recent cases. Some of these cases will be discussed here.

Transgenic Goats. Transgenic goats were successfully produced recently by groups headed by *John McPherson* and *Karl Ebert,* both from USA. These transgenic goats expressed a heterologous protein (a variant of human tissue-type plasminogen activator = LAtPA) in their milk. This protein is used for dissolving blood clots *i.e.*, for treatment of coronary thrombosis. A cDNA representing LAtPA was linked with either the murine whey acid promoter (WAP) or a β casein promoter in an expression vector and used for injecting early embryos obtained surgically from the oviducts of superovulated dairy goats. These injected embryos were either immediately transferred to the oviducts of recipient females (surrogate mothers) or cultured for 72 hours (blocked at 8 –16 cell stage), before transfer to the uterus or recipient females of 29 offsprings from 36 recipients, one male and one female contained the transgene. The transgenic female delivered five offsprings, one of which was transgenic showing expression of LAtPA at a low level of few milligrams per litre of milk. In another case of a transgenic goat, few grams of LAtPA per litre of milk could be obtained. At this concentration, the dairy goat may become an economically viable bioreactor for human pharmaceuticals.

Transgenic Cows. In most earlier attempts (in Canada) for the production of transgenic cows, embryos or fertilized oocytes produced *in vivo* were utilized. Fertilized oocytes or proembryos were surgically terrieved from superovulated and artificially inseminated cows. Microinjected zygotes were than transferred by surgery either directly into the oviduct of recipient cows or into temporary hosts like sheep or rabbits. In view of two surgical operations, this method is labour intensive and more expensive.

In 'the Netherland', recently (1991) a technique has been developed for *in vitro* embryo production. In this new procedure, oocytes obtained from the ovaries of slaughter house cows, were matured and fertilized *in vitro*. Their pronuclei were microinjected with a construct containing a bovine alpha – S1-casei promoter (bovine = ox) driving a cDNA encoding the antibacterial human iron binding protein, 'lactoferrin'. The embryos were cultured to morula/ blastula stage and then non-surgically transferred to recipient females. Two of the 19 calves born from 103 transferred zygotes were transgenic (one male and other female). This procedure may facilitate the use of cows as bioreactors at the commercial level.

Transgenic Fish. Attempts to produce transgenic fish started in 1985 and some encouraging results have been obtained. The genes that have been introduced by microinjection in fish include the following: (*i*) human or rat gene for **growth hormone,** (*ii*) chicken gene for **delta-crystalline** protein, (*iii*) *E. coli* gene for β-**galactosidase**, (*iv*) *E.coli* gene for neomycin resistance, (*v*) winter flounder gene for **antifreeze protein** (flounder–flat fish), (*vi*) rainbow trout gene for **growth hormone**.

The technique of microinjection has been successfully used to generate transgenic in many species such as common carp, catfish, goldfish, loach, medaka, salmon, Tilapia, rainbow trout and zebrafish. In other animals (*e.g.*, mice, cows, pigs, sheep and rabbits), usually direct microinjection of cloned DNA into male pronuclei of fertilized eggs has proved very successful, but in most fish species studied so far, pronuclei cannot be easily visualized (except in medaka), so that the DNA needs to be injected into the cytoplasm. Eggs and sperms from mature individuals are collected and placed into a separate dry container. Fertilization is initiated by adding water and sperm to eggs, with gentle stirring to facilitate the fertilization process. Egg shells are hardened in water. About 10^6 to 10^8 molecules of linearized DNA in a volume of 20 ml or less are microinjected into each egg (1–4 cells stage) within the first few hours after fertilization. Following microinjection, eggs are incubated in appropriate hatching trays, and dead embryos are removed daily.

Since in fish, fertilization is external, *in vitro* culturing of embryos and their subsequent transfer into foster mothers (required in mammalian systems) is not required. Further, the injection into the cytoplasm is not as harmful as that into the nucleus, so that the survival rate in fish is much higher (35% to 80%).

Human growth hormone gene transferred to transgenic fish allowed growth that was twice the size for their corresponding non-transgenic fish (goldfish, rainbow trout, salmon). Similarly antifreeze protein (AFP) gene was transferred in several cases and its expression was studied in transgenic salmon. It was shown that the level of AFP gene expression is still too low to provide protection against freeze. There is also a report (in 1991) of the production of transgenic zebrafish from an Indian Laboratory (Madurai Kamraj University). In this attempt, a plasmid containing rat growth-hormone gene was microinjected into fertilized zebrafish eggs, and its presence confirmed in adult fish.

Transgenic Pigs. The efficiency of the production of transgenic pigs is still very low compared to that of the production of transgenic mice. In mice, 2.5% to 6% of the microinjected eggs developed into transgenic mice, but in pigs this frequency was as low as 0.6% even when as many as 7000 eggs were injected. Despite this low frequency, transgenic pigs carrying growth hormone (GH) gene from bovine (of 'ox' origin) of human, and sheep globin gene have been produced (by V.G. Purse) at Agriculture Research Service, Beltsville, USA. The pigs carrying hGH gene showed different levels of expression and only 66% of these animals showed detectable levels of hGH and bGH in their plasma. The animals grew a little faster but did not become large; similarly pigs with sheep globin gene did not show any expression of the transgene for unknown reasons. In these transgenic pigs, however, a modest increase of 10–15% in daily weight and 16–18% in feed efficiency was observed, which is though lower to those in mice, but are comparable to those obtained due to daily injection with pig growth hormone. It was also observed that there was a marked reduction in the subcutaneous fat in some of these transgenic pigs suggesting the possibility of producing leaner meat with lower fat content. These results may have a significant impact on the 9.5 billion dollar annual pig industry in USA.

It is also reported that a long term elevation of growth hormone was generally detrimental to health. The pigs had incidence of gastric ulcers, arthritis and several other diseases. Therefore, techniques will have to be developed to manipulate better the transgene expression by a variety of methods (*e.g.*, changing genetic background or modifying the husbandry regiment).

Transgenic Sheep. The rate of transgenesis in sheep is very low (0.1 to 0.2%). This can be improved, if only transgenic viable embryos (after necessary checking) are transferred to surrogate ewes (female sheep). Embryos at 8–16 cell stage can be split into two parts, one for continued culture and the other for detection of integrated genes using polymerase chain reaction (PCR). Although microinjection is the most common method for DNA delivery, gene targeting may be increasingly used in future. In this approach, embryonic stem (ES) cells in culture are transferred with a vector which targets the gene to a particular site by homologous recombination (as discussed above). The technique, though successfully used in mice, has yet to be applied to sheep, where ES cells will have to be isolated first.

The first reports of transgenic sheep were published by *J.P. Simons* (1988) of Edinburgh. Two transgenic ewes were produced, each carrying about 10 copies of human anti-hemophilic factor IX gene (cDNA) fused with the 10.5 kb BLG gene (BLG = β-lactoglobulin). BLG gene was used, because it is necessary for specific expression of gene in mammary glands. Consequently, the gene had a tissue specific expression and ewes secreted human factor IX (or alpha-1 antitrypsin) into their milk; this human factor IX is active, even though the expression of transgenic is low. The transgenic ewes were born in early summer of 1986 and were successfully mated same year in December. In 1987, each ewes gave birth to a single lamb. Each lamb inherited BLG-F IX transgene and secreted factor IX in the milk. This programme of the production of transgenic animals by J.P. Simons at Edinburgh was funded by Pharmaceutical Proteins Ltd. (Cambridge, UK) due to its commercial appeal.

In another report (published in 1991), also from Edinburgh, five transgenic sheep were produced (*Alan Colman and colleagues*). In all these cases, transgene involved fusion of the ovine β-lactoglobulin gene promoter fused to the human α_1 antitrypsin (hα_1 AT) gene. Four of these animals were female and one male. In one female the protein hAT hα_1AT reached a level of 35 grams per litre of milk. The protein purified from milk had a biological activity indistinguishable from human plasma derived antitrypsin. The deficiency for hα_1AT leads to a lethal disease emphysema, (BC-22), which is a common hereditary disorder among caucasian males of European descent. Therefore any strategy giving high yield of this protein economically will be most welcome. In view of this, transgenic sheep with hα_1AT gene will prove very useful as a bioreactor.

Recombinant DNA technique can also be used to increase the ability of sheep for wool growth. For this purpose, genes essential for synthesis of some important amino acids found in keratin proteins of wool, have been cloned and introduced in embryos to produce transgenic sheep. For instance, genes (cysE and cysM) for two enzymes (serine acetyl transferase = SAT and O = acetylserine sulphydryase = OAS), involved in cystein biosynthesis, were isolated from bacteria and cloned in a vector. These genes were introduced in sheep cells, ultimately leading to the production of transgenic sheep, where these genes are expressed. Growth hormone (GH) genes have also been introduced and can be used to promote body weight. Other genes involved in wool production have also been cloned and well be used for transgenesis, thus increasing the potential of wool production through genetic engineering.

32 Isolation of Auxotrophs

INTRODUCTION

In order to study microbial mutants, one must be able to detect them readily, even when there are few, and then efficiently isolate them from the parent organism and other mutants. Some techniques used in mutant detection, selection and isolation.

MUTANT DETECTION

When collecting mutant of a particular organism, one must know the normal or wild type characteristics so as to recognize an altered phenotype. Usually, mutation are rare, about one per 10^7 to 10^{11} cells, therefore, it is necessary to have a very sensitive direction system and one has to look at perhaps thousands to millions of colonies or clones. Mutation detection systems in bacteria and other haploid organism are straight forward, because any new allele should be seen immediately, if it is a recessive mutation.

Sometimes detection of mutants is direct of albino mutants of a normally pigmented bacterium are being studied, detection simply requires visual observation of colony colour.

Other direct detection systems are more complex, for example, the replicate plating technique is used to detect auxotrophic mutants. It distinguish between mutants and the wild type strain based on their ability to grow in the absence for a particular biosynthetic end product.

MUTANT SELECTION

1. **Selection Technique:** An effective selection technique uses incubation conditions under which the mutant will grow, because of properties given to it by the mutation, whereas the wild type will not. It involves reversions, mutations or the development of resistance to an environmental stress.
2. **Resistance Selection Technique:** It follows a similar approach often wild type cells are not resistant to phage attack or antibiotic treatment, so it is possible to grow the bacterium in the presence of the agent and took for surviving organisms. Consider the example of a phage sensitive wild type bacterium when the organism is cultured in medium lacking the virus and then plated out on selective medium containing phages, any colonies that form will be resistant to phage attack and very likely will be mutants in this regard. Resistance selection can be used together with virtually any environmental parameter. Resistance to bacteriophage antibiotics or temperature are most commonly employed.

3. **Substrate Utilization Technique:** This technique is also employed in bacterial selection. Many bacteria catabolize only a few primary carbon source, with these bacteria, it is possible to select mutants by a method similar to that used in resistance selection. The culture is plated on medium containing an alternate carbon source. Any colonies that appear can use the substrate and are probably mutants.
4. **Mutant Detection and Selection Methods:** These methods are used for purposes other than understanding more about the nature of genes or the biochemistry of a particular micro-organism. Replica plating used for the isolation of auxotroph mutants.

CLASSIFICATION OF AUXOTROPHS

The classification of mutants of organisms can be done by replica plating technique. Here a suspension of bacterial cells treated with mutants is poured onto a master plate containing medium supplemented with those growth factors which will allow the desired mutants to grow. After incubation, the replication is applied to the plate and this picks up few colonies from the master plate, and it is pressed successively on series of plates containing minimal medium along and with various supplements. After further incubation, colonies appear on these replica plates in the same pattern as on the master plate, absence of a particular colony from a plate is an evidence that it requires that supplement missing from the plate, and so that mutant is classified according to the missing growth factor, *e.g.*, arginine requiring mutants, threonine requiring mutants, etc.

Auxotrophic Mutants

Many variants of bacteria have been isolated which are unable to synthesize one or more amino acids, or other factors required for growth. Such strains are called **auxotrophs**, while nutritionally independent, wild type strains are called **prototrophs**. In *E. coli* strains with lac^{+} are prototrophs and those with lac^{-} are auxotrophs. An auxotrophic strain may be identified by its inability to grow and multiply in a synthetic medium lacking its specific growth requirements, in which the wild type, prototrophic strain flourishes. If the synthetic medium is supplemented with the growth factor which the auxotroph cannot synthesize for itself, then auxotrophs and protrophs can grow equally well in it and can be indistinguishable. Similarly strains of bacteria which have become resistant to the lethal action of antibiotics or bacterial viruses may appear identical to the wild type unless the drug or the phage is present in the environment to detect the difference between them. Usually, auxotrophs are given the symbol (–) and prototrophs the symbol (+).

Auxotroph Selection

It is easy to pick out a prototrophic mutant from a mixture of auxotrophs and prototrophs because practically auxotrophs will not grow in a selective medium not containing the missing growth factor. But it is difficult to pick out an auxotroph from a medium because the prototrophs will grow pretty as well as the auxotrophs when the medium is supplemented.

We can solve this problem by the fact that penicillin kills only growing bacteria. If a suspension of *E. coli* prototrophs is treated with a mutagen and then grown in minimal medium containing a lethal concentration of penicillin, all the cells which are capable of growth will be killed. After a suitable delay the bacteria remaining including any auxotrophs present which cannot grow because of lack of supplements, are centrifuged down, washed, re-suspended in penicillin-free medium and poured on supplemented agar plates. The resultant colonies are tested and classified as requiring various amino acids or vitamin for normal growth.

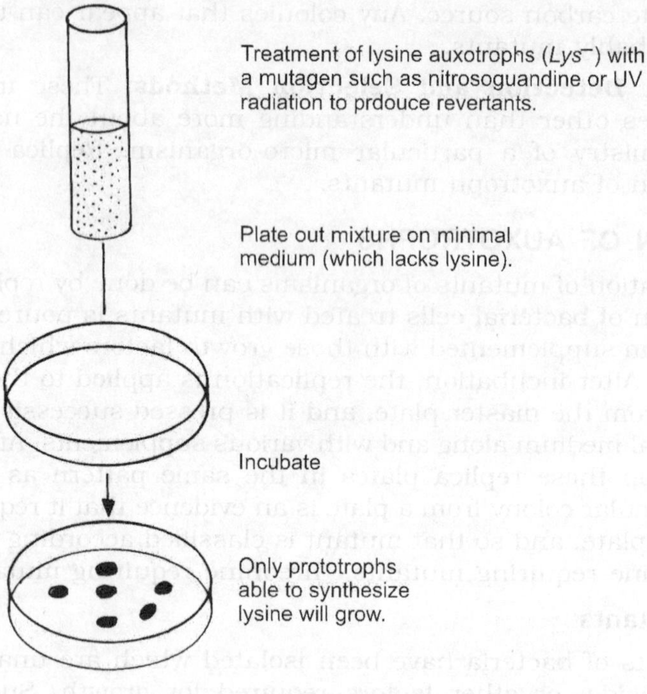

Fig. 32.1. Mutant Selection. The production and direct selection of auxotroph revertants. In this example, lysine revertants will be selected after treatment of a lysine auxotroph culture because the agar contains minimal medium that will not support auxotroph.

33 REPLICA PLATING TECHNIQUE

INTRODUCTION

In the 1940s, *George Beadle* and *Edward L. Tatum* set out to determine the nature of the products specified by genes. For their study, they used the pink bread mold *Neurospora crassa.* They recovered the auxotrophs at the cost of substantial labour. They analyzed about 2000 ascospores to recover the three vitamin-requiring mutants, all of the transfers being done one at a time. Current technology facilitates the recovery of auxotrophs in fungi; such as *Neurospora* and *yeast*, and in *bacteria.* One of the technique Replica Plating.

REPLICA PLATING

After studying nutritional mutants in *Neurospora. Edward Tatum* turned his attention to bacteria, in particular to *E. coli.* The first *E. coli* auxotrophs to be isolated were recovered in a laborious manner similar to the one originally used for *Neurospora* mutants; individual strains were transferred one at a time to different types of media. Subsequently, in 1952, Tatum's associates *Ester* and *Joshua Lederberg* introduced the procedure of replica plating. This technique is very effective in recovery of mutants in organisms, such as bacteria, that produce discrete colonies by division on solid agar medium.

The recovery of mutants by using replica plating starts; with treating the bacteria with a mutagen that is with an agent stressed against such as irradiation that cause mutations. A suspension of bacterial cells is then "plated" on a solid enriched medium in a petri dish, so that about 100 cells are randomly distributed across the medium. These cells divide to produce clonal colonies of genetically identical bacteria. The petri dish is then inverted on a circular block which is covered by a sterile velvet cloth, and some bacteria from each colony stick to the cloth. Several strile Petri dishes each containing a different medium, are then pressed against the surface of the cloth. Some bacteria adhere to the surface of the medium in each dish in a pattern identical that in the original dish, thus the original pattern of colonies is reproduced, or replicated on the different experimental media.

The different petri dishes contain media of different compositions; one is minimal others may be enriched with different vitamins, with amino acid, or with purines and pyrimidines. After a day or two the dishes are examined and the distribution of colonies recorded. If colonies are found on enriched media that are absent on the minimal medium, new auxotrophs have been identified, one of the auxotrophs might be growing on the medium enriched with amino acids, whereas another is present on the medium containing purines and pyrimidines.

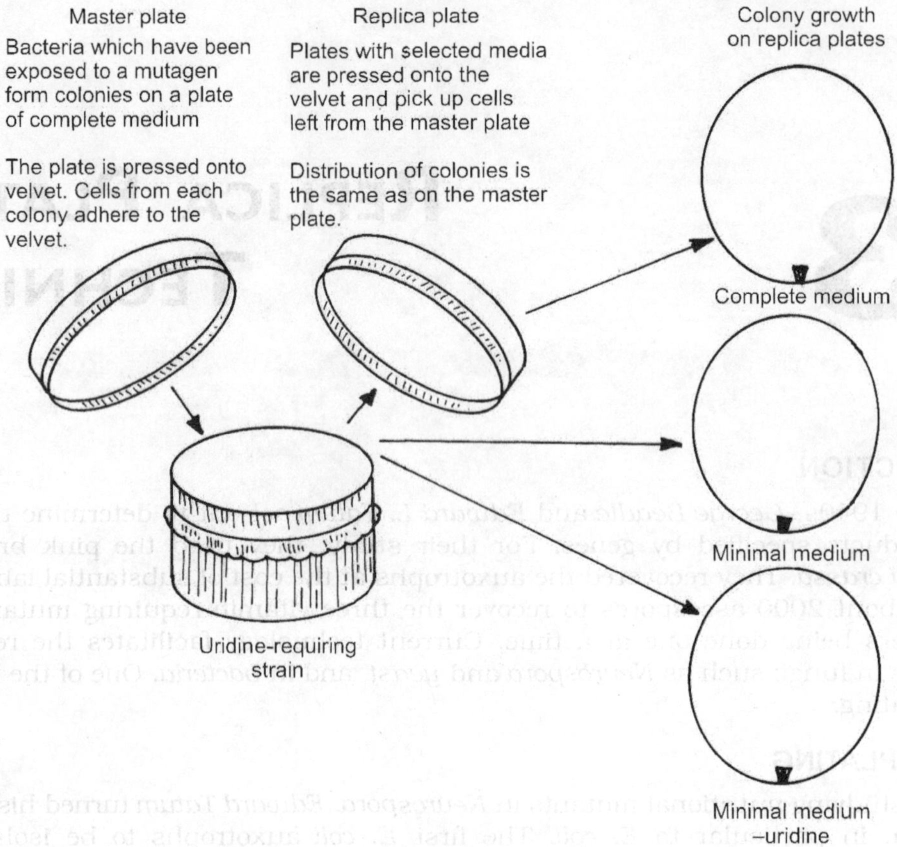

Fig. 33.1

The replica plating procedure allows rapid screening of hundreds of colonies for the possible existence of new mutants. Replica plating is now limited to the recovery of auxotrophs. It may also be used for the selective recovery of mutants that are resisted to antibiotics. (Different antibiotics may be included in the test plates). Similarly, phage resistant mutants and mutants sensitive to physical agents, such as ultraviolet light, can be recovered, as can conditional lethals, for example, by growing the bacteria at different temperatures. Indeed, the types of mutants that can be recovered by replica plating are perhaps limited only by the imagination of the experiment.

34 Analysis of Mutation in Biochemical Pathway

HIGH RESOLUTION DISSECTION OF BIOLOGICAL PROCESSES VIA MUTATION ANALYSIS

Mutations have been used extensively to elucidate the pathways by which biological processes occur. Metabolism occurs via sequences of enzyme-catalyzed reactions. By isolating and studying mutations in the genes coding for the enzymes involved, the sequence of steps in a pathway can often be determined. Morphogenesis frequently involves the sequential addition of proteins in the formation of specific three dimensional structures. Again, the sequence of protein additions can often be determined by isolating and studying mutant organisms with mutation in the genes coding for the proteins involved, Because an appropriate mutation will eliminate the activity of a single polypeptide, mutation provide an extremely powerful probe with which to dissect biological processes.

Consider, for example, the following simple pathway:

$$\begin{array}{l} \text{Precursor X} \\ \quad \downarrow \ \text{enzyme A} \leftarrow \text{gene A} \\ \text{Intermediate Y} \\ \quad \downarrow \ \text{enzyme B} \leftarrow \text{gene B} \\ \text{product Z} \end{array}$$

Intermediate Y is produced from precursor X by the action of enzyme A, the product of gene A. Intermediate Y, however, may be rapidly converted to product Z by enzyme B, the product of gene B. If so, intermediate Y may be present in very low quantities and be very difficult to identify biochemically. However, in a mutant organism that has a mutation in gene B, resulting in the synthesis of either an inactive form of enzyme B or no enzyme B, intermediate Y will often accumulate to much higher concentrations, facilitating its isolation and identification. Similarly, a mutation in gene A may aid in the identification of precursor X. Furthermore, the functions and mechanisms of action of individual gene products can often by deduced by comparative biochemical and biophysical analyses of mutant and wild-type organisms.

The resolving power of mutational dissection in biological processes has been emphatically demonstrated by the elucidation of the pathway of morphogenesis in bacteriophage T4. This complex process involves the products of about 50 genes. Each gene encodes a structural protein of the virus of an enzyme that catalyzes one or more steps in the morphogenetic pathway. By (1) isolating mutant strains of phage T4 with temperature sensitive and suppressor-sensitive conditional lethal mutations in each of the approximately 50 genes, and (2) analyzing the structures that accumulate when these mutant strains are grown under the restrictive conditions by electron microscopy and biochemical techniques, *R. S. Edgar, W.B. Wood, J. King* and *colleagues* have been able to work out almost the entire pathway of phage T4 morphogenesis (Figs. 34.1 and 34.2).

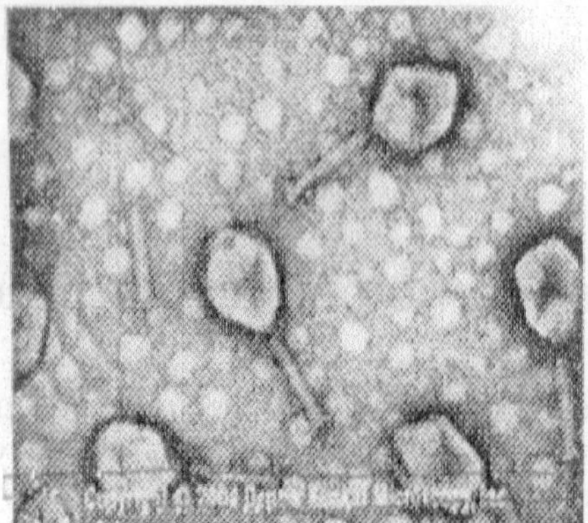

Fig. 34.1. Electron micrograph of negatively stained T4 bacteriophages, showing their intricate morphology. The DNA-filled heads, the tails, and the six-pointed baseplates with their attached tail fibers are easily recognized. On some of the phage (particularly the one at the lower right), the collar between the head and the tail is visible. The tail sheath of the phage at the lower left is contracted, revealing the distal portion of the inner core or tube of the tail through which the DNA is ejected during infection. The tail sheath functions like a little muscle. After the tail fibers and spikes on the baseplate have become firmly attached to receptor sites on the bacterial cell, the tail sheath contacts, pushing the tail core through the cell envelope, much like an injection needle. Magnification is × 37,000 (Courtesy *E. Boy de la four* and *E. Kellenberger*).

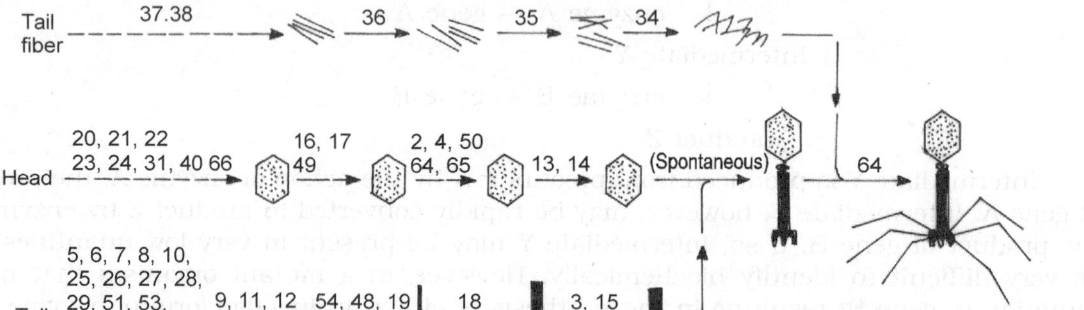

Fig. 34.2. Summary of the pathway of morphogenesis in bacteriophage T4, as revealed by mutational dissection and electron microscopic and biochemical analyses. The head, the tail, and the tail fibers are synthesized by means of separate branches of the pathway, which are then joined in terminal steps of the pathway. The numbers beside the arrows indicate the T4 genes whose products are required at that particular step in the pathway. The pathway is obligatorily sequential in all but a few steps, that is, the morphogenesis must proceed in the sequence indicated. The tail fibers will not attach to headless tails, for example, even though the structure of the baseplate appears complete prior to the joining of the tail to the head. (After *W.B. Wood, R.S. Edgar, J. King, I. Lielausis,* and *M. Henninger, Fed. Proc,* 27: 1160–116, 1968).

Other biological processes that have been successfully dissected by mutational analysis include the photosynthetic electron transport chains in *Chlamydomonas reinbardi* and maize and nitrogen fixation in bacteria. This approach is currently being used to dissect differentiation and development in higher plants and animals. *S. Benzer* and *colleagues* are using mutations to dissect behaviour and learning in *Drosophila.* In theory, mutational dissection should be applicable to any process that is under genetic control. The existence of a gene that does not mutate to a non-functional state seems extremely unlikely. Thus, mutational dissection of biological process is limited only by the ingenuity of scientists in identifying mutations of the desired types.

EXERCISE

(Based on Unit–5)

Q.1 Explain the different phenomenon of mutation.

Q.2 Write short notes on:

(*a*) Spontaneous mutation

(*b*) Induced mutation

Q.3 What are the mutagens ?

Q.4 Describe physical and chemical mutagens.

Q.5 Write short notes on:

(*a*) Isolation of auxotrophs

(*b*) Replica plating technique

(*c*) Analysis of mutation in bio-chemical pathway

(*d*) Induced mutation in microbes and animals for economic benefit of man.

INDEX